GERM-FREE BIOLOGY
Experimental and Clinical Aspects

ADVANCES IN EXPERIMENTAL MEDICINE AND BIOLOGY

Volume 1
THE RETICULOENDOTHELIAL SYSTEM AND ATHEROSCLEROSIS
Edited by N. R. Di Luzio and R. Paoletti • 1967

Volume 2
PHARMACOLOGY OF HORMONAL POLYPEPTIDES AND PROTEINS
Edited by N. Back, L. Martini, and R. Paoletti • 1968

Volume 3
GERM-FREE BIOLOGY – EXPERIMENTAL AND CLINICAL ASPECTS
Edited by E. A. Mirand and N. Back • 1969

Volume 4
DRUGS AFFECTING LIPID METABOLISM
Edited by W. L. Holmes, L. A. Carlson, and R. Paoletti • 1969

GERM-FREE BIOLOGY
Experimental and Clinical Aspects

Proceedings of an International Symposium on Gnotobiology held in Buffalo, New York, June 9-11, 1968

Edited by
Edwin A. Mirand
Roswell Park Memorial Institute
New York State Department of Health
State University of New York, Buffalo, New York
Niagara University, Niagara University, New York
Canisius College, Buffalo, New York

and

Nathan Back
Department of Biochemical Pharmacology
School of Pharmacy
State University of New York
Buffalo, New York

Springer Science+Business Media, LLC 1969

Library of Congress Catalog Card Number 69-16518

Originally published by Plenum Press, New York in 1969.

ISBN 978-1-4899-6234-8   ISBN 978-1-4899-6495-3 (eBook)
DOI 10.1007/978-1-4899-6495-3
Softcover reprint of the hardcover 1st edition 1969

PREFACE

The purpose of this publication is to provide a cross-sectional version of the present state of gnotobiology and its clinical applications by biomedical scientists and physicians. With contributions from abroad as well as from the United States, the current status of gnotobiology is amply explored. We wish to express our sincere appreciation to the participants for their contributions to this Symposium held in Buffalo, New York on June 9 - 11, 1968, sponsored by Roswell Park Memorial Institute, State University of New York at Buffalo School of Pharmacy, and the Association for Gnotobiotics, Inc. We owe a great debt to those pioneers who labored under great hardships to develop this area of science. Such people at the Lobund Institute at the University of Notre Dame, notably the late Professor J. A. Reyniers and Professor P. C. Trexler and their colleagues as well as others in Sweden, Japan and elsewhere, are to be recognized for their enormous effort to keep this field alive.

Who can determine the limits of the usefulness of gnotobiotic techniques in solving problems in a number of the sciences. Its usefulness in immunological research is becoming more clearly understood. The techniques in gnotobiotic research may be useful too in aerospace science. To those working in gnotobiology, Dr. Nathan Back and I would like to commend all of you who have contributed to so many important developments in understanding the biological relationships between animals including man and their microflora.

Edwin A. Mirand

December, 1968

CONTENTS

PATIENT CARE IN STERILE ENVIRONMENT

EXPERIMENTAL BIOLOGY

LEUKEMIA IN GERMFREE RESEARCH

CARCINOGENESIS

PHYSIOLOGY OF CECUM

DISEASE RESISTANCE OF GERMFREE ANIMALS

TECHNOLOGY

PAPERS RECEIVED TOO LATE

FOR INCLUSION UNDER PROPER SUBJECT HEADINGS

THE APPLICATION OF LAMINAR FLOW ROOMS TO PATIENT ISOLATION

Donald Vesley, George S. Michaelsen, & Alexander A. Levitan

School of Public Health, University of Minnesota

Minneapolis, Minnesota

I. INTRODUCTION

The need for providing ultraclean environments for hospital patients with reduced immune defense mechanisms is rapidly increasing. Cancer chemotherapy and organ transplantation are two medical procedures which require extreme measures to protect recipients from infection.

The use of plastic tent isolators for this purpose has been investigated by several research groups (Haynes and Hench, 1965, Schwartz, et al, 1965). The system shows considerable promise for providing patient protection but also has several inherent limitations. Specific limitations include the difficulty of giving nursing care to adult patients through gloveports and the adverse psychological impact on the patient of long duration treatment. Both of these difficulties can be minimized, theoretically, by the use of a full size room.

The development of laminar flow ventilation (Whitfield, 1962) provided a possible approach to ultraclean patient protection in a full size room. This system provides a very high level of filtration combined with minimum turbulence, unidirectionality, and a very high ventilation rate compared to conventional facilities.

Under contract from the National Cancer Institute, the School of Public Health at the University of Minnesota undertook to investigate the potential applicability of a laminar flow room to the problem of patient isolation. The facility is illustrated in Figure 1. It consists of an anteroom, a dressing and storage area, and a patient room. Only the patient room is laminar flow, but the other

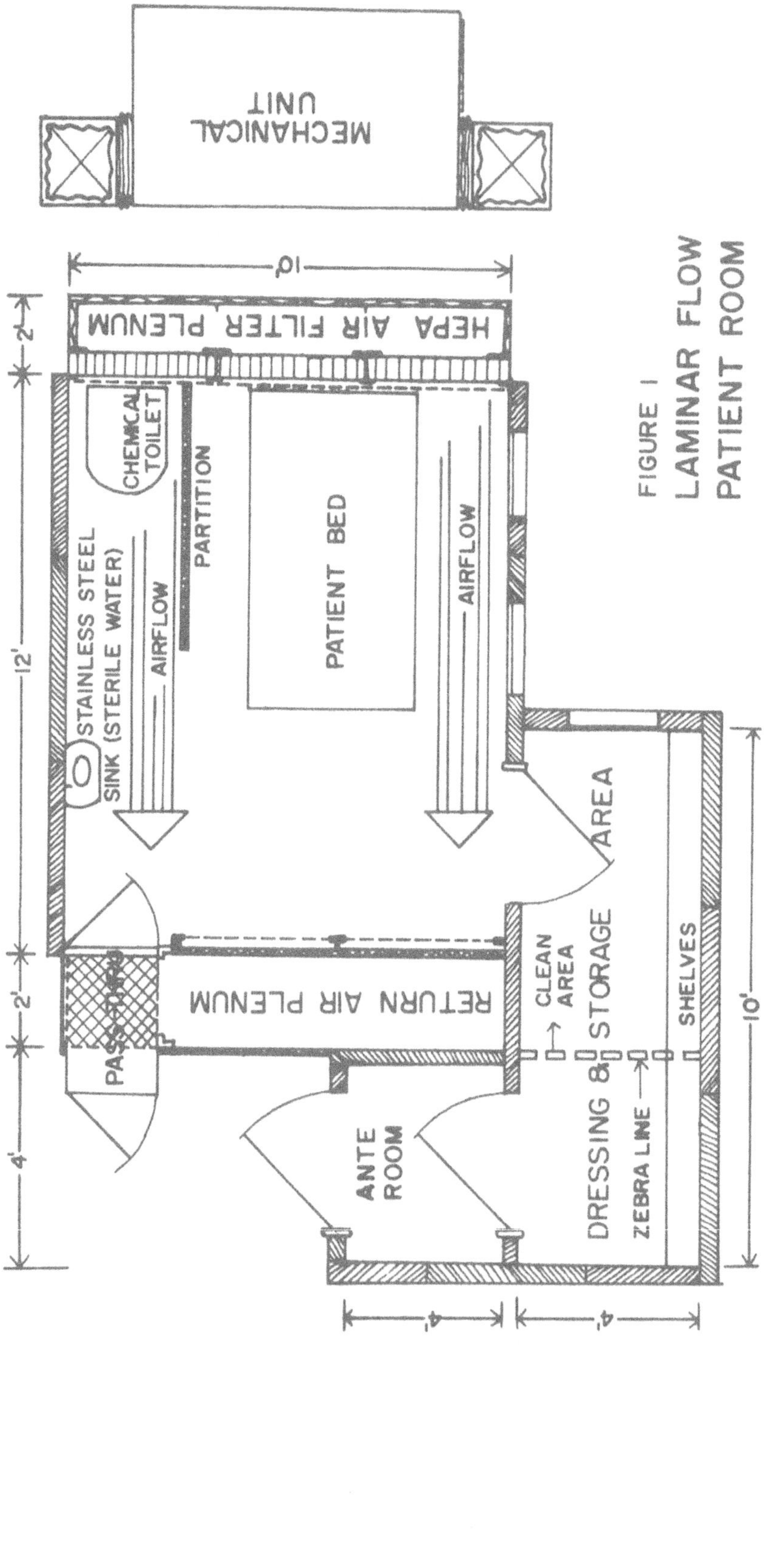

FIGURE I
LAMINAR FLOW
PATIENT ROOM

areas are pressurized in relation to the outside. A single air lock pass-through located in the return air wall serves for introduction and removal of supplies.

Temperature and humidity controls permit ranges of approximately 67°-80°F. and 40-65% relative humidity in the room. Mean air velocity is approximately 90 feet per minute at full operation. Later additions to the facility included a bucket-type toilet, without plumbing connections, set behind a partition to provide privacy and a foot operated stainless steel sink supplied with hot and cold running water, filtered through a Millipore HA filter.

Experiments were conducted with presterilized manikins and with healthy human volunteer subjects domiciled for various lengths of time. Each of these studies will be described in succeeding sections.

II. PRESTERILIZED MANIKIN EXPERIMENTS

The initial experimental approach was to use fiberglass department store type manikins to simulate the patient. The theory was that the manikins could be presterilized so that all microbial contaminants subsequently isolated would be known to be exogenous. The manikins were sterilized by ethylene oxide. First evaluations were made based on individual procedures involving maximum activity and patient contact. The nursing procedure tested was the bathing of the "patient" followed by a complete bed linen change accomplished with the dummy in bed. The medical procedure was a complete simulated physical examination which involved considerable handling of the "patient".

Later, emphasis was placed on sequential evaluations following routine nursing care over a two-day period. The activities included bathing, oral care, bedmaking, feeding, vital signs, physical examination, and bandage removal.

Detection of viable contamination during this phase of the investigation was divided into three categories; namely, airborne contaminants, contamination accumulating on personnel gloves, and contamination reaching (and surviving on) the dummy "patient". Microbiological sampling techniques for these purposes were as follows:

For air sampling, two Reyniers slit samplers equipped with two-hour mechanisms were set up at the upstream and downstream ends of the room. Samples were collected at one cubic foot per minute on pre-incubated Trypticase Soy Agar plates, then counted after 72 hours of incubation at 32°C. Samples were collected only during periods of activity as no contaminants could be detected when the room was unoccupied. Gloves were sampled before and after specific activities by having the wearer press first the fingers and then the heel of the gloved hand to a Trypticase Soy Agar surface in a 150 mm petri dish. The plates were then incubated at 32°C. for 72 hours before counting.

Considerable emphasis was placed on detection of viable contaminants on the manikin preceding and following specific nursing and medical procedures.

The procedure was as follows: The manikins were divided systematically into 23 anatomical areas. Each area was swabbed once with a sterile cotton ball moistened with 2 ml of M/15 pH 7 phosphate buffer. The ball was held with sterile forceps during sampling, then dropped into a screw cap bottle containing 50 ml of Tryptone Glucose Yeast broth plus 0.5% Tween 80 for neutralization of hexachlorophene. The broth was incubated for one week at 32°C. at which time results were recorded as growth (turbid) or no growth (clear). The sensitivity of the technique was tested on several occasions by using it on a presterilized manikin exposed for a few minutes in a conventional hospital room. Under these conditions 100% of the samples demonstrated growth.

The results of air sampling conducted during periods of activity related to simulated care of the presterilized manikins are summarized in Table I. The over-all contamination levels observed, even at the downstream end of the facility, were about one order of magnitude below results previously reported for conventional industrial clean rooms.

The full dress procedure resulted in lower contamination levels downstream by a factor of about two. The upstream results, showing two or three colonies per 1,000 cubic feet of air, approach the lower limit of detectability.

Table II summarizes contamination of personnel gloves before and following specific activities during the manikin experiments. More than 30% of the gloves sampled prior to activity showed some

Table I. Microbial Air Contamination in a Laminar Crossflow Room During Periods of Simulated Nursing and Medical Care Activities on a Presterilized Manikin

Experimental Conditions	Colonies/cu.ft.	
	Filter wall (upstream)	Return wall (downstream)
Manikin Experiments - Personnel in full dress (24 experiments)	0.002	0.007
Manikin Experiments - Personnel in laboratory coats (26 experiments)	0.003	0.016

Table II. Personnel Glove Contamination Before and After Simulated Nursing Activities on Presterilized Manikins

Subject	Before Activity			After Activity		
	No. observations	% contaminated	Col./glove	No. observations	% contaminated	Col./glove
Technician	80	36	0.7	133	82	13.5
Nurse	76	28	0.7	142	80	10.9
Doctor	34	26	0.6	34	77	13.7
All	190	31	0.7	309	81	12.4

contamination. While the number of colonies actually detected in these instances was usually small, there was a strong indication that "sterile gloving procedure" often resulted in some contamination even before the start of activities. Following manikin handling procedures the number of gloves showing some contamination increased to more than 80%. As the manikin and all items of furnishings or equipment were presterilized, it is probable that these contaminants originated from the experimenter.

Table III compares microbial evaluations obtained using two different dress procedures during the individual activity experimentation on the manikins. It can be observed that addition of exogenous contamination to the manikin during handling was relatively slight regardless of the dress procedure used. Approximately 90% of areas sampled before experimentation were sterile. The figure dropped to about 80% after the nursing procedure, then to about 70% after the physical examination. Thus, it appeared at this point

Table III. Effect of Two Barrier Techniques on Microbial Contamination on a Presterilized Manikin in a Laminar Crossflow Room

Barrier Technique	% Sterile Swabs ±1 Standard Error From Manikin (Out of 23 Anatomical Areas)		
	Before Experiment	After Bedmaking	After Physical Exam
Full Dress Procedure (1)	89.7 ± 3.4	82.1 ± 4.0	70.0 ± 5.8
Minimal Dress Procedure (2)	89.0 ± 5.2	77.6 ± 8.7	71.8 ± 2.8

(1) Includes boots, Dacron coveralls with tight ankles and wrists, full hood, face mask, and gloves; all sterile.

(2) Includes sterile boots, sterile gloves, paper hat, and standard cotton laboratory coat or nurse's uniform.

that, while some contamination was unavoidable in this setup, adherence to the principle of remaining downstream from the "patient" and handling only with "sterile" gloves minimized the transfer of contaminants from nursing personnel to the "patient".

During the longer period manikin experiments, relatively heavy glove contamination occurred on several occasions and the impact on manikin contamination was striking. Table IV summarizes the viable contamination measurements on the dummy "patient" during the simulated two-day care experiments. Very high percentages of negative observations were maintained throughout the two-day period when personnel glove contamination was avoided and full barrier uniforms were worn. The partial dress procedure yielded somewhat fewer negative observations. However, a marked difference was observed on those occasions when heavy glove contamination occurred, regardless of type of uniform worn.

The manikin experiments demonstrated that it was possible to utilize nursing personnel in the facility with relative freedom of movement and a minimum of contamination transfer to the "patient".

Table IV. Microbial Contamination Detected on a Presterilized Manikin Following a Set Routine of Nursing Activities During A Two-Day Period

Activity Prior to Sample	% Sterile Swabs		
	More than 10% of gloves with > 100 colonies (regardless of dress procedure)	Fewer than 10% of gloves with > 100 colonies	
		Full dress procedure for nurse and doctor	Partial dress procedure for nurse and doctor
Presterilized manikin unwrapped	58	96	81
pHisoHex bath, oral care	23	100	82
Bedmaking	10	94	66
Feeding, vital signs, physical examination	4	83	52
Overnight-no activity	13	90	70
Breakfast, oral care, and pHisoHex bath	37	89	79
Bedmaking, leg bandaging	28	93	78
Feeding, vital signs, bandage removal, physical examination	17	87	57
Mean of all observations	24	92	71

Furthermore, it had been shown that it was possible to retreat somewhat from a time consuming full dress procedure with relatively little additional contamination resulting.

III. HEALTHY HUMAN VOLUNTEER EXPERIMENTS

The second phase of evaluation of this system was the introduction of healthy human volunteer "patients" into the room in place of the presterilized manikins. Initially two male subjects aged 22 and 30 were selected. Each volunteer was domiciled in the laminar flow room on two occasions for periods of approximately 31 hours. During the identical period the other subject was domiciled in an ordinary patient room in University Hospitals. Both "patients" were subjected to skin disinfection by a series of five hexachlorophene soap tub-shower baths prior to the experimental period. Immediately before confinement each was subjected to an alcohol bath with a solution of 70% isopropyl alcohol, 0.5% cetyl alcohol, and 0.1% hexachlorophene. During occupancy both subjects were given identical routine nursing care, including vital signs, feeding, linen change, and daily pHisoHex bed baths.

Microbial samples from the "patients'" skin disclosed far greater differences between the two individuals than could be discerned between the laminar flow and conventional environments. However, the relative ease with which the subjects adjusted to and tolerated the laminar flow environment indicated the desirability of longer duration experiments.

Thus, a series of experiments involving 12-day continuous occupancy of the facility was initiated. According to the objectives of the project, information was sought for the fulfillment of four specific goals. They were as follows:

A) Determination of operational needs and problems arising from nursing care and ancillary services for a patient in a laminar flow facility.

B) Determination of physiological effects of laminar flow during 12 days of continuous occupancy by a healthy male subject.

C) Determination of factors influencing patient comfort and psychological reactions.

D) Determination of microbial patterns in the environment and microbiological sampling technique efficacy for a continuously occupied laminar flow room.

The general method of procedure for the experiments was as follows: Four subjects were selected. All were males ranging from 22 to 31 years of age. One of these subjects subsequently was selected for retrial in the fifth experiment.

Each subject was instructed to bathe daily with a germicidal soap for a period of two weeks prior to admission to the laminar flow facility. On the morning of admission, the subject was given a final pHisoHex shower followed by an alcohol bath with a solution of 70% isopropyl alcohol, 0.5% cetyl alcohol, and 0.1% hexachlorophene.

All items of equipment were sterilized by steam autoclaving or ethylene oxide and were introduced into the facility through the pass-through. Food was not sterilized because the "patient" retained his normal gastrointestinal flora.

The nurse wore a uniform consisting of a calf-length smock coat with full length front Velcro closure, a hood, and calf-length booties. These items were made from Bar-Bac cloth, a closely woven cotton material, and were presterilized. A sterile face mask and sterile surgical type gloves also were worn. It was decided that maximum barrier effect could be combined with convenient entry procedures using this configuration. The use of Velcro closures on the coat, hood, and boots greatly facilitated personnel entry in a relatively short time.

The basic format was not changed drastically through the five experiments. However, several experimental factors were introduced during the later trials, and microbiological monitoring techniques which evolved during the first two experiments were standardized for the last three.

Microbiological monitoring of the facility consisted of the following:

A) Reyniers air samplers were set up at the absolute filter wall and at the return air wall in a manner similar to previous experiments. Four 2-hour samples were run consecutively at each location on each day starting at about 7:30 a.m. and continuing until late afternoon.

B) Nurses' gloves were sampled routinely only during the last three experiments. On three occasions each day (following morning care, following afternoon backrub, and at the end of the nursing shift) the nurse pressed each gloved hand to a Trypticase Soy Agar surface before discarding the glove.

C) Room furnishings and surfaces were monitored using Rodac agar contact plates and following a predetermined pattern of sampling sites during the last three experiments.

All sampling plates were filled with Trypticase Soy Agar and were incubated for 72 hours at 32°C. before counting.

Table V summarizes air sampling data for all five experiments. It is again obvious that, even with continuous activity over an extended time period, the laminar flow facility provides virtually contamination-free air at the upstream end of the room. The mean colony count of one colony per thousand cubic feet at the filter wall location again represents the minimum level of detectability of the sampling apparatus. A comparison of downstream results for the five experiments indicates that this sampling device is a sensitive indicator of "patient" shedding potential.

Table VI summarizes nurses' glove contamination for the last three experiments. Of great interest is the fact that contamination levels associated with each "patient" are exactly correlated with downstream air sampling levels for these same subjects. This table also indicates a generally higher level of glove contamination following the period of greatest activity (morning care) as compared with other sampling times (158.3 col./glove vs. 87.8 col./glove).

Surface sampling results were generally consistent with the air sample and glove impressions in discerning between patients. Thus, the microbiological monitoring techniques employed all proved to be sensitive indicators of biocontamination problems.

Table V. Microbial Air Contamination in a Laminar Crossflow Facility During 12 Days of "Patient" Occupancy

Experiment	Colonies/cubic foot	
	Upstream	Downstream
1	< 0.001	1.235
2*	0.001	0.052
3	0.003	0.673
4	< 0.001	0.257
5*	< 0.001	0.132
All	0.001	0.436

* The same subject was involved in experiments two and five.

Table VI. Microbial Contamination on Nurses' Gloves Following Routine "Patient" Care Performance in a Laminar Flow Facility

Experiment	Col./glove (Mean of Two Nurses)		
	Following a.m. care	Other	All
3	291.6	170.7	212.2
4	144.0	71.3	97.0
5	34.1	21.4	25.6
All	158.3	87.8	112.1

Other conclusions reached for these experiments include the following:

A) The laminar flow system provides minimization of airborne contamination levels far below any conventional system.

B) Autogenous contaminants from the patient remain the most difficult source to control.

C) Contamination of personnel gloves is very difficult to prevent consistently and contaminants from this source can easily reach the patient.

D) The environment is comfortable and evokes no noticeable physiologic response from healthy individuals. However, the effect on seriously ill persons is yet to be demonstrated.

E) Logistics of providing patient care are manageable but include large staffing requirements and other difficulties also inherent in plastic tent isolation systems.

BIBLIOGRAPHY

Haynes, B.W. and M.E. Hench, 1965. Hospital isolation system for preventing cross-contamination by staphylococcal and pseudomonas organisms in burn wounds. Annals of Surgery, 162:641-649.

Schwartz, S.A., M. Colvin, K.C. Himmelsback, and E. Frei, 1965. The effect of bacterial suppression and reverse isolation on intensive cancer chemotherapy. Clinical Research, 13:48.

Whitfield, W.J., 1962. A new approach to clean room design. Sandia Corporation Publication SC-4673 (RK) Albuquerque, New Mexico.

This study was conducted under Contract PH 43-65-999 within the Special Virus Leukemia Program of the National Cancer Institute, National Institutes of Health, United States Public Health Service.

TECHNICAL ADVANCES IN PATIENT ISOLATION

R. P. Hummel, M.D., M. Maley, B.S., B. G. MacMillan, M.D.*

*Shriners Burns Institute and Department of Surgery

University of Cincinnati, Cincinnati, Ohio

Extension of gnotobiotic techniques from the animal laboratory to the clinical ward presents a number of problems and challenges. Previous publication by Levenson and Trexler, as well as Haynes and others, have reported some of the uses and advantages of reverse isolation technique with patient isolators both in the operating room and on the surgical ward. The use of these techniques in the clinical management of patients and in the clinical research of infection and immunologic response to infection has also been demonstrated. Much work has been done which indicates that clinical gnotobiotic research and application to patient care will develop rapidly and expand to new areas as the technical problem of patient care in an isolator are minimized.

Isolator design, methods of entry, air filtration systems, etc. are being developed and clinically tested. As experience increases with each patient treated, changes in the isolator design will inevitably be required. The isolator system currently in use at our institution does perform the intended task of protection and isolation of the burn patient while also requiring a minimum of extra time and work on the part of supporting medical and technical personnel.

<u>Basic Design</u>

The isolator currently in use is constructed of flexible plastic and produced by the Snyder Manufacturing Company of New Philadelphia, Ohio. This isolator is approximately 8 feet long, 6 feet wide and 6 feet high above the bed. Its rectangular shape is maintained by a frame of metal traction supports, which are mounted on a Hill-Rom electric bed and the isolator suspended by rings from four 9 foot parallel bars. With one exception all supporting bars are external to the isolator. The internal bar affords a site for the attachment of the hangers that hold storage pouches and a trapeze for patient use.

The mattress and side rails of the bed are external to the isolator and special sheets are used to protect the plastic from the bed components. A large shelf has been provided at the head of the bed which affords an area for storage of patient items and medical supplies. The plastic storage pouches mentioned above are of several designs and can be used for the storage of a large number and variety of items. The pouches are detachable and can be folded and stored if they are not needed at the time.

Several types of openings are present in the walls of the isolator. At the foot of the bed is a large oval port that measures 36" x 28" and which affords a site for entry of the patient. This port also acts as the attachment for the ultraviolet entry lock. The patient is introduced through a special plastic sleeve attached to the external lip of the port which can be detached after patient entry, thus, reducing the possibility of contamination. This sleeve is designed so that the air flow is from inside the isolator to the outside during the time the patient is being introduced or removed so as to prevent external contaminated air from entering the sterile isolator environment.

There are two twelve-inch diameter entry ports attached to the isolator. One of these is located at the foot of the isolator next to the patient entry port and serves as a site for attachment of a specially designed laundry and waste disposal bag. This port has a 90° firm, plastic elbow with two flares which are used when sterilizing the waste disposal port and attached bag. The remaining port is located at the head of the bed and serves as an attachment for the trap isolator (used primarily for food preparation and discussed below).

There are 44 flares attached singly and in clusters measuring either one or two inches in diameter. These openings provide attachments for electrical probes, extension cords, oxygen tubes, urine drainage tubes, intravenous and blood tubes, and similar equipment. Double split stoppers are used to hold the tubes and cords in place. Pass-out bags that have been sterilized prior to attachment are used when items are passed through the flares from the inside out. The combination of double stoppers, flares, and pass-out bags eliminates the possibility of contamination if procedures are properly performed.

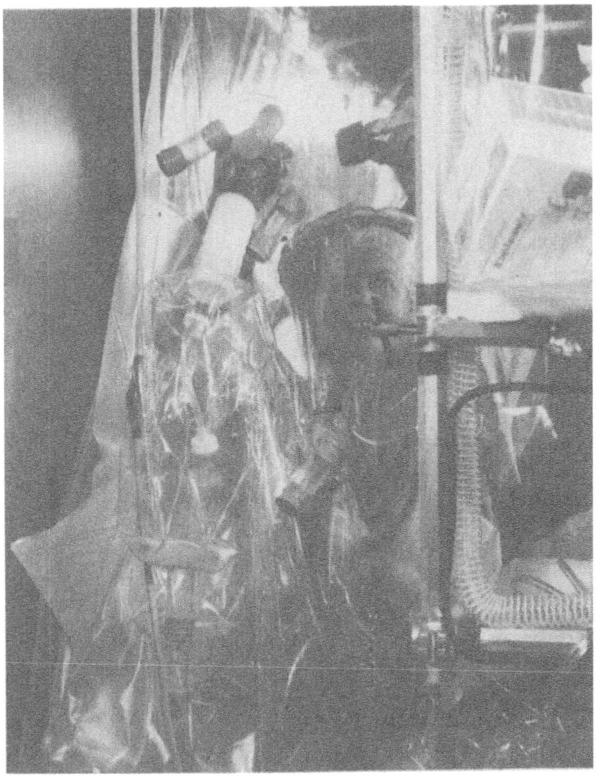

Air Filtration System

The inflow and outflow air blowers are individually controlled so that it is possible to create a positive or negative pressure within the isolation unit by regulating the volume of air produced by each blower. The air circulation system is usually regulated so that the inflow equals the outflow resulting in a neutral air circulation. The air the enters and leaves the unit first passes through a polyurathane pre-filter and then passes through a H. E. P. A. type of final filter. Since the blowers are situated on the upper portion of the isolator away from the floor, there is a reduction in the amount of dust and dirt that must be filtered.

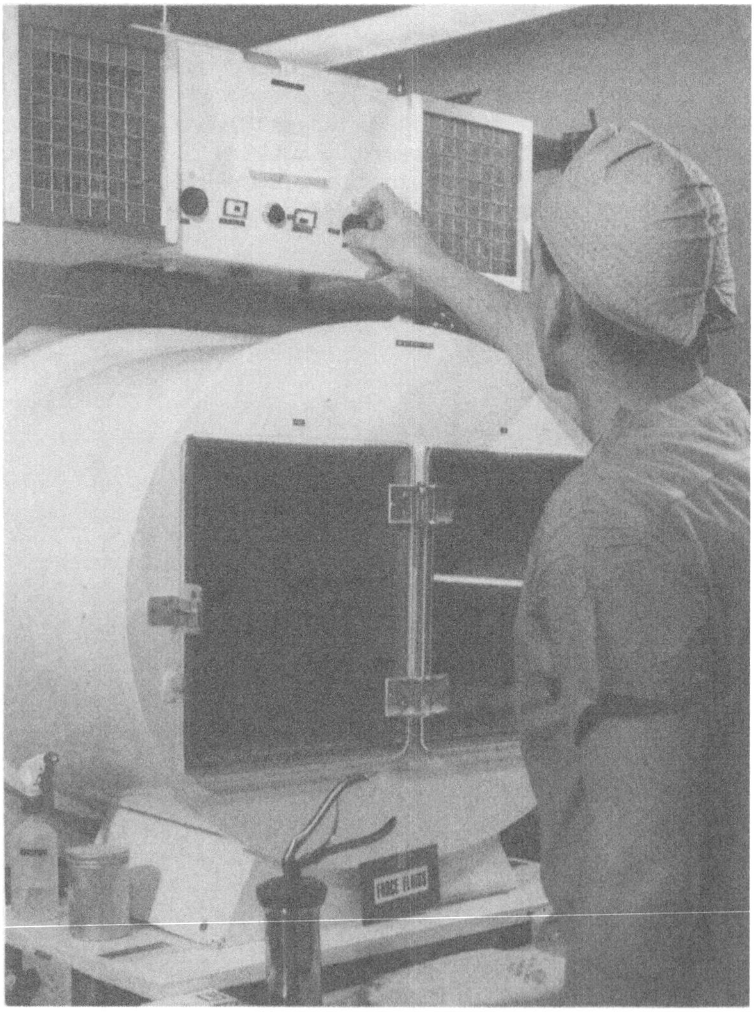

At the head of the bed there are two fiber glass filters similar to those used in animal isolators. These are used to exhaust the peracetic acid fumes after the isolator has been sterilized. These fiber glass filters also afford a safety factor since they can be used in place of the previously mentioned H. E. P. A. filters. A 12 volt direct current blower unit is available at all times and can be attached to these filters in emergency situation, if there is a power failure.

Ultraviolet Lock

The ultraviolet lock consist of two chambers each having two ultraviolet lights. One chamber is used to pass items into the isolator and the other to pass them out. The lock is the same shape as the oval entry port and fits snugly into the port when attached to the patient isolator. The cart that holds the ultraviolet lock has a small storage area and a flat surface upon which items may be placed prior to entry into the lock. The UV lock is not intended to function as a sterilizing chamber but only to act as a convenient method of introducing and removing material from the isolator. The materials passed into the lock are doubled wrapped at the time of sterilization. When they are brought to the isolator unit the outer wrap is removed. With its inner wrapper in place the sterilized material is then introduced into the input lock chamber by the nurse who is wearing a mask and using sterile gloves or sterile instruments. The lock furnishes a means of decontaminating any external air that may have entered when the package was introduced.

Personnel Entry Jackets

This patient isolator has been provided with six jackets, consisting of gloves, helmets with clear face pieces, adjustable head bands, and sufficient plastic to allow the attendant in the suit to move freely within the isolator and perform medical or technical duties almost as easily as one could at the conventional bedside. The manuverability is such that a person can contact any part of the bed from any of the individual personnel jackets. Each suit has its own individually controlled air supply. The separate blowers for each jacket have proved to be superior to one large air supply for all six jackets. The two jackets at the head of the bed are designed primarily for emergency use. The attending personnel can rapidly enter the isolator with these jackets should the need arise. The remaining four jackets are designed for less rapid entry, but more freedom of motion and comfort to the attending personnel over a longer period of time.

The Trap Isolator Unit

This separate isolator unit is a modification of the trap-type isolator used for Cesarean sections delivering gnotobiotic animals. It consists of a double doored, baffled germicidal trap;

a 36" x 24" x 30" plastic isolator hood; a twelve-inch entry port to attach the trap unit to the patient isolator; four 32" gloves for manipulating items within the isolator; and a metal frame inside the isolator to support the plastic hood when the connection between the trap unit and patient isolator is open.

This supporting frame is necessary, since there is a positive pressure inside the trap unit and a neutral pressure inside the patient isolator unit. When the connection between the two is open, the trap-hood will deflate as the positive pressure decompresses into the patient isolator. The trap unit is equiped with its own intake and exhaust filter and fans.

An electric skillet for heating and preparing food is kept inside the trap isolator and techniques for food processing have been developed so that the patient can be supplied with most any type of food he should desire or require.

Sterilization and Disinfecting the Isolator

After the unit has been checked for possible leaks with Freon leak detector, and prior to introduction of a patient into the

isolator, a 3.6% concentration of peracetic acid is used to sterilize the unit. After a 2-6 hour period has passed, the acid fumes are exhausted through a chemical scrubbing tower into the room atmosphere. Cultures are taken after the unit has been standing for 2 days. All cultures are examined aerobically and anaerobically and negative cultures are kept for 14 days before being discarded. All suspected positive cultures are subcultured and examined microscopically.

Because of the disagreeable odor of peracetic acid and its possible toxic effect, items attached and detached from the isolator, such as caps, stoppers, tubes, etc. are sprayed with a 1:600 concentration of benzyl ammonium cloride compound, which has been found to sterilize clean objects in approximately 15 minutes. This is the same solution in the germicidal bath of the trap unit. Items which cannot be steam sterilized or put through the liquid bath are packaged and gas sterilized with Ethylene Oxide prior to introduction into the unit.

Two basic types of chemical scrubbers have been used to neutralize the noxious peracetic acid fumes as they are evacuated from the isolator unit. One unit consist of a saturated solution of Sodium Carbonate distributed over coils of plastic. A blower draws air from the isolator and passes it through the plastic coils that are bathed in the carbonate solution. The second type of scrubber consist of alternate layers of activated charcoal, calcium carbonate, and pellets of calcium oxide.

SUMMARY

An attempt is being made to develop a patient isolation system that will protect and isolate the patient from his environment and secondary infection, while requiring a minimum of extra time and work on the part of the supporting medical and technical personnel. Although the unit currently in use has proved to be satisfactory, several modifications are in the process of being made which will further enhance the facility and feasibility of caring for a patient under strict isolation standards.

This type of patient isolator would seem to offer an excellent opportunity for the continued study of the role of infection and bacterial contamination on the course of the burn patient; the importance of endogenous and exogenous infection in burned and otherwise traumatized patients; as well, as controlled studies on wound healing and the immunologic response of the patient. These studies are currently underway and will be expanded as facilities permit.

BIBLIOGRAPHY

1. Haynes, B. W., and Hench, M. E.: Hospital isolation system for preventing cross-contamination by staphylococcal and pseudomonas organisms in burned wounds. Annals of Surgery 162:641-649 (October, 1965).

2. Hummel, R. P., MacMillan, B. G., Altemeier, W. A. and Hill, E. O.: Immune Response of Germ-Free and Mono-Contaminated Burned Animals. The Journal of Trauma. Vol. 6, p. 386-389, 1966.

3. Levenson, S. M., Trexler, P. C., Malm, O. J., La Conte, M., Horowitz, R. E., Moncrief, W. H. Jr.: A Plastic Isolator for Operating in a Sterile Environment. Am. Journal of Surgery 104:891-899 (December, 1962).

4. Levenson, S. M., Trexler, P. C., La Conte, M.,Pulaski, E. J.: Application of the Technology of the Germfree Laboratory to Special Problems of Patient Care. Am. Journal of Surgery 107: 710-722 (May, 1964).

5. Trexler, P. C.: Germfree Isolators. Scientific American 211:78-88 (July, 1964).

LAMINAR AIR FLOW UNIT FOR PATIENTS UNDERGOING CANCER CHEMOTHERAPY

Gerald P. Bodey, M.D.

The University of Texas M. D. Anderson Hospital and

Tumor Institute at Houston

Several investigators have demonstrated that one type of protected environment, the Life Island unit*, can be used for patients undergoing cancer chemotherapy. Several disadvantages have been noted with this type of unit. The patient is confined to the bed and a small adjacent area. Bathing and toilet facilities must be provided from outside of the unit. Patient procedures are time-consuming since they must be performed though plastic sleeves on the side of the unit. Because there is turbulent air flow, organisms may disseminate in the unit from sites of contamination on the patient.

Laminar air flow facilities have been utilized in industry for some time. Recently, medical investigators have become interested in applying these techniques to patient isolation facilities. The laminar air flow room provides a low level of environmental contamination without being as restrictive as other types of isolation units. Two horizontal laminar air flow rooms† have been installed at the M. D. Anderson Hospital and Tumor Institute for cancer patients undergoing chemotherapy.

A floor plan of the isolation unit is shown in Figure 1. The laminar air flow system consists of three wall modules, 3 feet wide by 8 feet high, located at the supply end of the room. These modules occupy the entire wall and include high efficiency particulate air filters, supply plenum and a protective-decorative screening. Each wall module is supplied by a remote blower unit suspended from the ceiling. Air circulates through the room at a velocity of 90 feet

* T. M. Matthews Research, Incorporated, Alexandria, Virginia.
† Envirco, Incorporated, Albuquerque, New Mexico.

per minute and returns through a plenum in the ceiling at the opposite end of the room. The pressure in the room is such that the flow of air is from the patient room to the access corridor to the hospital hallway. Each room has a separate air conditioning unit suspended from the ceiling in the corridor. The temperature can be regulated between 65° and 80° F with a relative humidity of 45-55%.

The sanitary facilities in the patient's room consist of a bedpan seat and a lavatory. This type of toilet facility eliminates possible contamination and prevents disposal of specimens needed for studies. The hot and cold water provided in the lavatory flow through a millipore filter. The water pipes can be purged with steam to prevent contamination of the water system. A portion of the room adjacent to the door is marked by darker tile, indicating an area where personnel may enter without dressing completely in sterile clothing. The furnishings in the room were selected to provide minimum obstruction to air flow. For example, there is no head or foot board on the bed, and the bedside tables were specially constructed without sides.

In the corridor between the rooms there is a panel which contains controls for the blowers, air conditioners, lights and steam purging system for the water supply. There is an observation window, and an intercom system. A large storage area is located in the corridor with two pass-through locks equipped with ultra-violet light. On the patient room side, the storage cabinets are constructed of perforated anodized aluminum which permits a continuous flow of sterile air across the storage area.

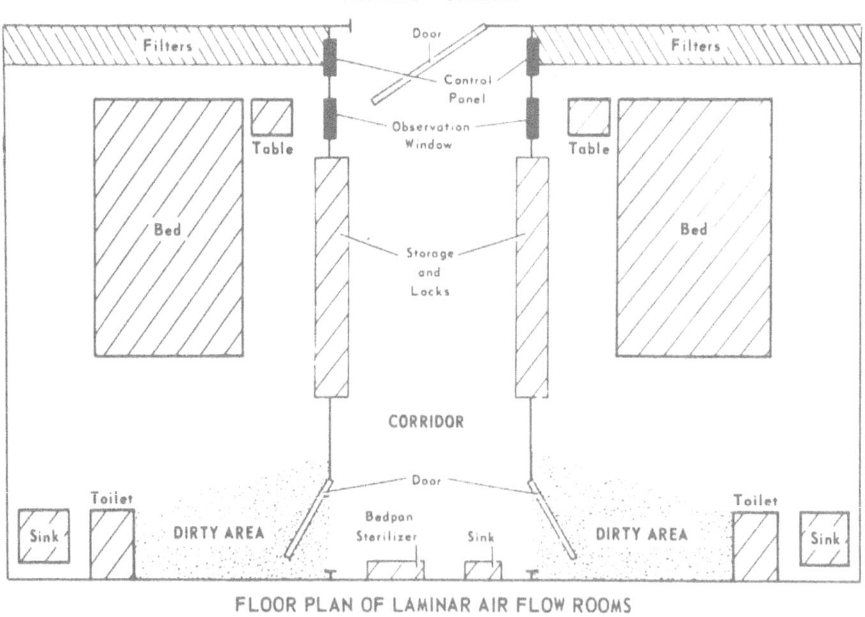

FLOOR PLAN OF LAMINAR AIR FLOW ROOMS

In order to protect the patient from contact contamination, personnel entering into the clean area dress in knee-length boots, gown, cap, mask and gloves. All items entering into the unit are specially wrapped and pre-sterilized. The patients receive specially prepared sterile food and are placed on an oral non-absorbable prophylactic antibiotic regimen. The two regimens of antibacterial agents are shown in Table 1. Regimen 1 consists of the three antibiotics, paromomycin sulfate, polymyxin B sulfate and vancomycin hydrochloride. Paromomycin is a broad-spectrum antibiotic similar to neomycin. Polymyxin B is added because of its antipseudomonal activity. Vancomycin is effective against Gram positive cocci, including Staph. aureus, and anaerotic bacteria. Regimen 2 substitutes gentamicin sulfate for paromomycin and polymyxin B. These antibiotics are administered in a flavored solution every 4 hours. Amphotericin B and nystatin are used as antifungal agents. Amphotericin B is administered in a flavored solution at a dose of 500 mg every 4 hours. Nystatin is substituted if amphotericin B fails to eliminate fungal organisms from the stool. The dose of nystatin used is 600,000 units of the suspension and 4 million units in tablets every 4 hours. A solution of neomycin, vancomycin and polymyxin B is sprayed into the nose and throat, 4 times daily. An ointment containing neomycin, vancomycin, polymyxin B and nystatin is applied to the anterior nares, external auditory canals, gums, groin and perianal area, 4 times daily. The patient bathes with hexachlorophene soap and rinses with 70% isopropyl alcohol. Cultures of the nose, ears, throat, urine, stool and skin are obtained prior to entry and once or twice weekly thereafter.

Prior to patient occupancy, the floors, walls and furnishings are scrubbed with an organic tin compound, Biomet 611‡. The unit is then fogged on two occasions with an organic tin solution. During patient occupancy the floors and furnishings are wiped once weekly with the organic tin solution and twice weekly with a solution of benzalkonium chloride.

Regimen 1	
Paromomycin sulfate	500 mg.
Polymyxin B sulfate	70 mg.
Vancomycin hydrochloride	250 mg.
Regimen 2	
Gentamicin sulfate	200 mg.
Vancomycin hydrochloride	250 mg.

Table 1. Oral Non-absorbable Antibacterial Agents.

‡ M & T Chemicals, Rahway, New Jersey.

Table 2 demonstrates the results of surface cultures within the laminar air flow room during patient occupancy. Five to seven floor sites, five to seven wall sites, one to seven furnishings and two to three sink sites were chosen for culturing. Cultures were obtained every 1 to 2 weeks just prior to cleansing with the organic tin solution. The majority of sites cultured remained sterile. Those organisms usually isolated were Bacillus sp., diphtheroids, Staph. epidermidis and Micrococcus sp. No Staph. aureus or Gram negative bacilli were recovered from these cultures.

Quantitative surface sampling has been performed during occupancy by two patients. Table 3 shows the results of one such study. Prior to the cleaning procedure, all sites were contaminated. After scrubbing and fogging no organisms were cultured from any site. During eight weeks of occupancy the commode top and the doors to the pass-through locks remained sterile. No organisms were cultured from 93% of 15 wall cultures even though no cleansing was done on these surfaces during patient occupancy. No organisms were cultured from 80% of 35 floor sites. The maximum contamination was 3350 colonies/1 square foot, found on one occasion.

Air sampling has been performed in the rooms during patient occupancy, utilizing the 2 hour Reynier slit sampler. Organisms usually isolated were Bacillus sp., Staph. epidermidis, Micrococcus sp., Streptomyces sp. and diphtheroids. No Staph. aureus or Gram negative organisms have been isolated from any air samples. The median number of colonies per 1000 cu. ft. of air in 103 samples was <8 colonies/1000 cu. ft. The highest number of organisms isolated was 200 on one occasion. In the access corridor between the patient rooms, there was a median of 130 colonies/1000 cu. ft. of air with a range of <8-706 colonies/1000 cu. ft. In the hospital rooms and hallways there was a median of 3800 colonies/1000 cu. ft of air with a range of 1600 to 8300 colonies/1000 cu. ft.

Four patients with acute leukemia have occupied the laminar air flow rooms for a median of 56 days and a total patient experience of 214+ days. Table 4 summarizes preliminary results of the cultures obtained from patient B.W. a 33 year old male who spent 56 days in the unit. By the end of the first week of his stay in the unit, most organisms isolated from cultures of the ears and nose had disappeared and the majority of subsequent cultures were sterile. Nine organisms were isolated from initial throat cultures, and although organisms were never completely eradicated from the throat, only Lactobacillus sp. and Micrococcus sp. persisted. All bacteria were eliminated from his stool cultures by the end of the first week, but Saccharomyces sp. persisted despite the daily administration of 21.6 million units of nystatin. Total body skin cultures were obtained from this patient weekly during his stay in the unit. Pre-treatment cultures revealed a median colony cound of 3250 organisms per area with a range of 150 to greater than 50,000 colonies per area. After

	(days)	Floor Sites		Wall Sites		Furnishings		Sink	
Patient	Duration	% Sterile	Total	% Sterile	Total	% Sterile	Total	% Sterile	Total
O.W.	60	69	48	89	46	87	38	96	23
S.M.	56	82	49	76	33	70	27	88	16
B.W.	56	79	34	95	22	100	5	89	9
J.P.	42+	70	20	92	12	67	3	60	5
TOTAL	214+	75	151	87	113	81	73	89	53

Table 2. Surface cultures in laminar air flow rooms. Organisms usually isolated: Bacillus, diphtheroids, Staph. epidermidis, Micrococcus.

	Floor Sites (7)		Wall Sites (3)		Locks (2)		Commode Top (1)
Time	No. Sterile	Maximum Cols./Area	No. Sterile	Maximum Cols./Area	No. Sterile	Maximum Cols./Area	Maximum Cols./Area
Dirty	0	Innumerable	0	200	0	50	300
After Scrubbing	7	0	3	0	1	5600	0
After Fogging	7	0	3	0	2	0	0
Week 1	6	800	2	100	2	0	0
Week 3	4	150	3	0	2	0	0
Week 5	7	0	3	0	2	0	0
Week 7	6	3550	3	0	2	0	0
Week 8	5	50	3	0	2	0	0

Table 3. Quantitative surface sampling - laminar air flow room #1.

Site	# of Cultures	% Sterile	Persistent Organisms
R.Ear	9	89	None
L.Ear	9	89	None
Nose	9	89	None
Throat	9	0	Lactobacillus, Micrococcus
Urine	5	100	None
Stool	8	0	Saccharomyces
Skin	94	67	Klebsiella, E.coli, Saccharomyces, Staph. epidermidis

Table 4. Cultures from Patient B.W.

two weeks of intensive skin cleansing, the median colony count was zero with a range of zero to 400 colonies per area. However, Klebsiella and E. coli persisted in low numbers on several body areas despite attempts to eradicate them. This patient achieved a complete remission and had no infectious episodes during his stay in the unit.

Table 5 summarizes the results of the cultures obtained from patient S.M., a 19 year old male who spent 56 days in the unit. The antibiotic regimen and cleansing procedures were not as effective in this patient. Staph. epidermidis and diphtheroids persisted in cultures of the ears and nose. Candida albicans persisted in the throat, stool and skin despite the administration of amphotericin B, and later, nystatin. A β hemolytic streptococcus was present in the throat and persisted despite the use of penicillin V and cloxacillin. This organism retained sensitivity to both of these agents. The patient achieved remission of his acute leukemia and had no infectious episodes while in the unit, despite the great amount of persistent contamination.

The results of cultures from patient O.W., a 52 year old male who spent 60 days in the unit, are shown in Table 6. No organisms

Site	# of Cultures	% Sterile	Persistent Organisms
R.Ear	15	7	Staph. epid., anaerobic diphtheroid
L.Ear	15	7	Staph. epid., anaerobic diphtheroid
Nose	16	0	Staph. epid.
Throat	16	0	β Strep., C. Albicans anaerobic Micrococcus
Urine	6	100	None
Stool	16	38	C. Albicans
Skin	123	13	Staph. epid., diphtheroid, C. albicans, anaerobic Lactobacillus, anaerobic Micrococcus

Table 5. Cultures from Patient S.M.

Sites	# of Cultures	% Sterile	Persistent Organisms
R.Ear	16	94	None
L.Ear	16	88	None
Nose	16	31	anaerobic Micrococcus
Throat	16	0	Lactobacillus, C. albicans
Urine	7	100	None
Stool	16	63	C. albicans
Skin	136	80	Staph. epid, diphtheroid, C. albicans, anaerobic Lactobacillus, anaerobic Micrococcus

Table 6. Cultures from Patient O.W.

were cultured from the ears after institution of the antibiotic regimen. An anaerobic Micrococcus sp. persisted in the nose and a Lactobacillus sp. in the throat. Although it remained sensitive to nystatin by in vitro testing, Candida albicans persisted in cultures of the throat and stool. Total body skin cultures were obtained at weekly intervals from this patient during his stay in the unit. There was a median of 25,000 colonies per area in the pretreatment skin cultures with a range of 1050 to greater than 50,000 colonies. After two weeks in the unit, the median number of colonies per area was zero with a maximum of 150 colonies isolated from the entire body surface. From the third week onward there were never more than three body sites from which organisms could be cultures. Diphtheroids were recovered persistently from the face and neck areas. This patient had already received extensive chemotherapy and failed to respond to cytosine arabinoside while in the laminar air flow room. He developed pneumonia three days after entering the unit and had a recurrence one month later.

Table 7 summarizes the preliminary results of cultures obtained from patient J.P., a 50 year old man who has spent more than 42 days in the unit. Staph. epidermidis persists in the ears, nose and skin. Candida sp. and Lactobacillus continue to be cultured from the throat. No organisms have been cultured from the stool since the second week of oral antibiotics, although budding yeasts have been observed on stool smears. This patient is not responding to antileukemic therapy. His course has been complicated by a jaw infection and septicemia.

In general, the results with the laminar air flow rooms have been similar to those with the Life Island unit. The antibiotic regimen eliminated 84% of the 21 strains of organisms originally isolated from ear cultures. Of the 12 strains of organisms originally isolated from nose cultures, 88% were eliminated and 81% of the 31 strains of organisms originally isolated from throat cultures were eliminated. One of the four patients had a consistently sterile

Site	# of Cultures	% Sterile	Persistent Organisms
R.Ear	5	80	None
L.Ear	5	0	Staph. epid.
Nose	5	0	Staph. epid.
Throat	5	0	Lactobacillus, Candida sp.
Urine	5	80	None
Stool	5	80	None
Skin	10	50	Staph. epid.

Table 7. Cultures from Patient J.P.

stool. All of the 39 strains of aerobic organisms, all of the 14 strains of anaerobic organisms and half of the 8 strains of fungi isolated from pretreatment stools were eradicated by the antibiotic regimen. Although the population of organisms constituting the skin flora was greatly reduced, some organisms persisted. Potential pathogens, such as Candida albicans, E. coli and Klebsiella were recovered consistently from two patients.

Despite the major reduction in the exogenous and endogenous microbial flora of these patients, infections still occurred. The initial infectious episodes occurred during the first week after entry into the unit, before the prophylactic antibiotics were maximally effective. Subsequent episodes of infection in these patients probably arose because organisms persisted at the original site of infection.

In conclusion, laminar air flow units provide a protected environment with adequate patient facilities and minimal confinement. Patients undergoing intensive cancer chemotherapy can remain in these units for extended periods of time. This type of program utilizing a laminar air flow room and prophylactic antibiotics results in a major reduction in the microbial flora of the patient and his environment. Further studies are necessary to develop the full potential of this program.

PSYCHOLOGIC ADAPTATION OF PATIENTS DURING TREATMENT OF ACUTE LEUKEMIA IN LIFE ISLAND ISOLATOR

Leslie Fine M.D., Morton Wachspress M.D., David N. Graubert M.D.
John Edmonson M.D., Howard Eisman Ph.D.

On staff of Coney Island Hospital affiliated with Maimonides
Medical Center, Brooklyn, New York

The patient treated for leukemia in the Life Island is subjected to the possibility of a complex set of severe stresses over a prolonged period of time. It is extraordinarily difficult to separate the stress of the leukemic process alone as distinct from the reaction to the unique treatment situation. These stresses can be catagorized into two major classes:

A. Factors influencing physico-chemical equilibrium.

B. Factors influencing psycho-social equilibrium.

Under Class A can be listed the following factors:

1. The toxic-metabolic effect of the disease process itself with its interference in normal brain function.

2. Leukemic infiltration of brain tissue.

3. Brain hemorrhages.

4. The effect of anti-neoplastic cytotoxic drugs.

5. The effect of severe infections.

6. The effect of large doses of antibiotic drugs.

Under Class B can be listed the following factors which could arise and provide additional burdens in the patient's adaptive struggle:

1. The patient's fantasies about his illness, e.g., as a punishment, as a loss of love, as a means of expiating guilt feelings, as a means of gratifying unfulfilled dependency needs, as a threat to his feeling of potency and power.

2. The patient's fantasies about the treatment itself, e.g., as a separation from love objects, as an assault on one's body, as an invasion of privacy, as the gratification of exhibitionistic and dependency wishes, as the loss of autonomy and control of one's destiny, or as a potential source of contamination and shame.

3. The patient's fantasies of death as a surrender and reunion with loved objects or as isolation from and loss of loved objects and prized functions.

4. In addition to the patient's fantasies of illness, treatment and death, as outlined above, there are very real losses which he is forced to undergo such as the severe curtailment of opportunities for the discharge of aggressive drives in motor activities, the curtailment of opportunities for gratification of sexual, social, intellectual and vocational needs, a very real loss of autonomy in which the patient is subjected to the rigid and demanding schedules necessitated by the complex treatment and observational procedures in a Life Island unit. There is also a potential danger from loss of adequate opportunity for sleeping and dreaming which is necessary for optimal psychic functioning. This can result from too frequent interruption of sleep patterns from pain or procedural activities. There is also a potential for diminution of sensory stimulation in units which are poorly designed and planned. The patient's adaptation to the possible stresses mentioned will be the resultant of his premorbid personality make-up with its built-in and acquired propensities for adaptive or maladaptive patterns of reacting as well as the particular combination of stresses that each individual is actually subjected to during his illness and treatment. The psychological changes which could occur can be classified in the following manner:

A. Changes in thinking:

1. perceptual disturbances

2. impairment of cognitive functioning such as orientation, memory, retention and recall, ability to sustain a mental or motor task and ability to utilize abstract concepts.

3. alterations in consciousness with

a. reduction in level of awareness

b. fluctuating attention

c. inattention

d. somnolence

e. coma

4. impairment in reality testing with the formation of delusions, illusions, hallucinations, changes in body image, etc.

B. Changes in affect (or altered ego states):

1. states of depression and/or withdrawal

2. hypomanic, manic and euphoric states

3. episodes of depersonalization

4. emergence of anxiety states

C. Behavioral changes:

1. irritability

2. restlessness and agitation

3. immobility

D. Institution of new psychological defense mechanisms and/or strengthening of old patterns:

1. Emergence of obsessive compulsive thinking and behavior with intensification of self-restraint, formalized behavior, obstinacy, orderliness, hesitation, doubting and a need for intellectual understanding.

2. Use of projective mechanisms leading to suspiciousness and withdrawal from staff and visitors, oversensitivity to slights, hurts and criticisms.

3. The emergence of infantile patterns of relating with demandingness, impulsivity, lowered frustration tolerance, need for constant attention, reassurance and affection.

4. Use of the mechanisms of denial either "in the

service of the ego" to ward off the awareness of painful events that would bring on excessive anxiety or denial that is maladaptive in that it prevents the patient from cooperating with his treatment.

5. The emergence of needs to be noticed and admired with teasing, seductiveness, jealousy, exaggerated manliness and courage.

6. The emergence of an increased "need to suffer" and feel martyred, to submit, to complain, which is comparable to the childlike expectation "you have to love me because I suffer so terribly."

7. The emergence of feelings of superiority, arrogance, smugness and vanity to compensate for feelings of anxiety, doubt and insecurity about the patient's self-image of perfection and invulnerability.

8. The emergence of remoteness, aloofness, reserve and lack of involvement with activities around him.

It is clear from the above that there is potentially a wide variety of psychological reactions of patients in a Life Island undergoing treatment for leukemia. The following case vignettes of patients studied in the Life Island unit at Coney Island Hospital will serve to illustrate some of the problems that can arise.

CASE NUMBER I

Mr. M. M., a 53 year old accountant, was put into the Life Island for treatment of acute myelogenous leukemia. Psychiatric examination and psychological testing prior to his entering the Life Island revealed him to be a fairly stable individual who tended to be passive and to suppress his aggressiveness. He seemed to handle anxiety by obsessive compulsive behavior.

During the first two weeks in the Life Island his psychiatric status was unremarkable except for occasional periods of irritability and preoccupation with physical symptoms, such as weakness, abdominal pains and diarrhea. During the following two weeks, when he had repeated severe infections of the kidneys and lungs which required vigorous treatment, he became more irritable and moderately depressed. His initial cheerfulness and denial of illness virtually disappeared during this time. During the fifth week he became withdrawn and complained of peculiar perceptions of his body - "like my legs are separated from the rest of my body or are unnatural extensions of my body." These were considered to be feelings of depersonalization and not somatic delusions. During the sixth week he developed urinary shut-down with uremia and recurrent infections. From then on he was more withdrawn, was reported to be quite confused and disoriented and at times comatose. This condition persisted until he expired a few weeks later.

CASE NUMBER TWO

Mr. H. E., a twenty-five year old junior executive at an import-export company, was treated for severe Hodgkins Disease in the Life Island unit. His mother died when he was fourteen of Myelogenous Leukemia. No one bothered to talk to him about his mother's condition, reason for hospitalization, or prognosis, although he had discovered his mother's diagnosis at age twelve and had harbored this knowledge without discussing it or sharing it with anyone. After his mother's death, he looked after himself with little supervision from his family and became something of a gambler who liked the excitement of crowds and the pleasure of competing and winning. He maintained a facade of pseudo-independence and insisted on doing everything for himself. In his treatment program in the hospital, he insisted that no relatives or friends be given any medical information about him - that he would do this himself. He made a bet with himself that he would get well and walk out of the hospital. Mr. H. E. was encouraged to do as much as possible in assisting in the details of his daily care. He was fully informed as to the nature of his diagnosis and treatment. He had an unbroken string of visitors every day of his hospital stay and viewed this as though it were a winning streak. He watched T.V. a great deal, especially sports events. Throughout his stay in the Life Island he remained alert, cooperative and generally cheerful. He won his bet and walked out of the hospital - greatly improved.

CASE NUMBER THREE

Mrs. A. W., a twenty-nine year old housewife, mother of two children, was suspicious, evasive, and uncommunicative prior to admission in the Life Island in an attempt to control her depression and feelings of hopelessness. Before her illness, she saw herself as a woman of great competence and self-determination who defied family tradition by running away from home at the age of eighteen and marrying a Negro man. She managed to have a reasonably successful life and re-established bonds with her family. At the time of admission, she felt that she had been left to deal with the fatal nature of her illness alone, since many of her questions had been met with evasiveness and denial. Initially, in addition to being depressed, she showed some tendency to perseverate with fluctuating levels of minimal impairment of memory, indicating the presence of an organic brain syndrome. During her first few days in the Life Island on anti-neoplastic cytotoxic drugs, she experienced painful abdominal distension which was accompanied by multiple complaints, crying and frequent demands for attention. This regression was supported and tolerated by the nursing staff who were encouraged to spend more time with her. Her general attitude at the present time is that of an easily hurt child with frequent, unfounded optimistic plans for the future. The picture is clouded by mood swings related to the administration of steroid medication.

CONCLUSIONS

It is important to understand the complexities of psychological adaptation of patients undergoing treatment for acute leukemia or related illnesses in a Life Island unit to minimize the amount of stress and discomfort he has to undergo, to increase his cooperation in the treatment procedures and to bolster his psycho-physiological mechanisms of coping with stress.

The following is an outline of how one might attempt to prevent undue stress and bolster defenses for patients in Life Island units:

1. Have the unit painted a bright, cheerful color, e.g., yellow, with posters or pictures on the walls rather than have the drab, sombre decor that one finds so often in hospitals.

2. It is advantageous if the patient can look out of the window or on to the ward when he wants to in order to "see life going on" and feel less isolated from the world.

3. A remote control television set and a telephone which the patient can use also contribute to his feeling of contact with the external world as well as providing him with an opportunity of controlling, to some extent, the amount of stimulation he receives.

4. Liberal visiting privileges for family and friends of the patient will help to maintain the necessary object ties and contacts with the outside world.

5. Permit the patient to have his watch, if there is no wall clock handy, to help him in his orientation in time and also to assist him in anticipating and preparing for procedures, medication, etc.

6. Permit the patient to use his glasses, hearing aids, false teeth, etc., so that he is not burdened by undue loss of visual, auditory or gustatory stimuli.

7. Many patients in Life Island units can benefit from some sort of planned recreational or physical activity program, especially those patients who badly need some outlet for motor activity or who react badly to long periods of enforced passivity. These activities could be in the form of occupational therapy, projects, exercise regimens, etc.

8. Early awareness of acute confusional states is important so that supportive measures can be promptly instituted, e.g., changing drug regimens, providing adequate sedation or tranquilization, leaving night lights on to assist patient in orienting himself, giving the patient repeated verbal clues as to where he is, what is happening, who people are, etc., to assist in reducing his confusion. One should avoid overtaxing the patient intellectually during confusional states, to minimize the anxiety reactions which can occur when patients are confronted with their inadequate functioning.

9. Relieve distressing symptoms such as excessive anxiety or depression, insomnia, undue pain, etc., so that the patient is reasonably comfortable, but do not over-medicate the patient so that he is somnolent or extremely apathetic. Often patients are over-medicated by treating physicians who find it difficult and painful to tolerate personal contact with dying patients.

10. In dependent patients, restrictions are better tolerated if there is no accompanying loss of interest on the part of the physician or if some kind of concession or gift is given as compensation, e.g., a desired change in the diet or helping the patient's family to visit by providing transportation, etc.

11. Compulsive individuals are helped by an intellectual understanding of their illness and treatment. They should react well to the orderliness and meticulous attention to detail that is necessary in a Life Island unit. Such patients should be encouraged to take part in details of their medical care.

12. Dramatizing, histrionic (hysterical) patients need the concrete appreciation of their attractiveness or courage and physicians should not be too reserved, but should have a measure of calmness and firmness. Reassuring explanations will often help distinguish reality from alarming fantasies with these patients. It is also helpful to allow such patients opportunities to discuss their fears repeatedly to release pent-up feelings.

13. With the guarded or suspicious patient, it is important to avoid getting excessively close or involved or to avoid situations where the patient may feel he is being forced or manipulated. Arguing or coercion only help to make such patients suspicious and querulous.

14. Patients with feelings of superiority, i.e., narcissistic

patients, may tend to feel more comfortable if their physician can acknowledge some aspects of the patient's superiority as an individual.

15. With the aloof or schizoid individual his "unsociability" has to be understood. One should make as few demands as possible for personal involvement with others and yet such a patient should not be permitted to withdraw altogether. This can be done by maintaining a considerate interest in him quietly and reassuringly without requesting a reciprocal effort on his part.

16. With long suffering or self-sacrificing patients, full accord should be given to such patient's pain and sacrifices. The physician should express his appreciation of the difficulties of illness as they are experienced by the patient. Such patients are better able to cooperate in their treatment out of a readiness to add to their burdens than for the personal relief that good health brings.

With time and a greater opportunity to study the effect of treatment in a Life Island unit on a multitude of different kinds of patients from different treatment centers, it is the authors' hope that the kind of information will be obtained that will permit the structuring and prescribing of an optimal therapeutic milieu for each patient having to undergo such intensive treatment. It is also hoped that information about the specific effect of such treatment, apart from the effect of hospitalization in general or the patient's reaction to the leukemic process, will be obtained.

FEVER AND INFECTION IN PATIENTS TREATED WITH INTENSIVE ANTINEOPLASTIC CHEMOTHERAPY IN AN ISOLATOR

John H. Edmonson, Wilfrido M. Sy,
Stephen J. Millian and Stanley L. Lee

The Coney Island Hospital affiliated with
Maimonides Medical Center, Brooklyn, N. Y.

During the first six months of 1968 three patients have been treated in an isolator.* This is a preliminary report of these initial experiences in the protection of patients receiving intensive cytotoxic drug regimens for acute leukemia and other neoplastic diseases. Two of the first three patients had drug resistant acute myelocytic leukemia and the third was suffering from Hodgkin's disease with bone marrow failure.

Prior to entering the isolator each patient received skin, bowel and pharyngeal prophylaxis and the female patient received vaginal cleansing. During this three to six day period the patients became accustomed to sterile food and used sterile linens and pajamas. After a final cleansing shower each patient was transferred to the isolator where the final phase of anti-microbial prophylaxis was carried out. Figure 1 outlines the prophylactic routine.

Case Summaries

M.M., a 53 year old accountant (Figure 2) with acute myelocytic leukemia had failed to achieve remission in a previous trial of daunomycin despite the induction of severe leukopenia. He had experienced a life threatening infection during that course and had recovered from a second infection (of the urinary tract) two weeks prior to entering the isolator. During 47 days in the isolator he received two courses of antileukemic treatment and intensive supportive care in addition to the microbial prophylactic regimen. Despite these efforts urinary tract infection, septicemia and pneumonitis developed and the urinary tract infection recurred soon after

*Life Island®, Matthews Research Incorporated, Alexandria, Va.

ANTIMICROBIAL PROPHYLACTIC REGIMEN

SITE	PRIMARY PROCEDURE	SECONDARY PROCEDURE
Gastrointestinal	Vancomycin 250mg. Paromomycin 500mg. Polymixin B 50 mg. Amphotericin B 100mg. } swallowed every 4 - 6 hrs.	Gentamycin, etc. for resistant bacteria; Nystatin for resistant Candida.
Nasal, Oral, Pharyngeal	Polymixin B Neomycin Vancomycin Nystatin } sprayed into nose and throat every 4 - 6 hrs.	being developed
Skin	Phisohex bath daily	Betadine Surgical Scrub bath N-P-B or Gentamycin ointment for resistant bacteria; Nystatin ointment for resistant Candida.
Nose and Ears	Neomycin, Polymixin, Bacitracin ointment every 4 - 6 hrs.	Gentamycin ointment for resistant bacteria; Nystatin for Candida.
Vagina	Betadine douche daily until clean, then twice weekly.	

Figure 1. - Prophylactic regimen in use January-June 1968.

systemic antibiotics were stopped. Various Gram negative bacterial species were recovered from the urine; E. coli bacteremia was demonstrated; and Serratia marcescens was found in the sputum during the episode of bronchopneumonia. The development of Citrobacter infection (E. freundii) at a time when the peripheral blood leucocyte count was 250 per mm^3 led to overwhelming sepsis, shock, and anuria. Despite peritoneal dialysis and complete control of the infection the patient developed uremic complications (including gastrointestinal hemorrhage) and died two weeks later.

H.E., a 25 year old exporter (Figure 3) with far advanced (Stage IV) Hodgkin's disease had ceased to respond to vinblastine and chlorambucil and was not a good candidate for further myelosuppressive treatment because of bone marrow failure. The initial resolution of his febrile course at the onset of intensive cytotoxic chemotherapy was maintained throughout the period of isolation until he began to take non-sterile food just before leaving the isolator. The only clinical infection observed during his period of isolation

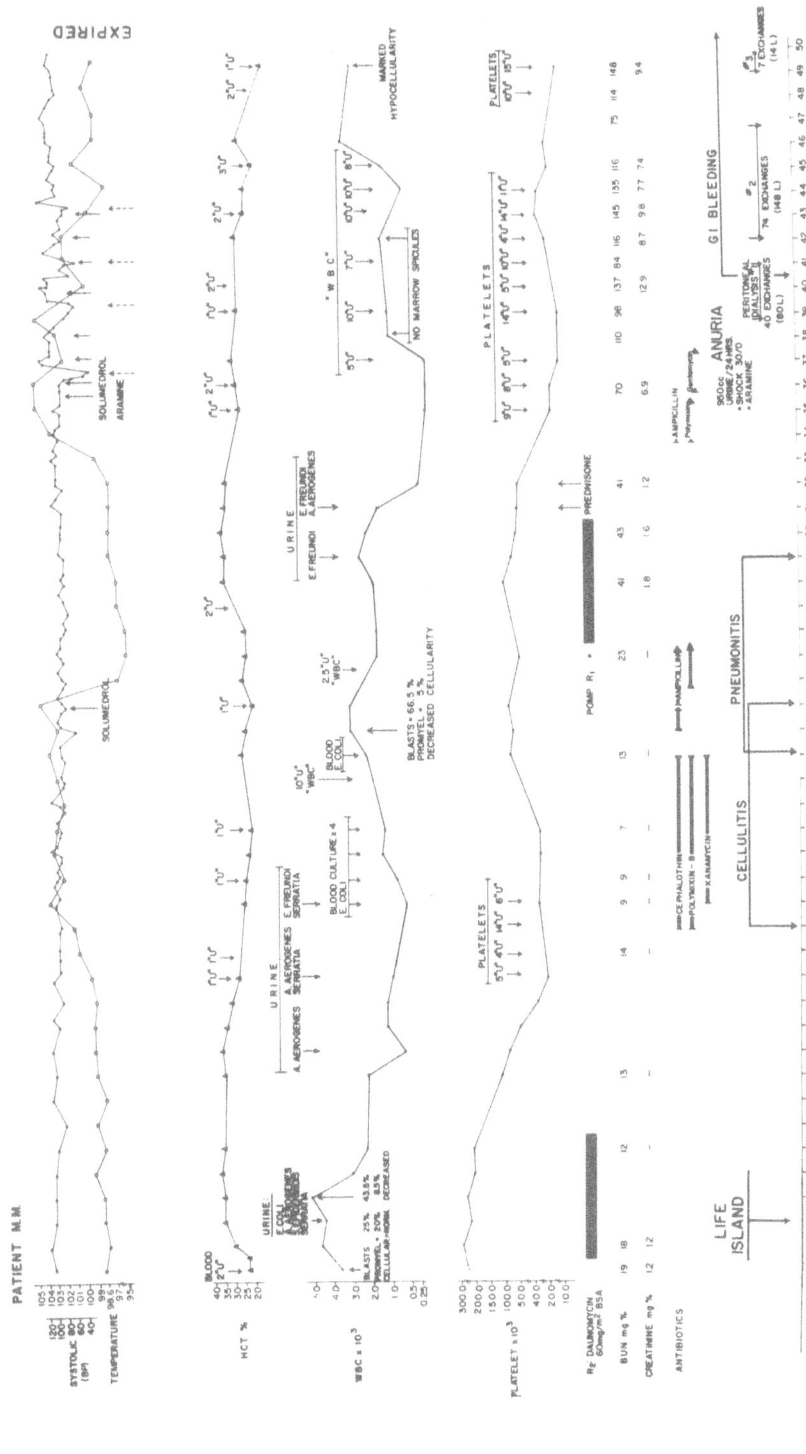

Figure 2. - Clinical course and treatment of patient M.M.
POMP R_1 = POMP regimen #1 (Prednisone, Oncovin, Methotrexate, and Purinethol).

was a recurrence of tinea cruris which promptly responded to local treatment with nystatin ointment. Candida albicans was isolated from these lesions.

A.W., a 29 year old housewife with acute myelocytic leukemia (Figure 4) had received several previous courses of antileukemic cytotoxic chemotherapy including cytosine arabinoside, daunomycin, cytosine arabinoside in combination with 1, 3-bis-2-chloroethyl-1-nitrosourea and hydroxyurea combined with prednisone. She had achieved transient remissions following treatment with cytosine arabinoside, but had not been in remission for several months prior to entering the isolator. At the time of entry she had a high circulating leukemic cell concentration and evidence of leukemic infiltrates in several soft tissue sites.

During her treatment with cytotoxic drugs in the isolator axillary hydradenitis recurred and she developed septicemia due to Aerobacter aeorgenes, followed by pneumonitis. She succumbed to pneumonitis despite vigorous systemic antibacterial chemotherapy specifically effective against the organism by in vitro testing. During the last week of life when mouth cultures were sterile she had lesions of the lips and tongue which resembled the lesions of Herpes simplex.

Microbial Screening Data

Stool cultures for bacteria and fungi rapidly became negative following onset of the antimicrobial regimen and were generally maintained sterile without difficulty. Anterior nasal cultures and ear cultures also were almost always sterile, except for a resistant Aerobacter aerogenes in the case of patient A.W. Bacterial flora of the skin were easily cleared from dry open areas, but persisted for three or four weeks in the moist protected areas (inguinal, gluteal, submammary, and to a lesser extent axillary). Mouth and nasopharynx were also resistant to cleaning, and in the case of H.E. a Lactobacillus strain persisted throughout the period of isolation. The two patients in whom oral and nasopharyngeal cultures became sterile also received systemic antibiotic therapy prior to this development.

Cultures of air from the isolator for a two hour period (Reyneirs slit sampler) during the period of the day of greatest nursing activity indicated a close parallel between the type and number of micro-organisms found in the isolator and those found in the patient occupying it. No resistant Gram positive cocci were encountered in any of the patients, however all three had strains of relatively resistant Gram negative bacilli and two patients had Candida albicans which was difficult to eliminate. No persistent anerobic organisms have been found. Patients M.M. and A.W. had six and four day periods respectively during which all patient and air cultures

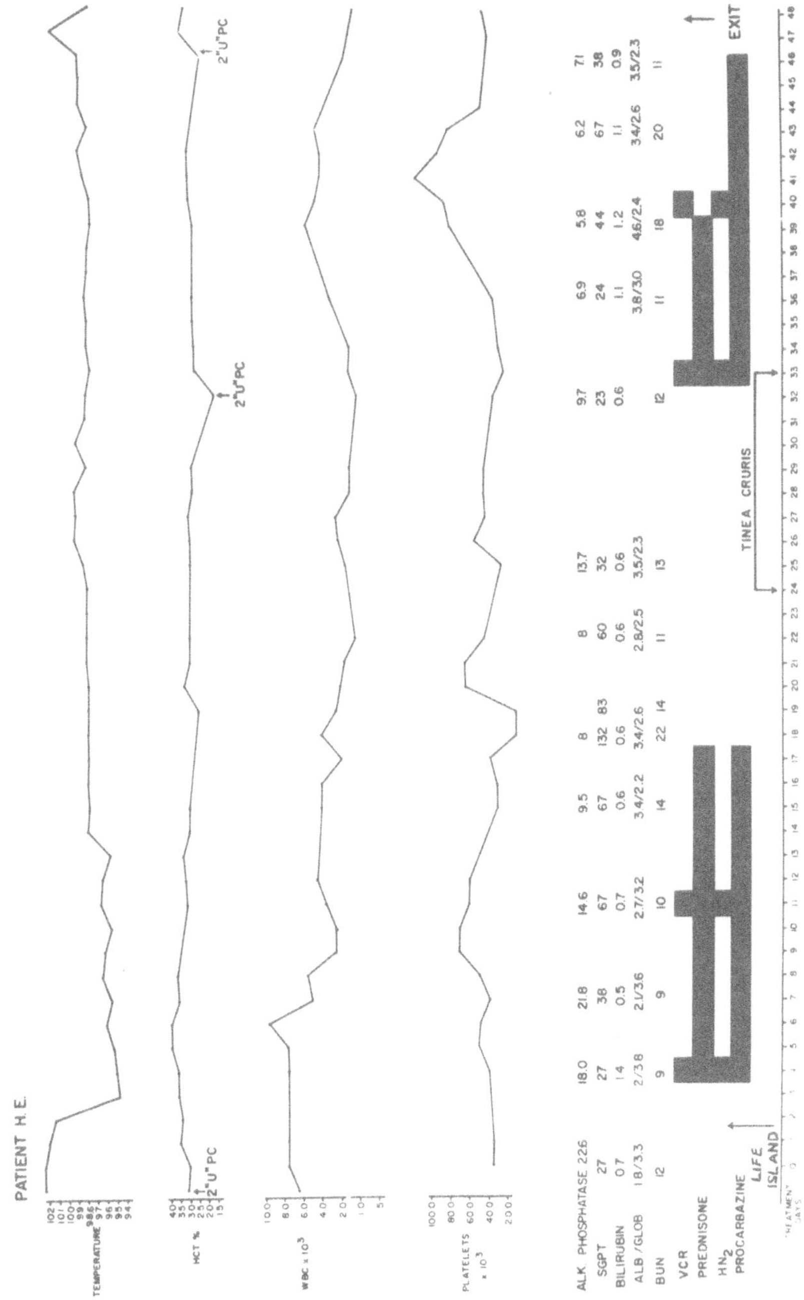

Figure 3. - Clinical course and treatment of patient H.E. VCR = Vincristine, HN_2 = Nitrogen Mustard.

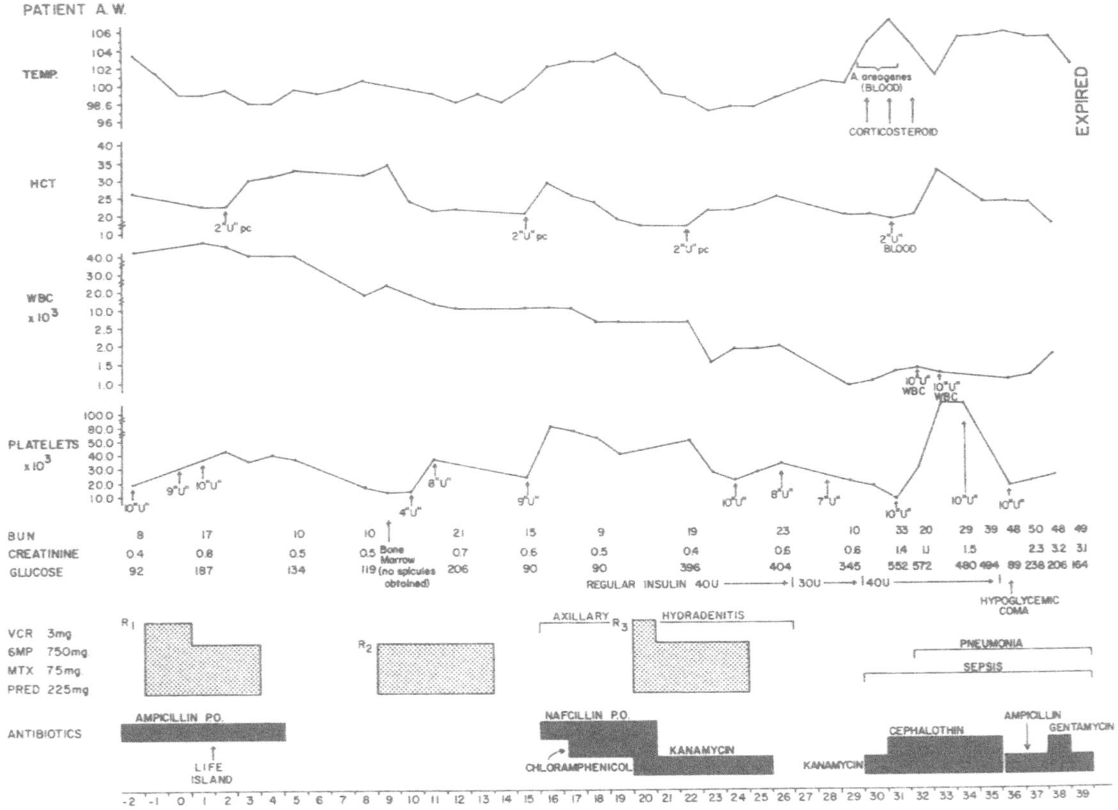

Figure 4. - Clinical course and treatment of patient A.W.
VCR = Vincristine, 6MP = 6-Mercaptopurine, MTX - Methotrexate, Pred = Prednisone.

for bacteria and fungi were negative, and even longer periods with sterile air samples have been recorded. While viruses and mycoplasmas have been recovered from these patients identification of these organisms is still in progress.

Discussion

Despite total isolation and nearly complete microbial decontamination of patients with acute myelocytic leukemia serious bacterial infection may develop. Such infections are difficult to control in these patients who are receiving intensive cytotoxic drug therapy, and as indicated by our experiences with patients M.M. and A.W. organisms ordinarily of low virulence may produce a fatal outcome despite intensive and specific antibiotic therapy.

In the case of M.M. the use of nephrotoxic antibiotics (especially polymixin B) might have aggravated the renal damage produced by the Gram negative (apparently Citrobacter) sepsis and shock. Unfortunately most of the antibiotic drugs of clinical value in the treatment of Gram negative bacillary infections are nephrotoxic in varying degree. Important exceptions are certain penicillin derivatives such as ampicillin and carbenecillin.

Although all of the significant infections observed in these three patients arose apparently from endogenous residual flora of the patients themselves other sources of contamination have been looked for. These are summarized in Figure 5. Contamination of the isolator atmosphere from outside has not been detected even though several leaks in the plastic enclosure have developed. Possibly the slight positive pressure maintained within the isolator prevented major inward air flow through these and the portals for I.V. tubing. Our observations tend also to confirm the effectiveness of the ultraviolet pass-through locks. No difficulty has been encountered in sterilizing equipment and supplies or in sterilizing various personal items for the patients. Vermin conceivably could enter sterile packs during storage after sterilization, but this has not been observed. Citrobacter and a few other bacterial species have at times been isolated in small numbers on screening fresh blood products, however none of these products or other infused or injected materials have been definitely identified as the sources of infection in these patients. Steam sterilization of food products has been generally effective. No infection has yet been definitely traced to food or oral medication sources despite the use of oral antibiotics and other oral medications directly from the manufacturer's bottle without additional sterilization.

The readaptation to non-sterile food and the reestablishment of normal flora has been observed in one of our cases. A detailed

POTENTIAL SOURCES OF INFECTION IN ISOLATOR

SOURCE	OBSERVATIONS	CONTAMINATION PROBABILITY
 ISOLATOR ATMOSPHERE	1. Air in unit sterile before patient entry. 2. Air has again become sterile or nearly so when patient flora is depleted. 3. Air flora parallels patient flora in kind and number generally.	—
 EQUIPMENT AND SUPPLIES	1. Time tested sterilization procedures used (steam or ethylene oxide). 2. No contamination has been identified. 3. Vermin (cockroaches) may enter cabinets.	±
 BLOOD PRODUCTS AND OTHER INJECTED MATERIALS	1. Citrobacter (E. freundi) isolated from several fresh platelet concentrates. 2. Alpha strep., B. megaterium, and Mima polymorpha also found in small numbers on screening fresh blood products.	±
 FOOD AND ORAL MEDICATION	1. No new organism in isolator has been definitely traced to food sources. 2. Oral flora has sometimes changed abruptly. 3. Stool sterility has not been difficult to maintain.	++
 RESIDUAL FLORA OF THE PATIENT	1. Total sterility difficult to attain in a previously contaminated individual. 2. Normal human flora is capable of producing infection and fever in the severly leukopenic host.	++++

Figure 5. - Evaluation of sources of infection in the isolator.

temperature chart during the process of recontamination of patient H.E. is illustrated in Figure 6. He had been afebrile during the six weeks preceding the cessation of antimicrobial prophylaxis, and except for Lactobacillus casei in his mouth and nasopharynx cultures were essentially sterile. Following blood transfusions on day 45 the oral temperature reached 99.6°F. Antibiotics were stopped at 2 A.M. on day 46. Non-sterile food (standard breakfast, lunch and dinner menus) was eaten on day 46 for the first time in more than seven weeks. Oral temperature reached 100°F by 4 P.M. and was 101.6° by 10 P.M. The patient was removed from the isolator on day 47 and continued to eat a regular non-sterile diet, and he was managed without any sterile precautions. His temperature reached 100° that night, but returned to normal on day 48 and rose only to 99.6° on the evening of day 48. This febrile episode occurred at a time when the patient was moderately leucopenic, but the fever quickly resolved during a period when the leucopenia was progressively worsening and did not recur despite the persistence of total leukocyte counts of under 1000 per mm^3 for more than a week.

Management of medical problems in this isolator has been more time consuming in some cases than might be expected in an open bed, however no other serious limitations to the ability of the physician

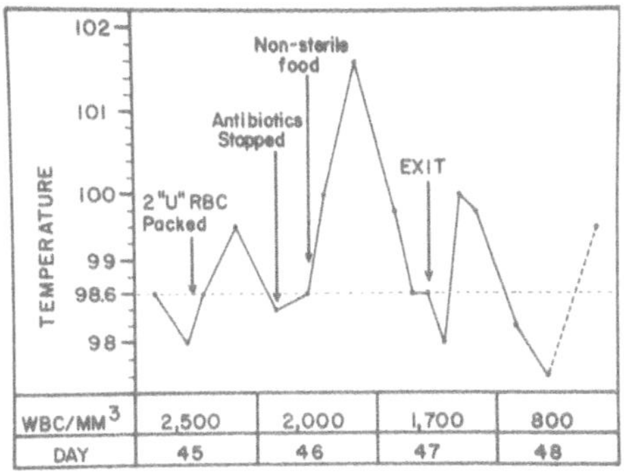

Figure 6. - Temperature elevation of patient H.E. following the ingestion of non-sterile food prior to exit from the isolator.

to treat medical problems are presented by the isolator. The effective management of hyperpyrexia, shock, renal failure requiring peritoneal dialysis, ileus requiring Levin tube suction has been accomplished, and intensive intravenous requirements for blood, platelets, drugs, and fluids have been satisfied. Venous cutdown has not been necessary despite the very poor condition of the veins in these patients. Three separate peritoneal dialysès were performed within a ten day period in one patient without difficulty, and 242 liters of dialysis solution were exchanged without contamination of the peritoneal space.

Summary

Some experiences with the treatment of patients in a Life Island isolator are reviewed. Two patients with acute myelocytic leukemia and one patient with Hodgkin's disease have been treated intensively with cytotoxic drugs. All three have developed clinical infection at some time during the period of treatment despite isolation from exogenous sources of contamination and great reductions of endogenous microbial flora. Both patients with acute myelocytic leukemia have died directly or indirectly as a result of Gram negative bacillary infections due apparently to residual endogenous flora. Other possible sources of infection have been reviewed. Most of the infections encountered in these three patients were recurrent at the sites of infection observed weeks or months previously. No serious difficulties have been encountered in the medical management of seriously ill patients in this isolator.

Aided by grants (CA10047 and CA05923)
from the National Cancer Institute.

GNOTOBIOTIC MAN

RONALD D. BARNES

LECTURER IN HAEMATOLOGY

INSTITUTE OF CHILD HEALTH, LONDON

In recent years the "Swiss" form of agammaglobulinaemia perhaps better known as the "combined immunity deficiency state" has become well defined as a congenital disease presenting in early childhood with recurrent and persistent infections (1). The familial nature of this disease is apparent from the case histories and some workers have proposed a genetic recessive inheritance (2). The prime characteristic of this disease is the deficiency of specific cellular and humoral immunity and in spite of attempts at treatment a fatal outcome has been to date inevitable. It is possible that treatment which has in certain cases included foetal tissue grafts has failed due to the poor clinical state of the infant at the time of treatment. Alternatively, therapeutic failure might be due to the treatment being delayed and given in a period where the infant's existing immunity mechanism was relatively "mature", reflecting in part the effect of the normal antigenic stimulus of infection. In view of this when the mother of such a child became pregnant again it was decided to deliver this infant into a germfree isolation unit for a period of investigation to establish whether this disease was in fact present. If the "combined immunity deficiency state" was present it was proposed to treat the infant with a graft of foetal tissue. Such treatment would be given in the germfree state to prevent infection and to ensure the relative immunological "immaturity" of the infant by preventing its exposure to the stimulus of foreign bacterial antigens.

TECHNIQUE

Various steps were involved in the delivery and subsequent care of the infant and these require to be described separately. Firstly a modified Caesarean section technique was used to deliver the child (3). Secondly the child was transferred to a connected aseptic area for resuscitation. Finally there was the transportation stage with the ultimate introduction of the child into the sterile maintenance isolator (4,5). Here the child received routine neonatal care during the period of her investigation (6).

a) Surgery

The mother was admitted to hospital for skin preparation several days prior to elective Caesarian section at 39 weeks gestation. Skin preparation involved the repeated use of povidone-iodine antiseptic solution. The treated area was then covered with either a sterile adhesive membrane or a conventional sterile linen towel until the next treatment.

Caesarian section was performed through an inflated polyvinyl chloride (PVC) surgical isolator overlying the patient (Fig. 1). Three pairs of PVC sleeves were welded as an integral part of the tent and each sleeve was fitted with a standard surgical rubber glove over a terminal rigid plastic cuff. The base of the surgical isolator was modified to form an adhesive incisional membrane to correspond to the abdomen of the patient.

A PVC tunnel from the surgical isolator was connected with a similar tunnel on the transfer isolator making both isolators into a single continous unit for sterilization. Surgical instruments and dressings for Caesarean section together with infant resuscitation requirements were enclosed in the unit prior to gamma irradiation used for sterilization. A total dose of 4 M rads from a CO^{60} source was administered to the isolators, contents, and the air filters attached to the isolators. Since the isolators had to be eventually parted and function individually, separate air supplies and filters were provided for each isolator. In each isolator air from a valved cylinder was sterilized through two

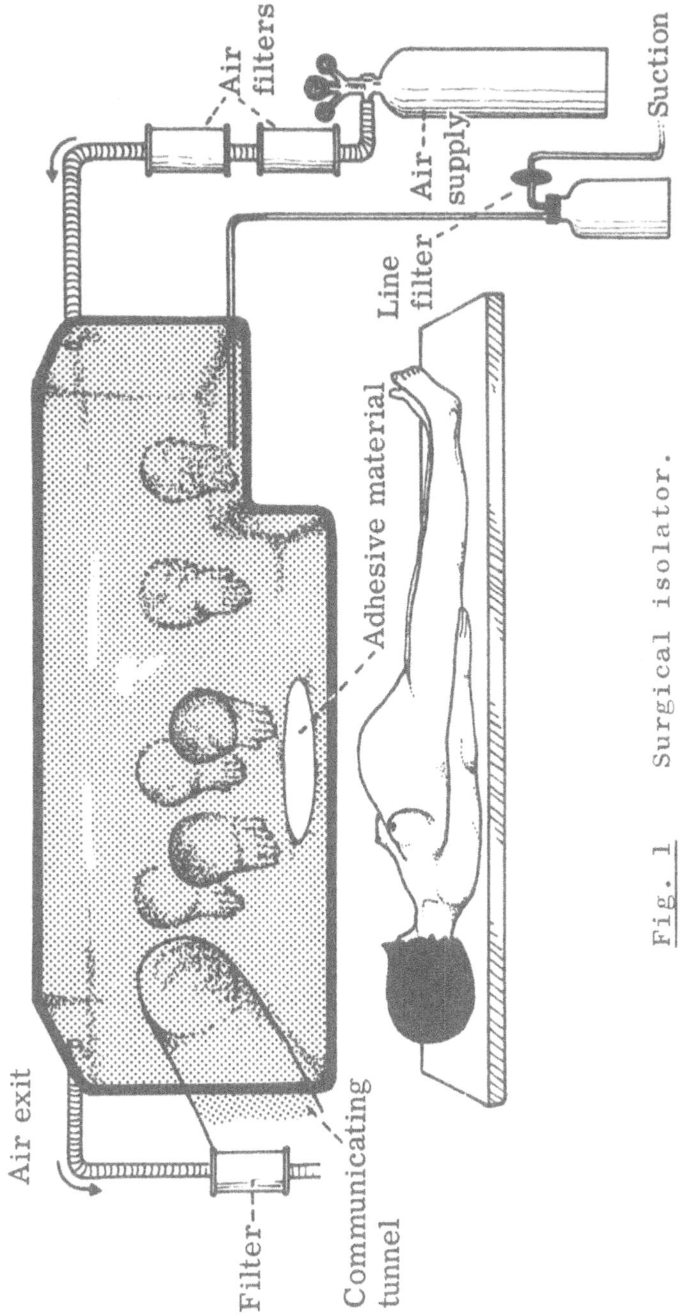

Fig. 1 Surgical isolator.

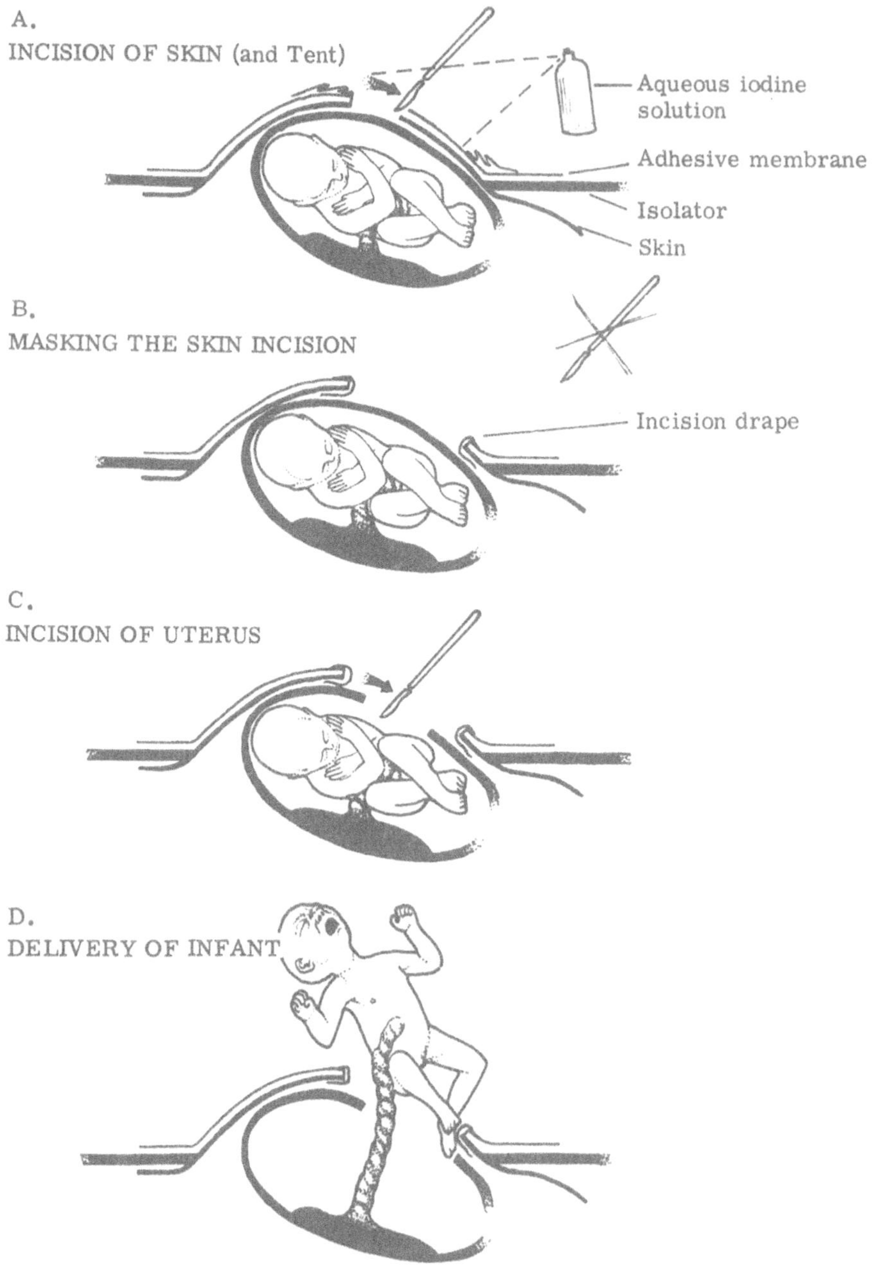

Fig. 2 Modified Caesarian section.

'high efficiency' air filters connected in series; each filter constructed of four layers of FM004 fibre glass filter media. A single similar filter was in addition fitted as a precaution to the air exit on each isolator. Sterilizing filters were also used for the oxygen line in the transfer isolator and were also fitted as a precaution on the suction lines incorporated in both isolators.

After induction of anaesthesia the patient was transferred to the operating theatre and the final skin preparation performed. Whilst the iodine antiseptic solution was drying, the adhesive surface of the base membrane in the now inflated sterile surgical isolator was exposed on removal of the protective backing paper. The antiseptic dry, the surgical tent was then carefully lowered and adhered to the patient's abdomen. The surgical team entered their gloves after a routine scrubbing up procedure, the surgical instruments were then unpacked in preparation for surgery.

The primary incision was made through both the plastic membrane and the skin. The scalpel used for this incision was then quickly removed from the operation site (Fig. 2). The incised skin edges (and the base of any exposed hair follicles) were immediately treated with the antiseptic iodine solution, and small plastic skin drapes were used to cover the incised skin. A standard lower segment Caesarean section was performed using a new scalpel and resulted in the delivery of the infant into the sterile surgical tent. As soon as the umbilical cord was clamped and cut the baby was passed through the connecting tunnel to the transfer isolator for resuscitation. In the surgical tent, the placenta was delivered and the Caesarean section repair was completed without difficulty and the tent removed.

b) Resuscitation and Transfer of the Infant

The resuscitation of the infant took place in the transfer isolator (Fig. 3) and at this time the tunnel with the surgical isolator was being clamped prior to their separation. Resuscitation with this infant required aspiration, the administration of oxygen and the passing of an endotracheal tube. The infant's temperature was maintained in a specially constructed incubator introduced in the isolator prior to sterilization. The incubator was

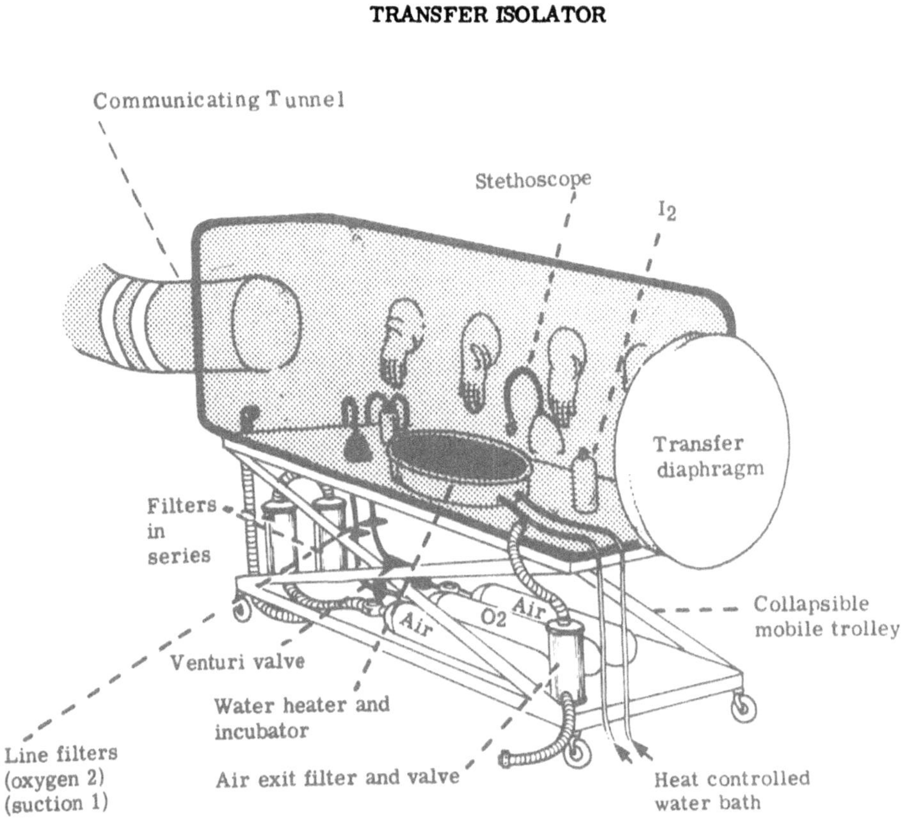

Fig. 3 Mobile transfer isolator

constructed with approximately 100-ft of thin walled plastic tubing lining the inside walls and floor of a simple plastic baby bath. This tubing was connected externally to a heat-controlled laboratory water bath fitted with a pump. The primary reason for the construction of this relatively simple incubator was that it was capable of withstanding the proposed dose of 4 M rad gamma irradiation without deterioration. The external connections of the incubator to the water bath naturally had to be separated prior to transportation but in practice its heat retaining capacity with a large jacket of warm water was found to be excellent.

Following the resuscitation of the infant and the separation of the two isolators, the other theatre supplies to the transfer isolator were disconnected. In the theatre the oxygen supply was from a special controlled positive pressure source in event of the necessity to inflate the infant's lungs. For the journey to the maintenance isolator (at another hospital) oxygen as a precaution was available from an oxygen cylinder fitted on the trolley supporting the transfer isolator. Suction, derived from an air cylinder activated venturi valve was also provided in the event of any emergency during the transportation. The pressure and flow of air in the transfer isolator was controlled from the valved air cylinder and an exit valve on the trolley. The infant was transported successfully in the transfer isolator to the vicinity of the large maintenance isolator, a journey that lasted about 30 minutes.

c) Maintenance Isolator

The "Cornercroft" type A human isolator (4) (Fig. 4) was designed for use with children at the Hospital for Sick Children, Great Ormond Street. The basic apparatus is constructed of separate head and foot consoles connected by adjustable rods. These rods are designed to support beds and cots of different sizes according to the age of the child. Over these connecting rods, the plastic tent was erected, and, when attached to both head and foot consoles, provided completely sealed unit (Fig. 5). The sealing of the tent to the consoles was achieved by inflatable pressure tubing connected and controlled from an air cylinder on each console.

The maintenance isolator incorporated the

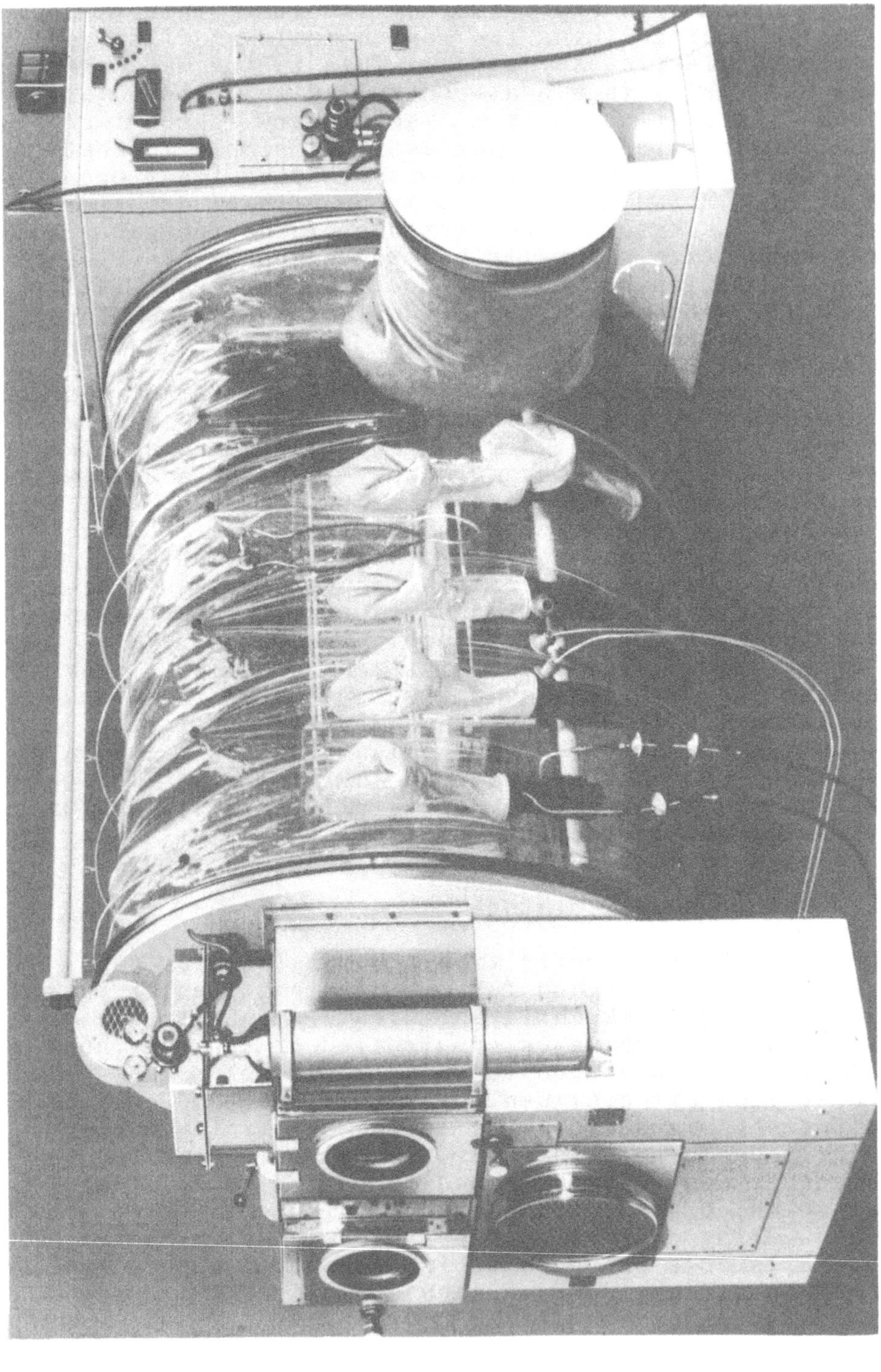

"Cornercroft" Type A Human Isolator

fail-safe device (4) fitted and tested with our germ-free animal isolation units (7). In the month before the isolator was used for this infant, the efficiency of the various components including the filters and tent had been confirmed. The assembled isolator was then tested as a complete unit after sterilization using perhaps the most sensitive and appropriate test system - namely "germ-free" animals. After an investigation period of one week, these animals were shown to be still germfree and this confirmed the microbiological integrity of the human isolator (5).

The "Cornercroft" isolator was assembled for use here with a nylon-coated metal cot and a PVC sealed mattress. An electrically heated incubator sealed in PVC and sterilized by ethylene oxide was also introduced prior to the sterilization of the isolator. The electricity supply for this (6v) and for a thermostatically controlled bottle warmer (230v) were provided with filtered emergency oxygen and suction lines through the wall of the tent. The tent was fitted with gloved sleeves necessary for nursing and medical care and a stethoscope was also provided.

In the preparation of the maintenance isolation system, considerable care was given to providing all the medical and nursing requisites likely to be required for the isolation period. Since it was considered that alternatively a period of one week or one month's isolation were likely, requirements for the latter period were prepared and sterilized in advance.

The method of sterilization varied and depended on the nature of the object and its ability to withstand the proposed method of sterilization. Primary sterilization of the articles always took place after first sealing in two layers of nylon or PVC film. To facilitate nursing care and to reduce the number of times that objects had to be introduced into the maintenance isolator, some items were sealed together in special packs, e.g. feeding packs. It only remained then to introduce the required sterile article into the isolator and this occurred through the specially designed entry lock after the aseptic removal of its protective plastic cover. In practice this cover was removed below the opened entry port door in a high velocity current of sterile air directed out from the port (Fig. 5). This with the subsequent timed exposure to ultra-violet irradiation

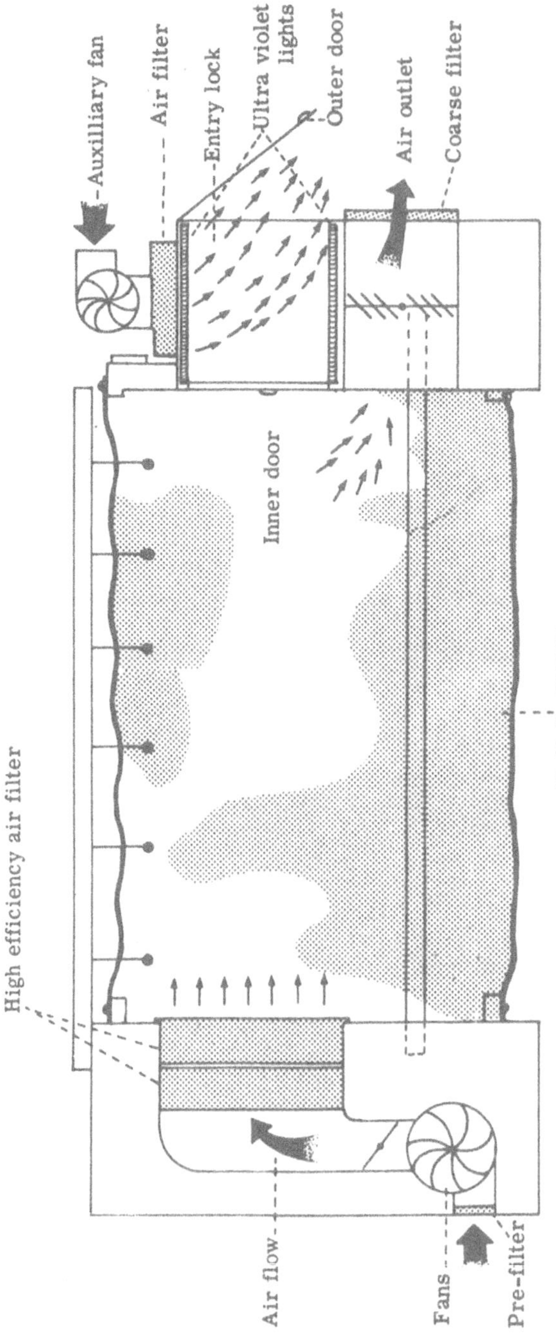

Fig. 5 "Cornercroft" Type A Human Isolator (diagram)

in the chamber on closure of the door, was incorporated to prevent and eliminate respectively any chance of air contamination during entry.

Gamma irradiation was the most frequent method of sterilization employed. Ethylene oxide was on occasions employed for objects such as instruments with glass components that are known to deteriorate with gamma irradiation. Ethylene oxide was also used to sterilize the external surface of the cans of evaporated milk required for feeding. Here the actual pasteurized milk was accepted as sterile upon the data supplied by the manufacturers and also its successful use in the care of germfree pigs (8). Drugs possibly needed in the event of any emergency were obtained in the parental and presumed sterile form; it remained only to sterilize the outside of the ampoule which was done with a 10% formalin solution.

The infant was introduced in the maintenance isolator from the mobile transfer isolator after the construction of a tunnel between the isolators. This tunnel was formed with two identical adhesive diaphragms fitted as an integral part of each isolator, and were constructed of single sided adhesive membranes stretched upon stainless steel cylinders. These membranes, with their outer adhesive surfaces covered with protective backing paper were sterilized with the individual isolators. The backing paper was removed and the membranes caused to bulge outwards by increasing the pressure in the isolators. The membranes were then joined, the union commencing from the centre of the bulge, preventing non-sterile air being trapped between the surfaces (Fig. 6).

As it was still possible that viable organisms were trapped in the adhesive union between the two membranes certain measures were employed against contamination from this source. The now common united adhesive membrane was first caused to bulge into the transfer isolator by reduction of the air pressure in this unit. The reduction of pressure in the transfer isolator also facilitated the spraying of this common membrane with the aqueous iodine antiseptic solution. Whilst wet the membrane was then incised with a scalpel from inside in a sawing manner (Fig. 7). This incision and subsequent removal of the incised panel equilibrated the pressure in both isolators. Further closure of the air exit

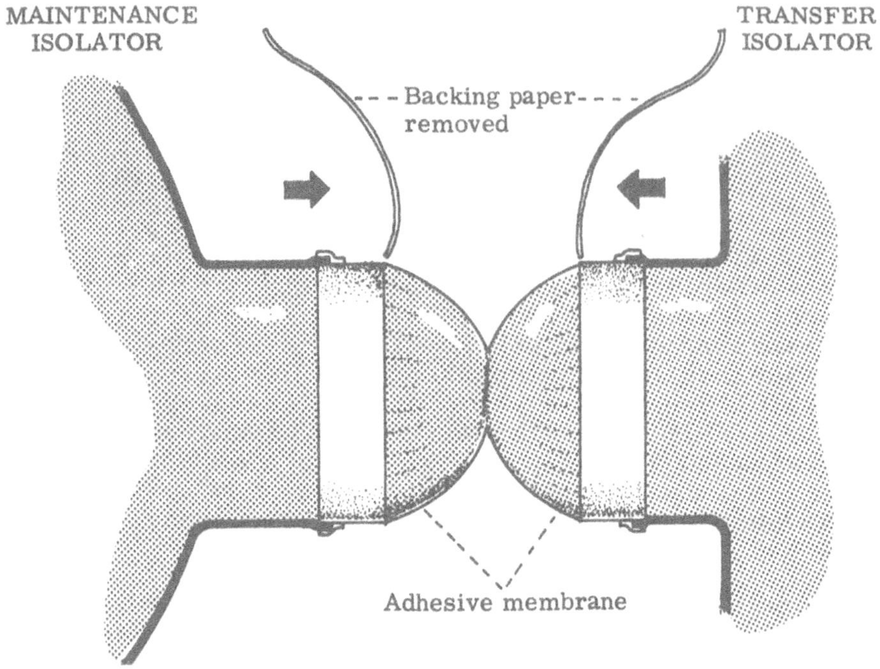

Fig. 6 Union of the membranes between the maintenance and transfer isolators.

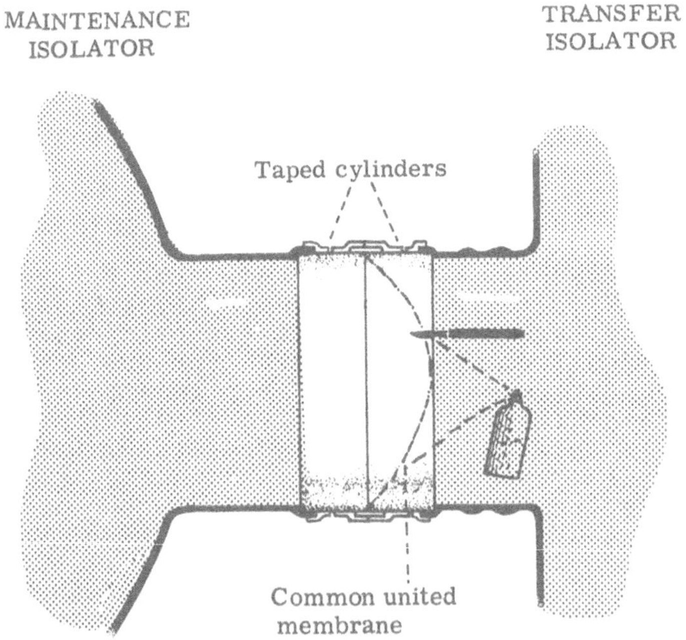

Fig. 7 Incision of membrane between the maintenance and transfer isolators.

valve in the maintenance isolator then caused the major part of the air supply to be directed out through the transfer isolator. In practice this meant that a large volume of sterile air was diverted at a high velocity from the maintenance isolator through the tunnel and out through the transfer isolator. In this situation, the baby was transferred against the gradient of the high velocity air flow into the maintenance isolator.

The baby safely in the maintenance unit, airtight clamps were applied, and the satellite transfer isolator removed.

Care of the Infant

During the period of isolation, nursing and medical care was carried out without undue difficulty and the details are described fully elsewhere (6). Physical examination revealed a clinically normal female infant and laboratory investigations excluded the presence of the "combined immunity deficiency state".

Laboratory investigation at this time also confirmed the germfree success of the procedure. Direct gram-stain films of her stools failed to show any microorganisms and repeated cultures from the infant and the isolator remained sterile (9). Psychological observations during the period of isolation confirmed that in spite of the physical barrier there were normal responses between the infant and nurse which are assumed to be essential to development (10).

The infant was removed from the isolation unit on the 7th day and special care was taken to ensure that her first exposure to outside (and infected) environment should be in the vicinity of other infants born by Caesarean section three days earlier.

A portable isolator for the removal and transfer of the infant back to the obstetric unit was essential since there was a minor epidemic of Salmonella typhi murium in the ward at this time. The effect of such an infection upon any newborn infant, and particularly this infant with time-reduced level in maternal transmitted humoral immunity was not underestimated and the return to the obstetric hospital was performed with strict aseptic technique. After

24 hours in the nursery in contact with other Caesarean derived infants, the infant was returned to her mother for normal care.

DISCUSSION

Various isolation units have been proposed and tested in recent years for patient care. Permanent bricks and mortar units have been described (11,12, 13,14), but suffer microbiologically by the necessary entry of medical and nursing personnel. Haynes and Hench (15) were the first to use a mobile plastic film isolation unit and showed a reduction in cross infection in the treatment of burns. Schwartz and Perry (16) later demonstrated the suitability of the same plastic isolator for cancer patients undergoing intensive chemotherapy. The maintenance isolation unit used here (4) was first tested with germfree animals (5). These animals were introduced into the assembled isolator after sterilization and the fact that these animals were still germfree after a period of one week confirmed the microbiological integrity of this particular isolator. The advantages of a plastic film unit compared with a walk-in bricks and mortar unit are not only microbiological; units constructed of plastic film are cheaper and mobile, and the fact that the isolator was quickly assembled in a normal hospital ward is very important.

The indications for germfree care of this particular patient have been discussed more fully elsewhere (9) as were the problems concerned with her derivation and introduction into the maintenance unit. Germfree patient care is now being considered routinely for three groups of cases. There is the group as the patient here, where there is a primary inability to combat infection. Cases with exposed tissue damage, e.g. burns would obviously benefit if infection could be prevented. Finally, and undoubtedly the largest group of cases are those that present with a secondary inability to combat infection. Here there are diseases such as aplastic anaemia, and leukaemia, who with cases receiving cytotoxic and immunosuppressive therapy, are known to have an increased susceptibility to infection. In such patients the problem of initial bacterial decontamination is of paramount importance. We have been fortunate here to evaluate an isolation unit with a germfree derived infant; however, the future and widespread use of such units is undoubtedly related to a successful bacterial contamination programme.

We have been fortunate also in using the maintenance isolation unit for the first time with a baby, and the nurses had no difficulty with the relatively easy common neonatal care procedures. Initially the nurses found that the routine procedures like feeding took rather more time than normal. However with practice this time was reduced and by the end of the week's isolation, it was considered that perhaps these procedures took less time than the routine barrier procedure, bearing in mind that the latter requires hand toilet and the donning of masks and gowns.

The neonatal period and early infancy was also a very suitable period to test the isolator in many other aspects apart from nursing. The relatively simple food requirements in early infancy was obviously advantageous and problems are anticipated in the future with the provision of a palatable, nutritious, yet sterile diet for children in other age groups. Not only was it easy to provide the infant here with its material needs but also it was considered that its emotional needs were supplied. The contact in terms of touch and sound between the attendant nurse and the baby differed very little from an infant nursed with barrier precautions. In certain respects there might be an advantage in the nursing of the young infant in the germfree isolator in comparison with conventional barrier nursing procedures. With the isolator the quick and easy access to the child was a distinct advantage. In addition the fact that the infant could see the attendant's face without a covering mask might be considered by some an advantage.

However, many problems will occur with the isolation of older infants, toddlers and pre-school children. Needs are more complex during these periods whether for movement, activity, stimulation or relationships - adult or children. The inevitable restriction of these through the isolation technique may lead to apathy, withdrawal and anxiety, alternatively to frustration or even sabotage of the isolator procedure. Isolation units have been used before for the care of children and although there have been considerable difficulties these in general have not been insurmountable and children of various ages have been, like adults, maintained in isolation units for periods in excess of one month (17). The difficulties with the care and occupation of children in isolation units may be overcome in a number of different ways; the introduction of appropriate play materials, providing adequate communication aids - T.V., radio, telephone, a "play room" extension to the isolator, play leaders, teachers and nursing staff to provide the sort of relationship through which the stress and frustration of the situation can be "played out" (18).

The infant here was not only in a very suitable age group for care in a germfree isolator but fortunately was in excellent clinical condition throughout. It might be anticipated and has been confirmed by others (19, 20), that considerable nursing and medical endeavour is necessary for the care of acutely ill patients requiring a lot of specialized attention. Here the only medical procedures apart from a routine clinical examination were minimal. However, the ease and success of the use of the human germfree isolation unit here has encouraged us with the future and major role of the isolation unit, the germfree care of patients with a secondary inability to combat infection, namely, those with aplastic anaemia, leukaemia and allied disorders, and cases receiving cytotoxic therapy.

SUMMARY

A mother having lost a previous child with the familially determined "combined immunity deficiency state" ("Swiss" agammaglobulinaemia) presented again pregnant. This infant was delivered by Caesarean section into a sterile germfree isolation unit for a period of investigation. The infant was maintained in this environment for seven days during which time laboratory investigations excluded the diagnosis of the "combined immunity deficiency state" and also confirmed that the infant in fact was germfree.

ACKNOWLEDGEMENTS

I am grateful to the many Government and commercial organizations that contributed to the success of this project. I wish also to thank the nursing staff at both the Great Ormond Street Hospital for Sick Children and University College Hospital for their enthusiastic efforts. I am particularly grateful to my colleagues Professor D. V. I. Fairweather,

Drs. E. O. R. Reynolds and Alina Piesowicz, Misses Maureen Tuffrey, Jean Holliday, Kate Mackay and Janet Anderson and Messrs. Darrock Mackay and David Pope for the tolerant but expert assistance. Finally, I acknowledge the expert advice of Mr. R. Cook and Dr. P. Trexler.

The work was supported by the Joint Research Board of the Institute of Child Health and Hospital for Sick Children with grants from the Camilla Samuel Trust Fund and the Wellcome Trust.

REFERENCES

(1) Hitzig, W. H. and Willi, H. 1961 Schweiz. med. Wschr., 91: 1625.
(2) Hitzig, W. H., Barandun, S. and Cottier, H. 1968 In Ergebnisse der Inneren Medizin und Kinderheilkunde. Springer-Verlag.
(3) Barnes, R. D., Fairweather, D. V. I., Reynolds, E. O. R., Tuffrey, M. and Holliday, J. 1968 J. Obstet. Gynaec. Brit. Cwlth., 75: 689.
(4) Barnes, R. D., Tuffrey, M. and Cook, R. 1968 Lancet, i:622.
(5) Barnes, R. D., Cook, R. and Holliday, J. 1968 J. appl. Bact., 31:
(6) Barnes, R. D., Bentovim, A., Hensman, S., Holliday, J., Piesowicz, A. and Tuffrey, M. J. Paediat. (in press).
(7) Cook, R., Tuffrey, M. and Barnes, R. D. 1968 Lab. Anim., 2: 51.
(8) Trexler, P. L. 1967 Personal communication
(9) Barnes, R. D., Fairweather, D. V. I., Holliday, J., Keane, C., Piesowicz, A., Soothill, J. F. and Tuffrey, M. 1968 Lancet (in press).
(10) Bentovim, A. 1967 Personal communication.
(11) Bagshawe, K. D. 1962 Brit. med. J., ii: 871.
(12) Mathe, G., Amiel, J. L. and Schwarzenberg, L. 1964 Ann. N.Y. Acad. Sci., 114: 368.
(13) Bowie, J. H., Tonkin, R. W., Robson, J. S. and Dixon, A. A. 1964 Lancet ii: 1383.
(14) James, K. W., Jamieson, B., Kay, H. E. M., Lynch, J. and Ngan, H. 1967 Lancet i: 1045.
(15) Haynes, B. W. and Hench, M. E. 1965 Ann. Surg., 162: 641.
(16) Schwartz, S. A. and Perry, S. J. 1967 J. Amer. med. Ass., 197: 105.
(17) Perry, S. 1968 Personal communication.
(18) Bentovim, A. 1968 Personal communication.
(19) Levitan, A. A., Seidler, F. M., Strong, C. W. and Herman, L. G. 1968 J. Amer. med. Ass., 203: 105.
(20) Levitan, A. A. and Perry, S. 1967 New Eng. J. med., 276: 881.

PENICILLIN PROPHYLAXIS OF POST-OPERATIVE CLOSTRIDIAL GANGRENE IN MONOCONTAMINATED GUINEA PIGS

Bruce P. Phillips

Laboratory of Germfree Animal Research, NIAID

National Institutes of Health, Bethesda, Maryland

The germfree guinea pig has shown a consistent inability to withstand the challenge of natural infection upon transfer to conventional animal quarters. The animals have tolerated quite well, however, induced infection, via the oral route, with a variety of single species of bacteria including *Bacillus subtilis*, *Bacillus cereus*, *Streptococcus faecalis*, *Staphylococcus aureus*, *Clostridium perfringens*, *Clostridium butyricum*, *Clostridium capitovale*, *Lactobacillus acidophilus*, *Lactobacillus bifidus*, 3 strains of *Escherichia coli*, *Micrococcus* sp., and *Bacteroides* sp. Of these only *B. subtilis* has shown any degree of pathogenicity for the animals when growing in the intact intestinal tract. Some of the animals harboring *B. subtilis* developed a moderate erythematous reaction in the cecum and adjacent colon. The most severe reaction to a single bacterial species, however, occurred in animals monocontaminated with *C. perfringens* and subjected to abdominal surgery. When laparotomies were performed on animals harboring *C. perfringens*, to facilitate intracecal inoculation with *Entamoeba histolytica*, the animals often developed a fatal gas gangrene. The gangrenous condition usually developed within 12 hours, and nearly always within 36 hours of surgery. The animals refused to eat, became moribund and bloated to nearly twice their normal size, and invariably succumbed. At autopsy the abdominal cavity was filled with gas, a serosanguineous fluid, and the dark-reddish gelatinous remains of muscle and connective tissue. Stained preparations of abdominal cavity debris showed an abundance of bacilli as evidence of its growth therein.

The use of penicillin therapeutically in animals which had already developed symptoms was without noticeable effects. The antibiotic was effective, however, when used prophylactically.

Penicillin G, 100,000 units subcutaneously in the back per animal at the time of surgery, has given the favorable results summarized in the table: during the specified periods of time gangrene did not develop in any animal which received the antibiotic.

TABLE I

Effect of penicillin on the development of fatal post-operative gangrene in guinea pigs monocontaminated with type A *Clostridium perfringens*

No. of Guinea Pigs**		Penicillin* Treatment	No. Guinea Pigs With Fatal Gangrene	Survival (days)
	10	None	6	1-3
	9	None	8	1-3
	9	None	6	1-4
	11	None	10	1-2
	10	None	7	1-4
Total	49		37	
	10	Yes	0	9-27 s
	10	Yes	0	12-21 s
	10	Yes	0	11-20 s
	11	Yes	0	16-19 s
	9	Yes	0	8-22 s
Total	50		0	

* 100,000 units penicillin G, subcutaneously, immediately following surgery

** Number of guinea pigs in the various experiments, all of which were subjected to laparotomy and intracecal inoculation with *Entamoeba histolytica*

s Sacrificed

INTESTINAL MONOCONTAMINATION IN THE NEONATAL PIG: MICROBIOLOGICAL AND MICROSCOPIC STUDIES

T. E. Staley, E. W. Jones, and L. D. Corley,

Departments of Veterinary Anatomy and Veterinary Medicine and Surgery, Oklahoma State University, Stillwater

At birth the porcine intestinal barrier to macromolecular protein is minimal but is rapidly established after feeding colostrum. Macromolecular protein absorption induces ultra-structural changes in duodenal[10] and jejunal[8,9] cells of the newborn pig. The neonatal intestinal epithelial barrier to intraluminal microorganisms has not been defined. The ease, by which macromolecules enter the newborn intestinal absorptive cell, suggest that bacteria may be similarly accepted. This study was designed to study the neonatal intestinal barrier to bacterial pathogens.

MATERIALS AND METHODS

Pigs were cesarean derived at 110-115 days of gestation. One gram of packed centrifuged Escherichia coli (055B5H7) organisms per kilogram body weight were suspended in 10 ml. sterile saline (subsequent plate counts 10^{10} organism per ml.) and, 1 to 2 hours after birth, were given via stomach tube. After dosing, the pigs were held in sterilized plastic isolators until sacrifice at 2, 6, and 20 hours later.

At sacrifice, piglets were anesthetized, prepared and draped by standard aseptic procedures and the abdomen opened. Samples

---

This investigation was supported in part by Public Health Service Research Grant AI-06461 from the National Institute of Allergy and Infectious Diseases, and General Medical Research Grant FR-05567.

were removed by electro cautery from the mesenteric lymph nodes draining duodenal, jejunal, and ileal regions, and from the liver and spleen.

The lumen of the intestine was cultured last to avoid contamination. Tissue samples were finely ground and suspended in 3 ml. of sterile saline, and each ml. was diluted with 10-12 ml. of Endo Agar and incubated in 37°C for 18-30 hours. Cotton swabs of the intestinal lumen were placed in thioglycollate, subsequently observed for turbidity, and smears were stained with Gram's stain. The Endo Agar plates were recorded positive when 3 or more colonies were present.

Tissues for light and electron microscopy were handled as previously reported.[9] Frozen and paraffin sections were examined for bacteria and for pathological changes after staining with acridine-orange, Gram's stain and hematoxylin and eosin (H & E).

RESULTS

Microbiological Studies

All samples from undosed control piglets were negative when cultured by the methods described. Two hours after intraluminal exposure to *E. coli*, positive cultures were obtained from duodenal lymph nodes in 5 of 6 piglets, from jejunal lymph nodes in 4 of 6 piglets, from no ileal lymph nodes in 6 piglets, from all 6 spleens and from but 1 of 6 livers. Intraluminal cultures from all 6 piglets dosed were positive. By 6 hours after intraluminal challenge, cultures from the duodenal lymph nodes of all 6 piglets sacrificed at this time were negative. Positive cultures were obtained from the jejunal lymph nodes of 3 of 5 piglets, from ileal nodes of 2 of 5 piglets, from all 6 spleens, but none of 6 livers. Intraluminal cultures indicated *E. coli* organisms in all 6 piglets. At 20 hours after intraluminal challenge, cultures were positive from jejunal and ileal lymph nodes of all 6 piglets, from duodenal nodes of 3 of 6 piglets, from 4 of 6 spleens and from 3 of 6 livers. Once again all intraluminal cultures were positive.

Light Microscopy Studies

Six micron sections stained with H & E were studied for tissue response to intraluminal *E. coli*. The brush border was prominent in the duodenal and jejunal epithelial cells, did not visibly change after exposure to *E. coli*, and at 20 hours after challenge was but sporadically observed in ileal epithelial cells.

Cytoplasmic vacuoles were not observed in duodenal cells during the 20 hour period studied. During this same interval, newborn jejunal and ileal cells contained large vacuoles (also observed in normal newborn piglets), and the ileum was infiltrated with neutrophils. Although villus to crypt ratios remained unchanged, villous connective tissue thickened only to approach normal once again in 20 hour exposed pigs.

Observations made with the aid of Gram's stain revealed abundant luminal bacteria 2 hours after infection. Bacteria were occasionally attached to the upper 1/2 of the villi of both jejunum and ileum. Six hours after infection, bacteria, although absent from the duodenal lumen, increased distally and in the ileum adhered to the upper 1/2 of the villi. Twenty hours after infection bacteria were absent from the duodenum and occurred only occasionally in the jejunum. Numerous bacteria were found both in the ileal lumen and adhering to the upper and lower portions of the villus. The core of the villus occasionally contained bacteria.

Electron Microscopic Studies

One micron sections were cut from the three representative areas of the intestine at 2, 6, and 20 hours after infection. Examination of duodenal and jejunal absorptive cells revealed neither intracellular organisms nor cellular damage. Bacterial invasion was only detected in the 20 hour exposed ileum in which numerous bacteria approximated the luminal surface of the absorptive cells. Once an invasion site was located, thin sections (800-1000Å) were cut for examination on the electron microscope. Intraluminal bacteria, prior to the 20 hour interval, were nonfimbriated, and bacterial degeneration was observed at the 6 hour time interval. In the 20 hour exposed ileum, all bacteria, both in the lumen and attached to the absorptive cells, had a well developed fimbriated coating. Sites of attachment were characterized by microulcers which were often associated with goblet cells and which resulted from exfoliation of the microvilli from 5-6 cells (Fig. 1). After loss of the microvilli, bacteria were closely associated with or adhered to the apical plasma membrane. The trilaminar structure of the plasma membrane was not apparent; it appeared only as a thickened single membrane. In the cells adjacent to the areas of attachment the microvilli were shortened and various degrees of swelling and budding occurred. The union between the bacterial cell wall and the absorptive plasmalemma was incomplete, due to a slight separation (250Å) which was occupied by the fimbria of the bacterial cell wall. The terminal web region immediately beneath the attachment site was increased in density. Bacteria appeared to enter the cell by invaginations of the plasma membrane where the terminal web is least dense. The invaginated plasma membrane enlarges to accomodate the organism, and the apical tubular structures appeared atrophic and in some cells were absent (Fig. 2). Once within the cytoplasm, the invaginated plasma membrane enveloped the bacteria in a membrane bound vacuole. Although these vacuoles originated from the plasma membrane, only a single limiting membrane was observed. Granular endoplasmic reticulum and mitochondria at the site of invasion were abundant. Degeneration of the mitochondria produced vacuolation and myelin figures. These alterations were, however, inconsistant and were observed occasionally in the ileal cell of undosed piglets. One to 3 bacteria occurred in the vacuoles and, due to the intervening fimbriae, did not adhere directly to the enveloping membrane. Some vacuoles with bacteria contained a dense osmophilic substance, the origin of which could not be established, but which was similar to vacuolar substance observed in unexposed ileal cells. Vacuoles filled with a dense osmophilic substance also occurred without bacteria in the exposed cells (Fig. 3,4). Some vacuolation and plasma membrane irregularities were observed within the microorganism, but it could not be established that these resulted from intracellular digestion. Although bacteria were observed within the connective tissue core of the villus by light microscopy, none were observed by electron microscopic techniques. Neutrophils were prominent in the blood capillaries and some had migrated into the gut lumen. In no instances were bacteria observed in the lateral intercellular spaces.

DISCUSSION

Bacterial uptake occurred within 2 hours after dosing in the duodenum and jejunum. Similarly rapid passage of macromolecular proteins has been reported. Although few bacteria were observed on the villus, recovery of the microorganisms in the mesenteric lymph nodes suggests that migration occurred via lymphatic vessels. Although E. coli were present in the intestinal lumen, it was neither possible to recover them from ileal mesenteric nodes 2 hours after infection, nor to detect them in the epithelium.

Fig. 1 - Apical ends of ileal cells after 20 hours exposure to E. coli. Fimbriated organisms collect on the absorptive cells adjacent to goblet cells. Microvilli are exfolicated and microorganisms attach to plasma membrane. x 5,000.

Fig. 2 - Apical plasma membrane of an ileal cell after 20 hours exposure to E. coli. Union between bacterial fimbriae and plasma membrane (arrows). Small invaginations of plasma membrane (*) enlarge to accommodate organism (X). Terminal web (TW) increases in density. Apical tubules (AT) atrophy. x 37,000.

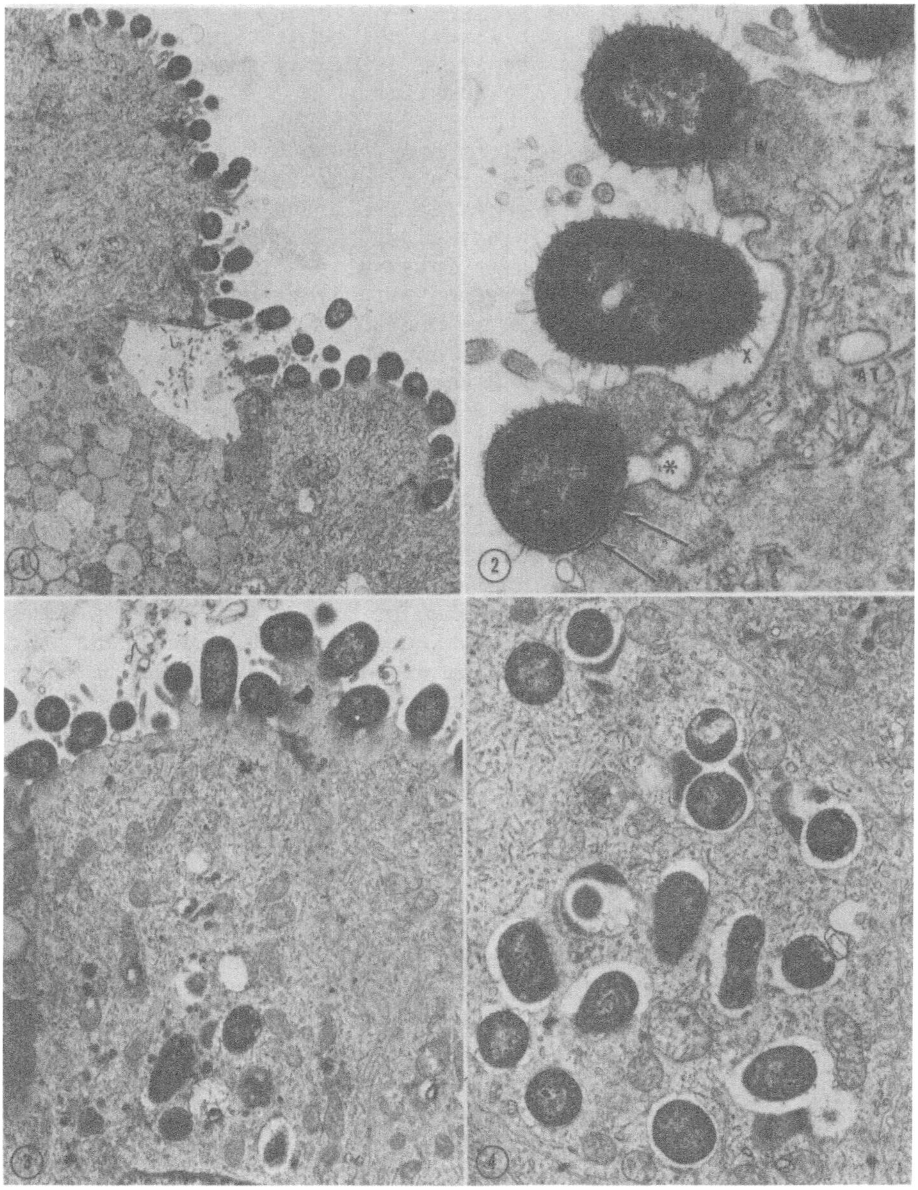

Fig. 3 - Apical ends of ileal cells after 20 hours exposure to E. coli. Microorganisms have attached to surface plasma membrane. Intracellular microorganisms are membrane bound near the nucleus. Small lysosome-like structures are also present. x 15,000.

Fig. 4 - Escherichia coli within ileal cell after 20 hours exposure to bacteria. All microorganisms enclosed in membrane bound vacuoles, some in combination with a dense osmophilic material. x 18,000.

Gram's stain revealed a few E. coli adhering to the apical regions of the villus. Assuming each intestinal segment received an equal exposure, penetration from the ileum evidently was slower than from other areas of the small intestine.

Tissue sections from piglets sampled at 6 hours after exposure indicated flushing of bacteria towards the ileum. Flushing has been demonstrated to be an effective means of preventing Shigella invasion in adult guinea pigs.[4] The duodenal lymph nodes were negative at the 6 hour time interval. Failure to detect bacteria in jejunal and ileal mesenteric lymph nodes of some piglets may reflect lack of exposure due to flushing. Positive spleen cultures in both the 2 hour and 6 hour E. coli exposed groups indicated bacteremia. In the 2 and 6 hour periods after exposure, there was no detectable tissue response to E. coli.

Twenty hours after exposure E. coli organisms were recovered from the mesenteric lymph nodes from all three areas of the small intestine, suggesting colonization and bacterial invasion. Twenty hours after exposure, electron microscopy revealed changes in the E. coli organism. Fimbriae appeared on the outer cell wall of the organism, such as those reported by Duguid, et. al.[3] A hemagglutinating function has been attributed to such fimbriae.[1,3] In the intestine, fimbriae may facilitate attachment of the microorganism to food particles. All piglets studied were unfed, hence, with the exception of sloughed cells or mucins secreted by the goblet cells, there was no readily available food supply. Mucins have been shown to be degraded by intestinal bacteria.[7] This may reflect the preponderance of E. coli attached near goblet cells. Alternatively mucin released from the goblet cell may facilitate attachment.

Cytopathic effects on the ileal cell are minimal at this early period. In the 20 hour E. coli exposed ileal cells, the microvilli were shorter than those of the newborn, despite absence of adjacent bacteria. Microvilli were observed in various stages of swelling and budding, a phenomenon shown to occur in normal intestinal cells[2,12] as well as in those exposed to pathogenic[12] and nonpathogenic[6] microorganisms. The microvilli were shed at sites, where 15-20 bacterial cells seemed attached. Single bacterial cells might induce exfoliation of 3-4 microvilli and lodge between the adjacent microvilli. In the 6 hour exposed pigs in which no attachment of organisms was observed, there was no exfoliation of microvilli despite non-fimbriated microorganism to microvilli distances less than 100-200Å. Takeuchi[12] reported that when Salmonella typhimurium came within 350Å, the microvilli degenerated. Seemingly microvillous degeneration and exfoliation requires both fimbriae on the E. coli and contact between these and plasma membrane. The absence of fimbriae prior to 20 hours may explain why organisms passed the microvillous border were absorbed by the epithelial cell and were recovered in the lymph node without inducing visible cellular pathology. When the exfoliated plasma membrane contacted the fimbriated E. coli, it thickened, formed a small invagination, and enlarged to accommodate the organism.

The terminal web increased in density beneath the site of organism attachment but this apparently offered little resistance to invasion. Hampton and Rosario[6] have described a plaque-like thickening of the plasma membrane subsequent to the attachment of Streptobacillus moniliformis to intestinal epithelial cells.

Intracellular E. coli organisms were in all instances enveloped by a surrounding membrane. No intercellular invasion was observed. A space occupied by fimbriae was often evident between the bacterial cell wall and the surrounding membrane. In some vacuoles a finely granular osmophilic substance occurred around the organism. This substance seemed to result from the fusion with a dense osmophilic vacuole. Similar osmophilic vacuoles were seen in ileal epithelial cells not exposed to bacteria and may be derived from[12] products of the vacuole, enclosing membrane, and bacterial metabolites. A similar material, presumably indicating intracellular digestion, occurred when E. coli were phagocytized by polymorphonuclear leukocytes.[5] There was, however, no evidence of digestion of the enveloped bacteria. Cellular response to invading bacteria was minimal, despite mitochondrial vacuolation and myelin figure formation. These were believed to be fixation artifacts.

Two routes of bacterial penetration through the intestinal epithelium seemed to occur. In the early hours after exposure (2 through 6 hours) non-fimbriated bacteria were absorbed or phagocytized by the immature epithelial cell, these passed rapidly to the regional lymph nodes. Phagocytosis by intestinal mucosal cells has been suggested by Trier[13], Takeuchi, et. al.[11] and Staley, et. al.[10]. The nodes were ineffective filters; organisms entered the general circulation and were recovered from the spleen. Penetration of fimbriated bacteria occurred by attachment to the absorptive cell with the help of mucin and enter into the cytoplasm via a vacuole. All areas of the small intestine apparently absorbed bacteria with equal ease in the early postnatal period. Passage of non-fimbriated organisms have not been observed to induce cellular pathology. After fimbriated forms predominated, cytopathic effects were evident. The early sites for uptake of bacteria have not been elucidated, however the villus tips would seem most likely, as this is where colostrum vacuoles first appear.[10] In the later stages organisms tended to collect around goblet cells. In no instance was invasion observed in the crypt cells.

REFERENCES

1. Brinton, C. C.: Non-flagellar Appendages of Bacteria. Nature, 183:782-786, (1959).

2. Donnellan, W. L.: The Structure of Colonic Mucosa. The Epithelium and Sub-epithelial Reticulohistocytic Complex. Gastroenterology, 49:496-513, (1965).

3. Duguid, J. P., Smith, I. W., Dempster, G. and Edmunds, P. N.: Nonflagellar Filamentous Appendages ("Frimbriae") and Hemagglutinating Activity in Bacterium Coli. J. Path. Bact., 70: 335-348, (1955).

4. Formal, S. G., Abrams, G. D., Schneider, H. and Sprinz, H.: Experimental Shigella Infections VI. Role of the Small Intestine in an Experimental Infection in Guinea Pigs. J. Bact., 85:119-125, (1963).

5. Fromme, H. G.: Studies on Intracellular Degeneration of Phagocytosed Microorganisms by Electron Microscopic Autoradiography. Lab. Invest., 18:211-215, (1968).

6. Hampton, J. C. and Rosario, B.: The Attachment of Microorganisms to Epithelial Cells in the Distal Ileum of the Mouse. Lab. Invest., 14:1464-1481, (1965).

7. Hoskins, L. C. and Zamcheck, N.: Bacterial Degradation of Gastrointestinal Mucins I. Comparisons of Mucus Constituents in the Stools of Germ-free and Conventional Rats. Gastroenterology, 54:210-224, (1968).

8. Mattisson, A. G. M. and Karlsson, B. W.: Electron Microscopic and Immunochemical Studies in the Small Intestine of the Newborn Piglets. Arkiv. für Zoologi., 18:575-589, (1966).

9. Staley, T. E., Jones, E. W. and Marshall, A. E.: The Jejunal Absorptive Cell of the Newborn Pig. An Electron Microscopic Study. Anat. Rec., (In Press).

10. Staley, T. E., Jones, E. W. and Corley, L. D.: The Fine Structure of the Duodenal Absorptive Cell in the Newborn Pig Before and After Feeding of Colostrum. Am. J. Vet. Res., (In Press).

11. Takeuchi, A., Sprinz, H., LaBrec, E. H. and Formal, S. B.: Experimental Bacillary Dysentery. An Electron Microscopic Study of the Response of the Intestinal Mucosa to Bacterial Invasion. Am. J. Path., 47:1011-1044, (1965).

12. Takeuchi, A.: Electron Microscope Studies of Experimental Salmonella Infection. I. Penetration into the Intestinal Epithelium by Salmonella Typhimurium. Am. J. Path., 50:109-136, (1967).

13. Trier, J. S.: Morphologic Alterations Induced by Methotrexate in the Mucosa of Human Proximal Intestine. II. Electron Microscopic Observations. Gastroenterology, 43:407-424, (1962).

THE PERIBRONCHIAL LYMPHOCYTIC TISSUE IN GERMFREE, DEFINED-FLORA, CONVENTIONAL AND CHRONIC MURINE PNEUMONIA-AFFECTED RATS

W. E. Giddens Jr. and C. K. Whitehair

Department of Pathology, Michigan State University

East Lansing, Michigan

Chronic murine pneumonia (CMP) is probably the most serious problem in maintaining healthy rat colonies. It has been estimated[6,9] that up to 75% of the rats from established colonies have lesions of this disease. It not only complicates routine experimental work using the rat, but also seriously diminishes the value of the rat for chronic toxicity studies.

Most workers are agreed that the first, and perhaps most significant lesion to develop in CMP is peribronchial lymphocytic infiltration.[4, 6, 14] Using histologic evidence of "excess" lymphocytic tissue as a criterion of CMP, Innes *et al.*[6] reported that over 91% of their rats were affected, including about 83% of rats which had grossly normal lungs. They reported that if a more strict yardstick had been utilized, the incidence would have been even higher.

Both research on and diagnosis of this disease are hindered by lack of knowledge of the amount of lymphocytic tissue in the normal rat lung. Much of the early work on the histology of the rat lung was done on rats affected with CMP. It was concluded[5, 13] that the rat had large amounts of pulmonary lymphocytic tissue as a normal species peculiarity.

A further complication is introduced by the present use of germfree rats for research on this and other respiratory diseases since there is evidence[3, 15] that germfree conditions produce a notable reduction in the number of lymphocytes in the lymph nodes and viscera. It has been reported that germfree rats[8, 10] and mice[12] have no peribronchial lymphocytic tissue, and that lymphocytic hyperplasia in the rat lung is a response to the etiologic agent of CMP.[7, 8]

One can readily appreciate the dilemma of a pathologist who observes and has to interpret peribronchial lymphocytic tissue in experimental rats.

The specific aims of this research were to examine the lungs of germfree, defined-flora, conventional and CMP-affected rats at selected ages, and to compare the amounts of peribronchial lymphocytic tissue in rats (a) raised under these different environments, and (b) at different ages.

EXPERIMENTAL PLAN

Source of Animals

Rats were obtained according to the schedule in Table 1. The germfree rats were a commercial colony of the Sprague-Dawley strain and were from cesarean-derived, isolator-maintained stock. They were considered germfree on the basis of the standards established by Wagner.[17] The defined-flora rats were from the same colony but were maintained in isolators to which organisms described by Schaedler and co-workers[16] (two strains of lactobacilli, one strain of anaerobic streptococcus Group N, two strains of bacteroides, and one coliform [SLF] strain) had been added. In addition, a micrococcus had contaminated the isolators. Except for the addition of these organisms, the rats were raised in isolators under the same conditions as the germfree rats.

The conventional rats originated from the same colony and stock. They were started by transferring rats from the defined flora isolators to rooms which had been thoroughly cleaned and disinfected with peracetic acid spray. These "subcolonies" were usually maintained for about one year, then destroyed, the rooms resterilized, and new stock added. Each room was supplied with filtered air under positive pressure, and the rats in it were cared for by an attendant who had contact with those rats only. The bedding was a wood product which was heat-treated during preparation. All feed was autoclaved before being used.

One of the CMP-affected colonies (CMP-1) was established by a chemical company in 1932 from rats of the Wistar strain and maintained as a closed colony. Rats from the colony were used for testing the toxicity of chemicals. The colony had a history of widespread CMP, and previous examination of rats in our laboratory about 6 months prior to this work had confirmed the presence of typical lesions.

Table 1. Origin, number and age of rats

Origin	Weanling Rats		Adult Rats		Aged Rats	
	No.	Age	No.	Age	No.	Age
Germfree	10	28 da.	10	8 wk.	7	12 mo.
Defined-Flora	10	28 da.	9	8 wk.	10	12 mo.
Conventional	10	21 da.	10	8 wk.	10	7 mo.
CMP-1	10	21 da.	10	9 wk.	10	12 mo.
CMP-2	-----		-----		10	11 mo.
CMP-3	10	21 da.	-----		10	16 mo.
Total	50	21-28 da.	39	8-9 wk.	57	7-16 mo.

The second CMP-affected colony (CMP-2) was of the Sprague-Dawley strain. It was a commercial colony which sold rats to academic and industrial laboratories. It had been affected with CMP for many years.

The third CMP-affected group (CMP-3) was from two rooms each containing approximately 500 breeding rats. The rooms were production rooms which had been disinfected and populated with defined-flora rats approximately one year prior to the time this study was made. About four to five months after they were populated, widespread CMP was noticed in the older rats.

Morphologic Examination of Tissues

Rats were euthanatized with ether. The trachea and lungs were exposed, and the trachea was infused with fixative until the lungs filled approximately 3/4 of the thoracic cavity. Tissues were fixed in 10% buffered formalin, glutaraldehyde, or Bouin's fixative. Three transverse sections of the lungs were made: the first included the right apical and cardiac lobes and the left lung at a plane just caudal to the heart; the second and third sections were approximately 5 and 10 mm. respectively caudal to the first. Morphologic and microbiologic examinations were also made of the nasal cavity, middle ear, trachea and lungs; the results will be reported at a later date.

All tissues for histologic examination were washed with tap water, dehydrated in ethyl alcohol, cleared in xylene, and embedded in paraffin. Sections were cut at 7 μ and stained with hematoxylin and eosin for general observations. Gomori's trichrome stain for collagen and smooth muscle, Wilder's stain for reticular fibers, Verhoeff's stain for elastin, and the Giemsa and Gram stains for microorganisms were occasionally employed, using standard procedures[1].

The three transverse sections of lung from each rat were examined to determine the amount and distribution of pulmonary lymphocytic tissue. Each bronchus or bronchiole in the plane of the 7 μ section was classified according to the number of lymphocytes (including occasional plasma cells and reticular cells) in peribronchial and peribronchiolar locations, using the scheme in Table 2. The groups were: slight (0-5 peribronchial lymphocytes), moderate (6-15 cells), severe (16-50 cells), and extensive (more than 50 cells).

The total number of bronchi and bronchioles in each group was multiplied by the natural logarithm of the approximate median number of that group. The natural logarithm of the average number of lymphocytes per bronchus or bronchiole, hereafter referred to as the Peribronchial Lymphocytic Index (PLI), was computed by adding these figures and dividing by the total number of observed bronchi and bronchioles.

Table 2. Computation of the Peribronchial Lymphocytic Index in a One-Year-Old Rat Affected with CMP

Classification of Peribronchial Lymphocytic Infiltration	Number In Each Group	Approx. Median In Each Group	Natural Logarithm of Median	Natural Logarithm of Number of Lymphocytes in Each Group
Slight (0-5 Cells)	42	3	1.09861	46.142
Moderate (6-15 Cells)	17	10	2.30259	39.144
Severe (16-50 Cells)	9	33	3.49651	31.468
Extensive (> 50 Cells)	22	100	4.60417	101.292
Total	90			218.046

$$\text{Peribronchial Lymphocytic Index} = \frac{218.046}{90} = 2.423$$

The PLI was computed for each rat and each age group from all the colonies sampled. Using the analysis of variance with least significant differences and 1% significance level[11], comparisons were made between rats of similar ages from the different colonies.

RESULTS AND DISCUSSION

Peribronchial lymphocytic tissue was present in all rats regardless of age or colony, the differences being more quantitative than qualitative. These aggregations of lymphocytes were surprisingly large in some germfree and defined-flora rats and were almost always located at the bifurcation of the bronchi or between a bronchus and an adjacent blood vessel (Fig. 1). The cells were almost entirely small and medium lymphocytes in germfree and defined flora rats (Fig. 2), whereas in conventional and CMP-affected rats, plasma cells and reticular cells were occasionally seen. The lymphocytes appeared to collect first outside of the peribronchial smooth muscle layer and, as their numbers increased, to infiltrate between the smooth muscle and the bronchial epithelium. This was also observed, although much less frequently, in germfree and de-

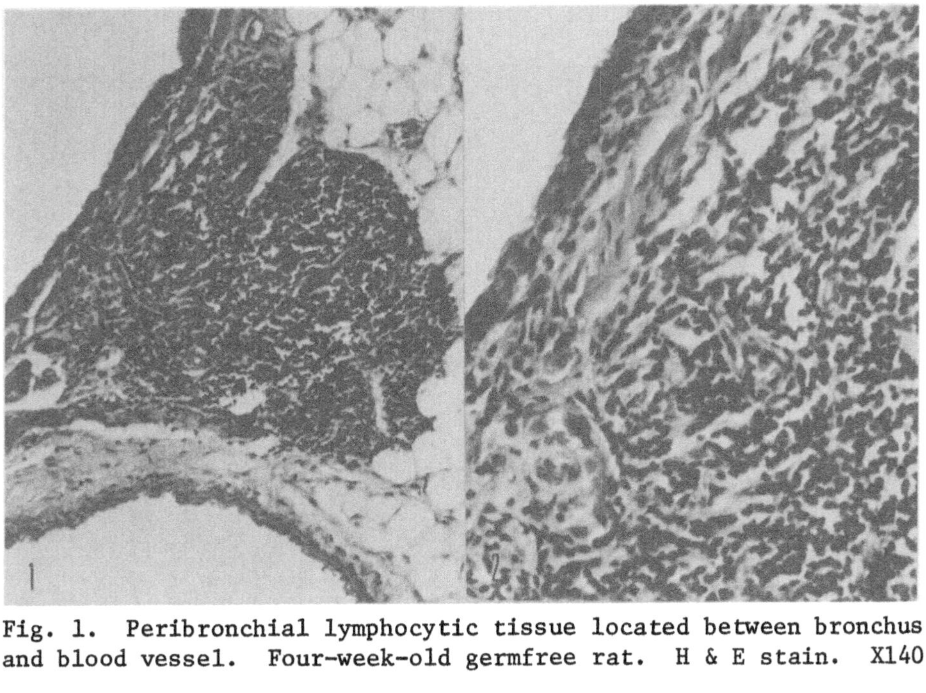

Fig. 1. Peribronchial lymphocytic tissue located between bronchus and blood vessel. Four-week-old germfree rat. H & E stain. X140

Fig. 2. Magnification from Figure 1. Small and medium lymphocytes in peribronchial location. Four-week-old germfree rat. H & E stain. X350 fined-flora rats, and it raised questions about the view, advocated by some, that infiltration of lymphocytes through the peribronchial smooth muscle layer is a diagnostic lesion of CMP. Occasionally, in CMP-affected rats (and rarely in conventional rats), lymphocytic tissue completely surrounded the bronchus (Fig. 3). This was in contrast to peribronchial lymphocytic tissue in germfree and defined-flora rats, which extended no more than a third of the distance around bronchi.

As peribronchial lymphocytic tissue increased in quantity and infiltrated closer to the bronchial epithelium, a decrease in the height of epithelium from columnar to cuboidal or even to simple squamous type was observed. This was accompanied by a disappearance of the cilia (Fig. 4). This change was common in the extensive peribronchial lymphocytic cuffing of CMP-affected rats, occasionally seen in conventional rats, but rarely seen in germfree and defined-flora rats.

Bronchiectasis occurred frequently in CMP-affected rats (Fig. 5). The dilated bronchi usually contained mucin, neutrophils and alveolar macrophages and were surrounded by lymphocytic tissue. The dilatation appeared to be a sequel to the peribronchial lymphocytic infiltration.

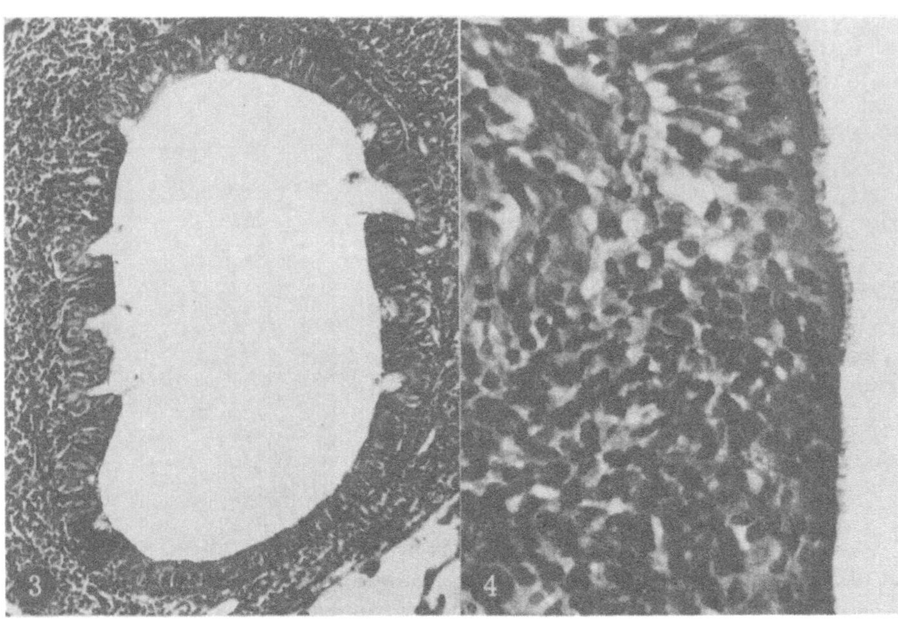

Fig. 3. Peribronchial cuffing with lymphocytes in one-year-old CMP-affected rat. H & E stain. X160

Fig. 4. Encroachment of epithelium by lymphocytic infiltration. Loss of cilia and decrease in height of epithelium. One-year-old CMP-affected rat. H & E stain. X560

Occasionally, clusters of alveolar macrophages characterized by a vacuolar cytoplasm were observed in the alveolar spaces of older rats from all groups examined (Fig. 6). These cells have been described previously in germfree and conventional rats[2] and the condition has been called "multifocal histiocytosis"[18], a somewhat misleading name since the cells do not resemble ordinary histiocytes. They were far more numerous in conventional and CMP-affected rats than in germfree and defined-flora rats. Both vacuolar as well as nonvacuolar macrophages and some mucin were occasionally observed in bronchial lumens, even of germfree rats (Fig. 7).

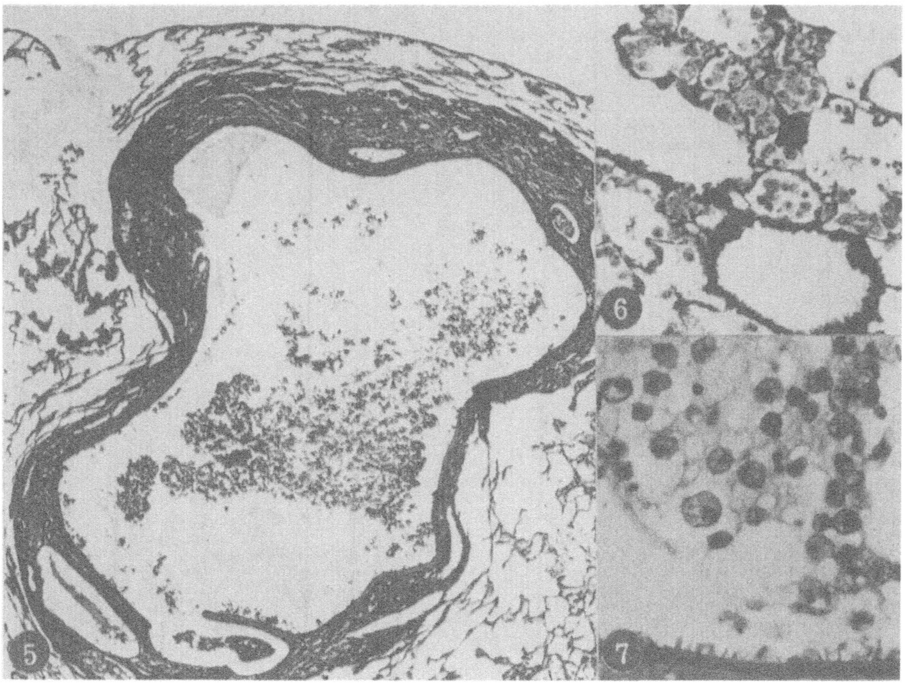

Fig. 5. Bronchiectasis in one-year-old CMP-affected rat. H & E stain. X56

Fig. 6. Vacuolar macrophages in one-year-old germfree rat. H & E stain. X140

Fig. 7. Macrophages and mucin in bronchial lumen. One-year-old germfree rat. H & E stain. X350

The means and standard deviations of the PLI are shown in Fig. 8. There was a significant difference between germfree or defined-flora rats and all CMP-affected rats in all three age groups, the difference tending to increase markedly in older rats. The values for conventional rats at all ages were between values for germfree and defined-flora rats on the one hand and CMP-affected rats on the other. Only in the adult group did conventional rats have a significantly different score from germfree or defined-flora rats. They were significantly different from CMP-1 rats at 8-9 weeks of age, and CMP-2 and CMP-3 rats at 7-16 months of age.

One can readily see the difference between the within-group variation in conventional and CMP-affected rats as opposed to germfree or defined-flora rats. The two latter groups were far more uniform.

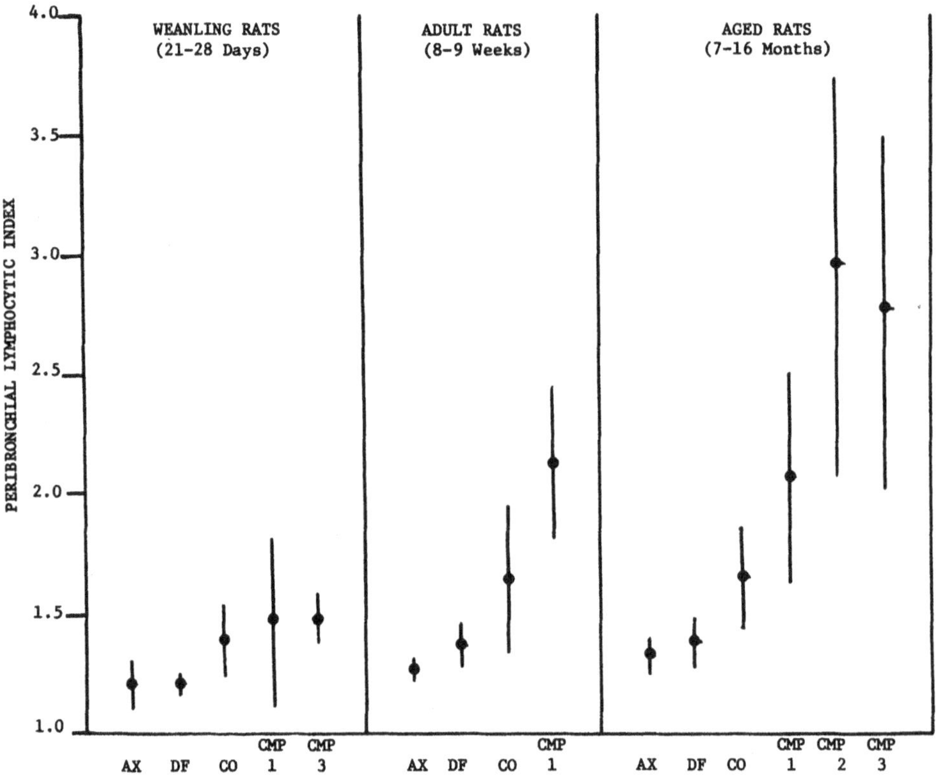

Fig. 8. Means and standard deviations of the Peribronchial Lymphocytic Index in germfree (AX) defined-flora (DF), conventional (CO), and three chronic murine pneumonia-affected colonies (CMP-1, CMP-2, CMP-3).

SUMMARY AND CONCLUSIONS

Laboratory rats were selected from each of the following: germfree, defined-flora, conventional and three chronic murine pneumonia-affected colonies. The rats generally consisted of three age groups: weanling (21-28 days old), adult (8-9 weeks old) and aged (7-16 months old). Peribronchial lymphocytic tissue was observed histologically in all rats regardless of age or source. In germfree and defined flora rats, the lymphocytic tissue was located at bronchial bifurcations or between a bronchus and an adjacent blood vessel, did not circumscribe more than one-third of the bronchus, and consisted almost entirely of small and medium lymphocytes. Using quantitative histologic techniques, an index of peribronchial lymphocytic tissue was determined for each rat and each age group. Significant differences were observed between germfree or defined-flora rats and all chronic murine pneumonia-affected rats at all three age groups. Differences were also observed between the conventional and some of the chronic murine pneumonia-affected rats in the adult and aged groups.

ACKNOWLEDGEMENTS

This study was supported by Grant No. FR 0038-01 from the National Institutes of Health.

Acknowledgement is given to Dr. J. L. Gill, Experiment Station Statistician, for suggestions regarding the methods for statistical analysis.

REFERENCES

1. Armed Forces Institute of Pathology: Manual of Histologic and Special Staining Technics. Washington, D.C. (1957).

2. Beaver, D. L., Ashburn, L. L., and McDaniel, E. G.: Lipid Deposits in the Lungs of Germfree Animals. Arch. Path., 76, (1963): 565-570.

3. Gordon, H. A., and Wostmann, B. S.: Morphological Studies on the Germfree Albino Rat. Anat. Rec., 137, (1960): 65-70.

4. Gray, J. E.: Naturally Occurring and Sulfonamide-Induced Lesions in Rats During a 1-year Toxicity Study. Amer. J. Vet. Res., 24, (1963): 1044-1059.

5. Guieysse - Pellisier, A.: L'Organe Lymphoide du Poumon. Arch. Anat. Micr., 23, (1927): 347-395.

6. Innes, J. R. M., McAdams, A. J., and Yevich, P.: Pulmonary Disease in Rats. A Survey with Comments on "Chronic Murine Pneumonia". Amer. J. Path., 32, (1956): 141-159.

7. Innes, J. R. M., Garner, F. M., and Stookey, J. L.: Respiratory Disease in Rats, in Pathology of Laboratory Rats and Mice, ed. Cotchin, E. and Roe, F.J.C., F.A.Davis Co., Philadelphia, 1967.

8. Innes, J. R. M., p. 291, in Pathology of Laboratory Rats and Mice, ed. Cotchin, E. and Roe, F. J. C., F. A. Davis, Philadelphia, 1967.

9. Joshi, N. N., Blackwood, A. C., and Dale, D. G.: Chronic Murine Pneumonia: A Review. Canad. J. Comp. Med., 25, (1961): 267-273.

10. Joshi, N., Dale, D. G., and Blackwood, A. C.: Etiology of Murine Endemic Pneumonia. Rev. Can. Biol., 24, (1965): 169-178.

11. Li, J. C. R.: Statistical Inference, Vol. I, Edwards Brothers, Inc., Ann Arbor, Michigan, 1964.

12. Lutsky, I. I., and Organick, A. B.: Pneumonia due to Mycoplasma in Gnotobiotic Mice. I. Pathogenicity of *Mycoplasma pneumoniae, Mycoplasma salivarium,* and *Mycoplasma pulmonis* for the Lungs of Conventional and Gnotobiotic Mice. J. Bact., 92, (1966): 1154-1163.

13. Miller, W. S.: The Distribution of Lymphoid Tissue in the Lung. Anat. Rec., 5, (1911): 99-119.

14. Newberne, P. M., Salmon, W. D., and Hare, W. V.: Chronic Murine Pneumonia in an Experimental Laboratory. Arch. Path., 72, (1961): 224-233.

15. Olson, G. B., and Wostmann, B. S.: Lymphocytopoiesis, Plasmacytopoiesis and Cellular Proliferation in Nonantigenically Stimulated Germfree Mice. J. Immunol., 97, (1966): 267-274.

16. Schaedler, R. W., Dubos, R., and Costello, R.: Association of Germfree Mice with Bacteria Isolated from Normal Mice. J. Exp. Med., 122, (1965): 77-82.

17. Wagner, M.: Determination of Germfree Status. Ann. N. Y. Acad. Sci., 78, (1959): 89-101.

18. Yang, Y. H., Yang, C. Y., and Grice, H. C.: Multifocal Histiocytosis in the Lungs of Rats. J. Path. Bact., 92, (1966): 559-561.

BACTERIAL BURDEN AND WOUND REPAIR IN CONVENTIONAL AND GERMFREE RATS

David T. Rovee, Carole A. Kurowsky, Patricia A. Flanagan

Johnson & Johnson Research

New Brunswick, New Jersey

It is generally agreed that gross infection is detrimental to the wound repair processes, however, the significance of wound contamination with bacteria without the ensuing clinical signs of infection is largely unknown. It is likely that most cutaneous wounds in conventional animals are inevitably contaminated at operation or shortly thereafter. Bacterial contamination does not guarantee that infection will follow, but infection does not occur in the absence of bacteria (Casten *et al*, 1964).

Since a correlation between contamination and incidence of infection probably exists, it is logical to attempt to avoid undue bacterial population in a wound by using aseptic technique, antiseptics or antibiotics. Ryan (1967) showed that the topical use of penicillin in surgical incisions was an effective prophylaxis to clinical infection. It has been shown that the large doses of penicillin prescribed by Ryan can reduce or completely eliminate the recoverable bacteria in experimental laparotomies in rats (Rovee *et al*, 1968). Ryan suggested that few organisms are resistant to the levels of antibiotic which can be delivered at the wound site (i.e., 50,000 units/cc saline), and since bacterial populations are small at operation the prophylactic use of penicillin could be effective.

There are reports that the introduction of penicillamine or penicillin in the diets of animals has a deleterious effect on the tensile strengths of cutaneous incisions (Geever *et al*, 1967), and that the penicillamine probably interferes with the normal cross-linking of collagen molecules in the repairing dermis (Seifter *et al*, 1968). Scapicchio *et al* have shown that sulfamylon acetate retards epithelialization in rabbits. Other effects of these or other antibiotics on wound repair are largely unknown. The germfree animal presents an ideal system for the study of antibiotic effects in the absence of bacterial interaction.

The few investigations on wound healing in the germfree animal contain conflicting interpretations, necessitating the establishment of base-line information for our own use. Miyakawa (1959) stated that epidermal repair in GF guinea pigs preceded dermal repair, and that epidermal repair in the GF animals did not show as great a lag phase as in CVN animals. Tipton and Dingman (1966), however, reported a greater lag phase in the healing of wounds in GF guinea pigs than in CVN animals. Rovin et al (1965) found no major histologic differences in the healing of tongue wounds in GF and CVN mice. Brody et al (1966) showed that wound collagen production in GF mice was no different from that in CVN animals. A review of both published and unpublished data by Luckey (1963) describes slower repair in GF goats to faster repair in GF rodents.

The present report centers on the repair of wounds in the presence or absence of bacteria. A topical antibiotic has been employed to eliminate or reduce wound flora in conventional (CVN) animals, and germfree (GF) animals were studied to detect possible tissue effects of the antibiotic in the absence of bacterial interaction. Wound repair in untreated CVN and GF animals has been studied with special attention to epithelialization and gain in incisional breaking strength.

MATERIALS AND METHODS

One-hundred GF (ca. 100 gm each) and 100 CVN (ca. 200 gm each) rats were laparotomized. Animals were anesthetized with chloral hydrate (36 mg/0.1 kg), and the ventral surfaces were clipped closely with scissors. A 5-cm longitudinal incision slightly to one side of the mid-line was made through skin and body wall. Closure was effected in two layers with 5-0 or 4-0 silk with careful placement of sutures ca. 5 mm apart and 4-5 mm from the cut edge of the wound. Fifty GF and 50 CVN rats received no further treatment and were divided into groups of 10 for study at wound ages 3, 6, 9, 12 and 20 days.

The remaining 50 GF and 50 CVN rats received local penicillin G. After closure, the penicillin was injected at all wound levels in an attempt to wash all cut surfaces with the antibiotic. Each wound was irrigated with 5 cc saline containing 250,000 units penicillin (50,000 u/cc), a dose exceeding by far maximum levels attainable with systemic therapy. Animals were divided into groups of 10 and killed for study at wound ages 3, 6, 9, 12 and 20 days.

At sacrifice GF animal wounds were excised along with ca. 2 cm tissue lateral to the incision. After removing the sutures, 1-cm-wide strips across the central portion of the skin incision were cut for tensile measurements. Skin samples were studied in the INSTRON[R] extensometer with a cross-head speed of 3 cm/min and a gage length of 1 cm. Forces required for break of the incision were recorded. Samples were also collected for histologic preparation (hematoxylin and eosin staining).

The same procedure was followed for wounds of CVN animals, and in addition, the quantitative microbiology of the untreated and antibiotic treated wounds was studied as follows.

Thirty-two CVN animals from the untreated and antibiotic treated groups representing 3, 6, 9 and 12 day wounds (4 animals/interval/group) were prepared for study of bacterial populations within the body wall tissue. Only 2 animals were studied at the 20-day interval. They were anesthetized with chloral hydrate, and the incision was swabbed with 70% ethanol. Sutures were removed and the skin incision was excised for the previously mentioned studies. The body wall incision was treated aseptically. Sutures were removed and the incision was excised along with 2-3 mm tissue on either side. The excised body wall wound was cut into ca. 3 strips, immersed in 70% ethanol, and flamed until the alcohol burned off in order to reduce the surface bacteria. Tissue was weighed, minced with fine iridectomy scissors and ground by a motor-driven pestle in 9 cc trypticase soy broth for 5 min in a 10 ml grinding tube. Appropriate serial dilutions of the homogenates were made and plated on trypticase soy agar. After 24-30 hours colonies could be counted and related to the weight of tissue studied.

To study epithelialization, wedge-shaped excisions of the rat ear were made by two radial cuts at the lateral margin with fine iridectomy scissors. The epithelium grows across the cut surfaces of the ear in a dorso-ventral direction rather than filling in the tissue loss. This accounts for uniformity in the wound studied, as the size of the defect to be bridged is determined by the thickness of the ear. Six excisions in GF and 6 in CVN rats were inflicted for histologic study at 1, 2 and 3 days.

Under sodium pentobarbital anesthesia, full thickness skin excisions 1 cm^2 were made on the dorsal flanks of 5 GF and 5 CVN mice. Daily measurements of wound areas were made until complete closure. Contraction in GF and CVN wounds was compared.

RESULTS

The breaking strengths of the untreated and penicillin treated skin incisions of the CVN rats did not differ significantly

($p > 0.5$, Figure 1 and Table I). Histologic study also revealed no differences in the two groups.

Within the GF treated and untreated incisions significant differences could not be seen ($p > 0.5$, Figure 2 and Table II). Histologic study revealed no differences in the two groups.

Both GF and CVN incisions were epithelialized by day 3, no differences existing. Cellular response to injury in the CVN animal was slightly greater than in the GF wound.

Table I. Breaking Strengths of Penicillin-Treated and Control Skin Incisions of CVN Rats
S.E. = standard error , N = 100

Wound Age In Days	Mean Breaking Strengths in Grams ± S.E. Skin	Treatment
3	32 ± 3	None
3	31 ± 3	Penicillin
6	115 ± 16	None
6	93 ± 8	Penicillin
9	242 ± 14	None
9	275 ± 32	Penicillin
12	317 ± 21	None
12	318 ± 30	Penicillin
20	628 ± 77	None
20	725 ± 76	Penicillin

Table II. Breaking Strengths of Penicillin-Treated and Control Skin Incisions of GF Rats. N = 100

Wound Age In Days	Mean Breaking Strengths in Grams ± S.E.	Treatment
3	37 ± 5	None
3	57 ± 8	Penicillin
6	68 ± 8	None
6	72 ± 8	Penicillin
9	166 ± 10	None
9	105 ± 28	Penicillin
12	183 ± 17	None
12	298 ± 25	Penicillin
20	322 ± 25	None
20	370 ± 27	Penicillin

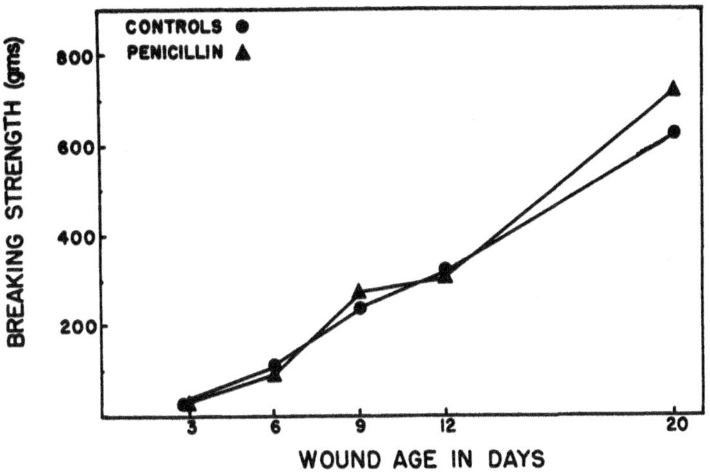

Figure 1. Breaking strengths of untreated and antibiotic treated incisions of CVN rats as a function of time.

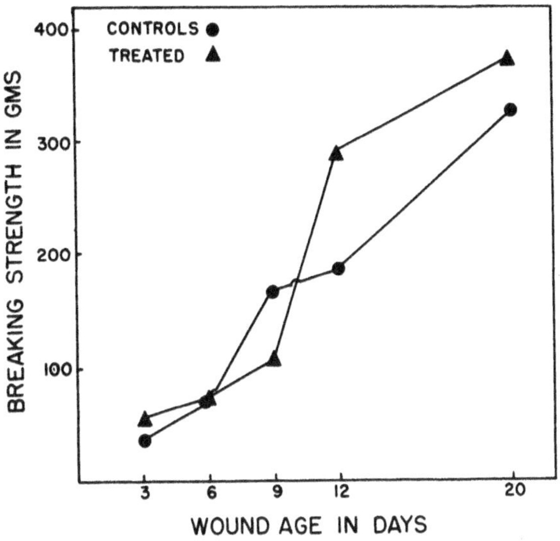

Figure 2. Breaking strengths of untreated and antibiotic treated incisions of GF rats as a function of time.

Untreated CVN wounds contained large bacterial populations at days 3 and 6 post operatively (10^3 - 10^6 bacterial/gm tissue, Table III). CVN wounds irrigated with 250,000 units penicillin G showed reduced to eliminated bacterial populations (Table III). The 3 and 6 day irrigated wounds contained no recoverable bacteria. The untreated wounds and penicillin treated wounds showed incidences of contamination of 0.706 and 0.117, respectively. These differences are significant at $p < 0.01$.

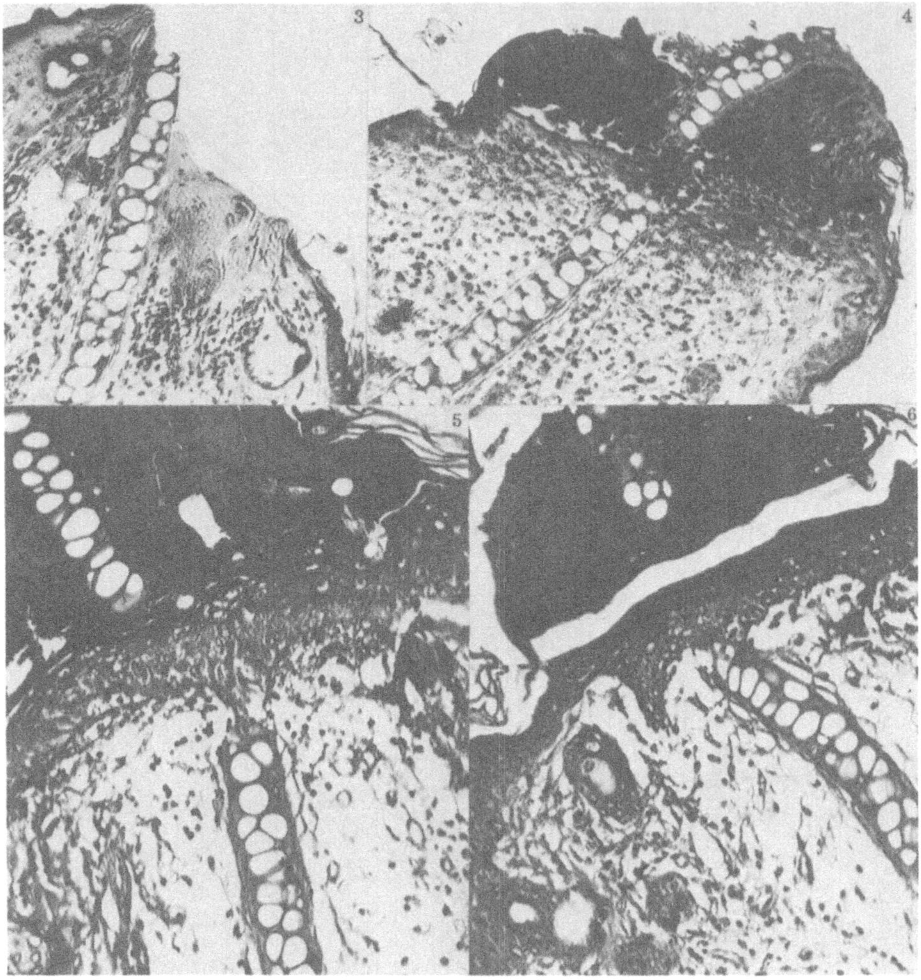

Figure 3. Ear excision in CVN rat at 1 day post wounding. Epithelium has not begun migration. Distal ear has desiccated (scab).

Figure 4. Ear excision in GF rat at 1 day post wounding. Epithelium has moved under scab and nearly covers the defect.

Figure 5. 3-day ear excision in CVN rat showing complete epithelialization.

Figure 6. 3-day ear excision in GF rat showing complete epithelialization.

Table III. Recoverable bacteria resident in body wall on control and penicillin-treated wounds. of CVN rats.

Wound Age	No. Bacteria/gm Tissue: Untreated Controls	Topical Penicillin
3	3.6×10^3	0
3	6.4×10^4	0
3	6.0×10^4	0
3	7.0×10^3	0
6	1.5×10^3	0
6	3.2×10^4	0
6	5.4×10^5	0
6	4.3×10^4	0
9	0	2.6×10^3
9	4.6×10^6	0
9	1.1×10^4	0
9	0	0
12	0	0
12	2.2×10^5	0
12	0	1.4×10^2
12	0	0
20	7.0×10^3	0

Epithelialization of ear excisions in GF rats proceeded with a shorter lag phase than in the CVN rat (Figures 3 and 4). By 3 days post wounding, both CVN and GF excisions had epithelial covers (Figures 5 and 6).

Daily measurements of healing full-thickness excisions in GF and CVN mice revealed no differences in closure rates with the exception of one CVN mouse which manifested a gross infection at the wound site at day 11. This animal is plotted individually from that time in Figure 7. Mean times for closure of the wounds were 17 (CVN) and 18 (GF) days. The infected CVN wound closed at day 22.

The migrating epithelium of a 6-day excision can be seen in Figure 8; in this instance, epithelial repair is well in advance of dermal repair (cf., Miyakawa, 1959).

DISCUSSION

The reduction or elimination of bacteria in the CVN wound did not result in detectable healing differences from the heavily

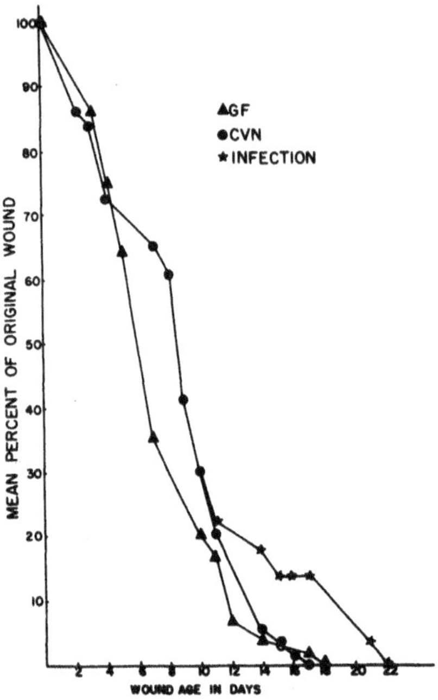

Figure 7. Contraction of excisions in GF and CVN mice. CVN animal showing infection at day 11 is plotted separately.

contaminated group. It could be assumed, however, that in a large group of a more susceptible species the higher incidence of contamination in the untreated wounds would result in a higher incidence of infection. The lower incidence of clinical infection reported by Ryan (1967) in wounds treated in the same manner can be explained on the basis of the data reported herein. The irrigation of the wound with the antibiotic allows (1) the use of massive doses (2) at a time when relatively few organisms are present in the wound. Ryan (1967) indicates that most of the so-called resistant organisms would be affected by the 50,000 unit/cc dose of penicillin at the site even it there is some tissue dilution of the antibiotic. Since antibiotic resistance is determined by maximal doses attained systemically, irrigation of the wound area requires a new interpretation of the efficacy of the antibacterial. Both the anticoccal and bacteriocidal properties of the antibiotic used are considered to account for the results obtained.

The presence of the antibiotic did not affect gain in incisional breaking strength or histology in the CVN rats. It could be interpreted that bacterial interaction with the antibiotic may alter any effect on repair processes, but the dose employed tends to argue against this at the time the antibiotic was employed (<u>i.e.</u>, when few bacteria were present). However, either the presence of bacteria <u>or</u> the antibiotic could conceivably interfere with repair, so the GF animal wound was also studied.

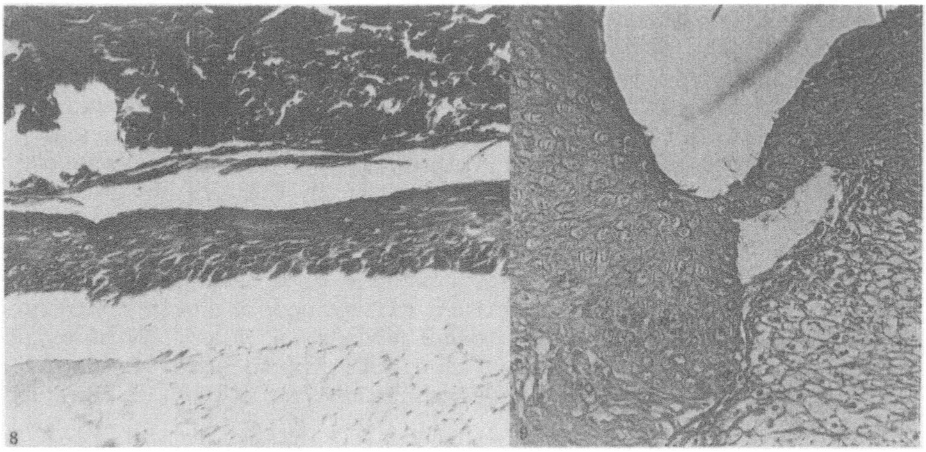

Figure 8. 6-day excision in GF mouse. Note migration of epithelium over fibrin clot (cf. Miyakawa, 1959).
Figure 9. 1-day incision in CVN guinea pig healing at 100% relative humidity. Note similarity to Figure 8.

Antibiotic treated and untreated incisions in GF rats did not differ with respect to breaking strength or histology. There appeared to be more variability in the breaking strengths of the GF animal, but this could be related to experimental error as surgical procedures within the isolator are somewhat more difficult than in the conventional environment.

Although incisional strength of GF and CVN animals can not be usefully compared because of the thinner skin in the GF rat used, histologic comparison indicated repair processes in the incision to be the same in both GF and CVN animals.

Closure of excisions in the GF mouse paralleled closure in the CVN animal. Contraction, a dermal phenomenon, is not affected by the presence or absence of bacteria. Miyakawa (1959) reported that the regenerative epidermis in skin incisions in GF guinea pigs moved under the wound slough, the cut edges fusing forming an epidermal bridge with a space beneath it. Incisions and excisions in GF rats have not shown this phenomenon in our laboratory, however, excisions in GF mice do frequently epithelialize as reported by Miyakawa. Epithelium of CVN animals will sometimes look similar in the presence of moisture (Figure 9, cf. Rovee and Miller, 1968).

Differences in epithelialization of rat ear wounds in CVN and GF animals were noted. The slightly shorter lag phase in the GF rat was a striking consistency in the wounds studied. It has been shown that epithelial repair occurs at a higher rate in the presence of moisture (Winter, 1962; Hinman and Maibach, 1963; Rovee and Miller, 1968), and that the lag phase may be shortened in the moist wound environment (Rovee and Miller, 1968). Relative humidity within our GF isolators runs from 50% to 60%+ whereas the CVN animal rooms are considerably lower (20%-40% R.H.). Temperature may also affect epithelial repair (Gimbel and Farris, 1965; Filston and Vennes, 1968; Culbertson and Tilstone, 1967); our intra-isolator temperature runs ca. 3°C higher than that in the CVN environment. Either humidity or temperature or both may effect the differences seen in the GF rat ear wound. The studies reported by Rovin et al (1965) are ideal in that intraoral wounds were examined thereby precluding humidity and temperature differences. Rovin found no differences in the healing of tongue wounds in GF or CVN mice, an observation which may have been made possible because of the similar wound environments (temperature and moisture) in the buccal cavities.

Ear excisions in CVN rats housed in a plastic isolator at ca. 60% R.H. and 3°C above conventional colony temperature (25°C instead of 22°C) epithelialize slightly earlier than CVN rats in the conventional environment (Rovee, 1968). The effects of temperature and humidity on repair in CVN and GF animals is under study at the present time. We believe it is important to carefully control both these parameters in any epithelialization study. The early epithelial migration in GF rat excisions is probably due, in part, to the physical environment.

CONCLUSIONS AND SUMMARY

(1) Irrigation of incisions in CVN rats with the antibiotic was effective in reducing or eliminating the recoverable wound bacteria.

(2) Tensile measurements and histology of cutaneous incisions treated with 250,000 units penicillin G showed no major differences from the untreated wounds at the intervals studied in either CVN or GF rats. Any effect of the normally expected flora of the experimental wounds could not be shown by these methods.

(3) Presence or absence of bacteria did not affect the rate of contraction in cutaneous wounds of mice. One wound in a CVN mouse showed gross infection and delay in closure.

(4) Epithelialization of ear excisions in GF rats precedes that in the CVN rat. This difference is due, in part, to humidity and/or temperature differences in the plastic isolators.

REFERENCES

1. Casten,D.F., R.J. Nach, J. Spinzia, Surg.Gyn.Obst. 118,783 (1964).
2. Ryan, E.A., Brit. J. Surg. 54, 324 (1967).
3. Rovee, D.T., C.A. Kurowsky, P.A. Flanagan, In Preparation (1968).
4. Geever, E.F., S. Youssef, E. Seifter, S.M. Levenson, J. Surg. Res. 7,160 (1967).
5. Seifter, E., J.J.Morris, S.M. Levenson, Nature 217, 1055 (1968).
6. Scapicchio, A.P., J.D. Constable, B. Opitz, Plast. Reconstr. Surg. 41, 319 (1968).
7. Miyakawa, M., in Recent Progress in Microbiology (Charles C. Thomas, Publisher, Springfield, Illinois, 1959), p.299.
8. Tipton, J.B. and R.O. Dingman, Plast. Reconstr. Surg. 38, 499 (1966).
9. Rovin, S., E.R. Costich, J.E. Fleming, H.A. Gordon, Arch. Path. 79, 641 (1965).
10. Brody, G.L., J.E. Bishop, G.D. Abrams, Arch. Path. 81,268 (1966).
11. Luckey, T.L., Germfree Life and Gnotobiology (Academic Press, New York, 1963).
12. Winter, G.D., Nature 193, 293 (1962).
13. Hinman, C.D. and H.I. Maibach, Nature 200, 377 (1963).
14. Rovee, D.T. and C.A. Miller, Arch. Surg. 96, 43 (1968).
15. Gimbel, N.S. and W. Farris, Arch. Surg. 92, 554 (1966).
16. Filston, H.C. and G.J. Vennes, Surg. Gyn. Obst. 126, 572 (1968).
17. Culbertson, D.P. and W.J. Tilston, Quart. J. Exp. Physiol. 52, 249 (1967).

EFFECT OF TYPE OF DIET ON CHOLESTEROL ABSORPTION AND BILE SALT EXCRETION IN GERMFREE AND CONVENTIONAL CHICKS

H. Eyssen, G. Van Messom, and J. Van den Bosch

The Rega Institute, University of Leuven
Minderbroedersstraat, 10, Leuven, Belgium

It has been a rather common observation that serum cholesterol levels of animals fed commercial or practical type diets are lower than those of animals fed purified diets. Although several studies (1, 2) support the hypothesis that the hypocholesterolemic effect of practical type diets could be due to acceleration of the turnover of bile acids, it is not clear whether the intestinal microflora play a role in these phenomena. In the experiments of Portman (3), addition to the diet of succinylsulfathiazole reduced the effect of Purina rat chow on bile acid turnover by approximately 50 %. On the other hand, Wostmann and Kan (4) concluded that the cholesterol-lowering effect of Purina chow was independent of the microflora since it also occurred in germfree rats.

To obtain more information on the nature of the hypocholesterolemic compounds in practical type diets, we studied the effect of various dietary compositions on cholesterol and bile salts in germfree and conventional chicks.

MATERIALS and METHODS

All animals were Hyline broiler chicks. The composition of the semi-synthetic casein-gelatin-starch diet has been described previously (5). The formula of the practical type diet is given in Table 1.

Table 1

Composition of the Practical Type Diet

	g
Milk powder	1
Casein	3
Gelatin	1
Extracted soya meal	7
Extracted sunflower meal	4
Ground oats	5
Ground milo	15
Ground barley	26
Ground corn	15
Wheat flour	5
Fish meal	10
Vitamin mixture	1.8
Choline hydrochloride	0.1
Methionine	0.1
Corn oil	3
Salt mixture	3

Germfree chicks were reared in the germfree facilities at the Rega Institute (5). Unless otherwise stated, the diets used throughout these experiments were steam sterilized for 20 min. at 121°.

Cholesterol was determined by the method of Abell et al. (6). Fecal bile salts were measured by the method of Evrard and Janssen (7).

RESULTS

1. Cholesterol Pools in Germfree and Conventional Chicks.

The first experiment was concerned with the influence of the germfree state on serum plus liver cholesterol pools.

Table 2 shows that no significant difference in cholesterol pools was observed between germfree and conventional animals on a diet without supplemental cholesterol. These results are in agreement with the findings of Coates et al. (8).

The data from Table 2 also indicate that cholesterol pools of chicks given supplements of dietary cholesterol are influenced

Table 2

Cholesterol Pools in Germfree and Conventional Newborn Chicks Fed the Experimental Diets for 2 Weeks

Diet	Cholesterol in diet	Pool* Serum + liver cholesterol mg/100 g body weight	
		Conventional	Germfree
Semi-synthetic	0	22	23
starch diet	0.25 %	66	115
Practical	0	20	20
type diet	0.25 %	27	68
	0.50 %	39	161

* The serum volume was assumed to be 6 % of body weight.

by two variables: the microflora and the type of diet.

Irrespective of the dietary formula, cholesterol pools of germfree animals fed exogenous cholesterol were substantially higher than those of the conventional control groups. This indicates that microflora interfere with uptake of dietary cholesterol. Similar results have been reported in rats (9).

On the other hand, cholesterol absorption was profoundly affected by the composition of the diet. In conventional chicks, feeding of 0.25 % cholesterol resulted in a threefold increase of the serum and liver cholesterol pools when the animals were on a semi-synthetic diet. Feeding of the same amount of cholesterol to animals on a practical type diet failed to affect the serum and liver cholesterol levels. In the latter animals the cholesterol supplement had to be increased up to 0.5 % before any significant rise of the cholesterol pools could be observed.

Since analogous effects were noted in experiments on germfree chicks, it can be concluded that at least a major percent of the hypocholesterolemic activity of the practical type formula was not mediated through modifications of the microflora.

Table 3

Excretion of Cholesterol and Bile Salts in Newborn Germfree and Conventional Chicks on Diets without Added Cholesterol

Diet	Fecal cholesterol mg/animal/12 days		Fecal bile salts mg/animal/12 days	
	Convention.	Germfree	Convention.	Germfree
Semi-synthetic starch diet	68	28	40	10
Practical type diet	70	35	75	40

2. Fecal Output of Bile Salts and Cholesterol.

The effect of the type of diet on fecal excretion of cholesterol and bile salts was studied in germfree and conventional newborn chicks fed dietary formula's without added cholesterol.

Table 3 shows that, irrespective of the type of diet, conventional chicks excreted 2 to 4 times more fecal cholesterol and bile salts than germfree chicks. These results reflect the influence of the gut flora on bile salt reabsorption.

The data from Table 3 also demonstrate that germfree chicks lost 4 times more bile salts when fed a practical type diet than when given a semi-synthetic diet. This indicates that the practical type diet contained certain compounds which promoted the excretion of bile salts independently of the effects of the microflora.

3. Cholesterol-Lowering Effect of Fish Meal.

Several experiments were performed to obtain more information on the nature of the compounds in the practical type diet which lowered cholesterol pools and promoted fecal bile salt output.

In a preliminary study the individual components of the practical diet were added to a semi-synthetic diet. Although it was difficult to keep the food composition isocaloric and comparable with respect to protein and fiber concentrations, the results were clear. Fish meal was the only additive which effectively prevented the rise in cholesterol pools of animals

Table 4

Effect of Fish Meal on Cholesterol and Bile Salts in Germfree and Conventional Chicks Fed a Semi-Synthetic Diet for 2 Weeks

Treatment	Basal* diet	Basal + 0.25 % cholest.	Basal + 0.25 % cholest. + 10 % fish meal**
Conventional Chicks			
Serum cholesterol mg/100 ml	227	330	198
Liver cholesterol mg/100 g	500	1312	503
Fecal bile salts mg/animal/12 days	21	44	79
Germfree Chicks			
Serum cholesterol mg/100 ml	230	324	241
Liver cholesterol mg/100 g	370	2656	1525
Fecal bile salts mg/animal/12 days	12	33	69

* Casein 20 %, gelatin 8 %, sucrose 34 %, starch 20 %.
** At the expense of 5 % casein and 2 % gelatin.

fed the starch diet with 0.25 % cholesterol. Further experiments showed that omission of fish meal from the practical type diet resulted in loss of hypocholesterolemic activity and concomitant reduction of the fecal bile salt output.

Table 4 gives an example of the effect of 10 % fish meal on serum and liver cholesterol levels of germfree and conventional chicks fed a semi-synthetic diet in which 34 % sucrose was substituted for 34 % starch. The diet of the conventional animals was not sterilized but contained 100 ppm of the antibiotic virginiamycin to assure uniform growth and nutrient absorption.

Addition of 10 % fish meal to the semi-synthetic diet resulted in a twofold increase of the fecal bile salt output. Liver cholesterol concentrations were 2 to 3 times lower in the animals receiving 10 % fish meal than in the control groups.

Since no significant difference in response to fish meal was seen between germfree and conventional animals, it can be concluded that modifications of the microflora were not a major cause of the hypocholesterolemic effect of fish meal.

4. Properties of the Fish Meal Factor.

The experiments with the sterilized diet already showed that fish meal can be autoclaved without substantial loss of hypocholesterolemic activity.

Since Wostmann and Kan (4) reported that compounds of Purina chow, which induced hypocholesterolemia in rats, could be extracted into hot hexane, fish meal was subjected to continuous extraction by n-hexane for 12 hr at 60°. In another experiment fish meal was stirred for 3 hr at room temperature with a mixture of chloroform/methanol (3 : 1). No hypocholesterolemic effect could be demonstrated in any of the solvent fractions. Continuous refluxing with water at 100° for 8 hr also failed to extract the active compounds. In addition the activity resisted digestion by trypsin. Following these various treatments, at least 90 % of the original activity was recovered from the insoluble residue. Only after digestion by papain could part of the active compound be liberated in the supernatant.

DISCUSSION

The present investigations have shown that germfree chicks respond to dietary cholesterol by demonstrating much higher serum plus liver cholesterol pools than conventional chicks. It was also observed that germfree chicks excreted considerably less fecal sterols and bile acids than their conventional controls. Comparable results have been reported for rats (9, 10, 11). These observations suggest that the intestinal microflora interferes with bile acid and cholesterol absorption from the intestine. No attempts were made to isolate the causative microorganisms.

The results of these studies also reveal that a practical type diet prevents the increase of serum and liver cholesterol concentrations of chicks fed low levels of cholesterol (0.25 %).

This formula also promoted the fecal bile acid output. According to the results, these effects were mainly due to the fish meal fraction since addition of 10 % fish meal to a semi-synthetic 0.25 % cholesterol diet resulted in similar effects. When fish meal was omitted from the practical diet, most of the cholesterol-lowering activity of this formula was lost and fecal bile acid output was significantly reduced.

The mode of action of fish meal as a cholesterol-lowering agent in chicks remains to be clarified. In chicks on a diet without supplemental cholesterol, fish meal promoted fecal bile acid output without significantly affecting cholesterol levels or cholesterol output. This suggests a specific effect on bile salts. An action via the bile salts could also explain why the effect of fish meal was most pronounced when supplements of cholesterol were added to the diet. Cholesterol uptake from the intestine is limited by the amount of bile acids available. Under extreme conditions, such as after overloading with dietary cholesterol, a small reduction of the bile salt pool could result in a significantly lower cholesterol absorption.

The nature of the compounds in fish meal responsible for the hypocholesterolemic effect are unknown. Our experiments suggest that they are not lipidic in nature and differ from the cholesterol-lowering agent in Purina rat chow. Moreover, the activity could not be extracted in water and remained in the insoluble residue after digestion by trypsin. It seems unlikely that the cholesterol-lowering action could be due to an aspecific effect of the insoluble non-nutritive bulk. In experiments not reported here, it was found that addition to the semi-synthetic diet of 5 % non-nutritive fiber, which is in large excess of the quantity of bulk in 10 % fish meal, did not significantly influence cholesterol levels. Moreover, omission of fish meal from the practical type diet resulted in loss of hypocholesterolemic effect; yet the bulk content was not appreciably affected by the omission.

Whether the fish meal factor shares properties in common with the atherosclerosis-preventing factor found by Howard et al. (12) in soya meal remains to be established. It should also be mentioned that certain mucilaginous polysaccharides exhibit cholesterol-lowering activity in chickens (13). The properties of the fish meal factor, as far as they are known, are not incompatible with those of mucopolysaccharides.

SUMMARY

Germfree chicks exhibited lower fecal cholesterol and bile salt output than conventional chicks, and responded to feeding of cholesterol by substantially higher serum plus liver cholesterol pools.

Practical type diets promoted fecal bile salt excretion. In addition, the rise in serum and liver cholesterol of chicks on a cholesterol diet could be alleviated. These activities were not related to modifications in the microflora since they were also observed in germfree chicks.

The cholesterol-lowering activity of the practical type diet was found mainly in the fish meal fraction. The active compounds were found to be resistant to heat and tryptic digestion. They were not extractable in n-hexane, chloroform/methanol (3 : 1), or water. Following these different treatments, at least 90 % of the original hypocholesterolemic activity could still be demonstrated in the insoluble residue.

REFERENCES

1. Portman, O. W., and Murphy, P., Arch. Biochem. Biophys. 76, 367 (1958).
2. Hellström, K., Sjövall, J., and Wigand, G., J. Lipid Res., 3, 397 (1962).
3. Portman, O. W., Am. J. Clin. Nutrition, 8, 462 (1960).
4. Wostmann, B. S., and Kan, D. F., J. Nutrition, 84, 277 (1964).
5. Eyssen, H., and De Somer, P., Poultry Sci., 46, 323 (1967).
6. Abell, L. L., Levy, B. B., Brodie, B. B., and Kendall, F. E., J. Biol. Chem., 195, 357 (1952).
7. Evrard, E., and Janssen, G., J. Lipid Res., 9, 226 (1968).
8. Coates, M. E., Harrison, G. F., and Moore, J. H., Ernährungsforschung, 10, 251 (1965).
9. Eyssen, H., Sacquet, E., Evrard, E., and Van den Bosch, J., Life Sci., in print.
10. Gustafsson, B. E., Bergström, S., Lindstedt, S., and Norman, A., Proc. Soc. Exp. Biol. Med., 94, 467 (1957).

11. Gustafsson, B. E., Norman, A., and Sjövall, J., Arch. Biochem. Biophys., 91, 93 (1960).

12. Howard, A. N., Gresham, G. A., Jones, D., and Jennings, I. W., J. Atheroscler. Res., 5, 330 (1965).

13. Fahrenbach, M. J., Riccardi, B. A., and Grant, W. C., Proc. Soc. Exp. Biol. Med., 123, 321 (1966).

INFECTIVITY OF AVIAN REOVIRUSES FOR DAY-OLD CHICKS UNDER GERM-FREE AND ISOLATION CONDITIONS

D. R. Deshmukh and B. S. Pomeroy

Department of Veterinary Microbiology

and Public Health, Univ. of Minn., St. Paul, Minn.

There has been considerable difficulty in associating reoviruses with human and animal infections. They have been isolated from healthy children, as well as children with mild febrile upper respiratory and intestinal infections, and from a number of animal species. The known pathological manifestations of reovirus infections in animals are as follows: Reovirus 2 is said to have been etiologically associated with an epidemic of coryza among chimpanzees in a laboratory and a strain of this type was isolated from the lung of a cynomolgus monkey which died of pneumonia. The prototype strain of reovirus type 1, and simian strains of types 1 and 2 produced a meningitis and necrosis of the choroid plexus when inoculated intracerebrally in monkeys. Some human strains of all these serotypes were pathogenic for suckling mice when inoculated parenterally in large doses, lesions occurred in organs including the brain, heart, and liver. Adult guinea pigs or calves inoculated intranasally with strains of human origin did not show any illness. No illness which could be definitely attributed to reoviruses was observed in a longitudinal study of three herds of dairy cattle from which numerous reoviruses were isolated.

The present report is regarding the infectivity of avian reoviruses 24 and 25 for day-old chicks under isolation as well as germ free conditions. The characterization and classification of reoviruses 24 and 25 as well as their relationship to human reoviruses were reported earlier by Deshmukh and Pomeroy (1967) and Deshmukh et al. (1967).

MATERIALS AND METHODS

Two major techniques were used for these studies, the first involved the use of germ-free isolators, whereas, the second involved the use of Horsfall-Bauer units. One-half millileters of a supernatant suspension of the eighth CAM passage of viruses 24 and 25 with an EID_{50} titer of 5.7 $logs_{10}$/ml was used to inoculate each bird orally. The flocks from which the birds and the eggs used for the infectivity studies were obtained did not have any previous history of cloacal pasting. For the germ-free isolators 30 eggs from a White Leghorn flock were set for hatching into each of the three isolators after disinfection. The number of chicks was adjusted to 20 in each isolator. The chicks in one isolator were inoculated with virus 24, and in another with virus 25. The chicks in the third isolator served as uninoculated controls. For the Horsfall-Bauer units day-old White Leghorn chicks hatched at our laboratory were divided into three groups of 25 chicks each. All the chicks from each group were weighed and the chicks from the first and the second groups were inoculated with viruses 24 and 25 respectively, whereas, the third group served as controls. The above experiment was repeated with day-old White Rock chicks in the germ-free isolators as well as the Horsfall-Bauer units.

The birds from both the germ-free isolators and the Horsfall-Bauer units were examined daily for 21 days for the presence of cloacal pasting or any other clinical signs of illness. At various intervals post-inoculation, three chicks from each group were killed and examined for the presence of lesions. Their intestines were collected and stored at -20 C. for virus isolation and to carry out second serial chick passages. Virus isolation was attempted in embryonating chicken eggs via the CAM route. On the twenty-first day post-inoculation the remaining birds were bled and the sera were pooled and stored at -20 C. for the determination of neutralizing antibody level to the virus inoculated.

EXPERIMENTAL RESULTS

None of the chicks inoculated with viruses 24 or 25 under germ-free conditions or those used for serial chick passages showed any cloacal pasting or any other signs of illness. Also they did not exhibit any gross pathology. In the Horsfall-Bauer units during the first trial approximately 28% of the chicks inoculated with both viruses showed cloacal pasting, whereas, during the second trial approximately 10% of the chicks inoculated showed cloacal pasting. The chicks that showed cloacal pasting were depressed, off feed, and lost weight. From all the infected chicks (in germ-free isolators and in Horsfall-Bauer units) virus isolations were made at various time intervals up to the seventh day post-inoculation. No virus isolations were made from the intestines collected on the twenty-first day post-inoculation. Control chicks showed neither signs of illness nor yielded any virus in samples collected at various time intervals. Control chicks showed no significant antibody levels, whereas the infected chicks both under germ-free and isolation conditions had between 1.6 and 2.3 $\log_{10}$ neutralizing antibody levels.

SUMMARY

In summary the infectivity of viruses 24 and 25 for day-old chicks on oral inoculation under germ-free and isolation conditions was shown by the presence of neutralizing antibodies in the sera of the inoculated chicks and by the recovery and serial passage of the viruses in chicks. Virus infection was associated with the presence of cloacal pasting in certain percentage of chicks under isolation conditions, but not under germ-free conditions. This indicated that biological and/or environmental factor (s) that were absent from the germ-free conditions compared to the isolation conditions were required in addition to a virus for producing cloacal pasting.

REFERENCES

1. Deshmukh, D.R., and Pomeroy, B.S. Avian Reoviruses. II. Physico-chemical Characterization and Classification. Avian Diseases (In Press).

2. Deshmukh, D.R., Sayed, H.I., and Pomeroy, B.S., Avian Reoviruses. IV. Relationship to human Reoviruses. Avian Diseases (In press).

ERYTHROPOIETIC RESPONSE OF GERMFREE MICE TO ERYTHROPOIETIN

Edwin A. Mirand*, Albert S. Gordon+, and Gerald P. Murphy*

*Roswell Park Mem. Inst., N. Y. State Dept. of Health, and Dept. of Biochemical Pharmacology and Urology, State University of New York at Buffalo, Buffalo, N. Y.; and +Lab. of Experimental Hematology, Graduate School of Arts and Sciences, New York University, New York, N. Y.

It is known that germfree and conventional animals possess differing hematological properties. In the germfree animal total leucocyte counts are lower, and there is a general diminution in the size of the lymph nodes, spleen, liver, and other organs comprising the reticuloendothelial system (1,2). Since differences in hematological properties between the germfree and conventional animals have been observed, the question arose whether there were differences in the humoral mechanism which regulates erythropoiesis between the two types of animals. Keighley _et al_ (3) have described differences in the erythropoietic responses to erythropoietin (ESF) in the fasted, specific pathogen-free rat and the conventional rat. Therefore, ESF, a hormone activated or produced by the kidney, was studied in these animals and this report describes particularly the differences in erythropoietic response to exogenous ESF between the germfree and conventional mice.

MATERIALS AND METHODS

Adult male germfree mice (Hauschka-Mirand/ICR Swiss strain) were obtained from the Charles River Breeding Laboratories, Boston, Massachusetts. They were kept in flexible plastic isolators of the Trexler type (4) to maintain the axenic environment. The conventional mice were of the same strain but raised under ordinary laboratory conditions. These were also placed in Trexler isolators to preclude the possibility that any effects occurring in the germfree animals were due to confinement in the isolators. Routine microbiological tests were made for bacteria, fungus, PPLO and other parasites (5). At the termination of the experiments, the blood, fecal contents, lung, liver and spleen from the germfree mice were cultured as a final check on the maintenance of the germfree state.

Half of the germfree animals were transfused intraperitoneally on 2 successive days with 1 ml. of homologous germfree packed red cells (6). These cells had been washed in sterile saline under aseptic conditions prior to injection. Half of the conventional mice were similarly transfused but with washed red cells from non-axenic mice. Experimental treatment was begun 3 days after the last injection of red cells. In these plethorized groups, mice with hematocrit values less than 60% at the termination of the experiments were not included in the data.

Four groups of mice were established: (A) axenic hypertransfused, (B) axenic non-transfused, (C) conventional hypertransfused, and (D) conventional non-transfused. Half of the animals in each of these 4 groups were injected subcutaneously with Step III sheep ESF (lot #K147137 supplied by the National Heart Institute, NIH) at a dose level of 1 unit per day in 0.2 ml. sterile saline for 3 days. Prior to use, the ESF was passed through a Millipore filter (HA-.45 μ) to ensure sterility. The remaining 50% of mice in each of the 4 groups received subcutaneous injections of 0.2 ml. sterile saline, the vehicle for the hormone, daily for 3 days. All mice were injected intramuscularly with 2 mg. iron dextran (Imferon, Lakeside Laboratory, Wisconsin) on the day that ESF or vehicle injections were begun to preclude changes in circulating iron pool that could falsify interpretation of the RBC Fe^{59} incorporation values. At 24 hours after the last injection of ESF or the vehicle, the mice received intravenous injections of 0.5 μc Fe^{59} in 0.2 ml. sterile saline; 48 hours later blood was drawn from the dorsal aorta and hematocrit and % RBC Fe^{59} incorporation values determined (6). Peripheral red, reticulocyte and white cell counts, and hemoglobin concentrations were also recorded at this time.

RESULTS

Table I enumerates the characteristic hematological responses in conventional and axenic mice, plethorized or non-plethoric, with saline injections or following ESF administration. The peripheral reticulocyte levels in these two groups of mice paralleled closely the radioiron incorporation values. No significant effects were produced in the four experimental groups on hemoglobin levels or red cell counts in response to ESF administration in these short term experiments. As noted by others (1,2), the total leucocyte counts were significantly lower in the axenic than in the conventional mice. These values (Table I) were not significantly altered by the administration of ESF or saline.

Figures 1 and 2 illustrate the augmented response in Fe^{59} 24 hour uptake in the axenic animals. This ESF value is most apparent in the hypertransfused (plethoric) axenic animals (61.6%)(Fig. 1). There was less difference in response to ESF injection in non-transfused axenic (50.9%) or conventional hypertransfused (30.3%) and non-transfused mice (34.0%)(Fig. 2).

Table I. Hematological Response of Axenic and Conventional Hypertransfused and Non-hypertransfused Mice (Means ± S.E.M.)

Group	No. Mice	Body Wt. (gram)	WBC/mm^3	Hb (gram)	RBC (mill/mm^3)	Hct %
(A) Axenic Hypertransfused						
1. ESF	18	29.0 ± .57	4,405 ± 320	18.7 ± 1.16	11.1 ± .71	63.6 ± 1.24
2. Saline	18	28.9 ± .32	3,759 ± 287	20.2 ± .92	11.4 ± .62	65.3 ± 1.46
(B) Axenic Non-transfused						
1. ESF	18	27.8 ± .29	3,299 ± 184	14.3 ± .84	8.7 ± .54	48.2 ± .93
2. Saline	18	27.6 ± .22	2,821 ± 112	14.1 ± .72	8.5 ± .66	44.8 ± .88
(C) Conventional Hypertransfused						
1. ESF	17	29.2 ± .41	6,275 ± 330	22.9 ± 1.03	12.6 ± .75	70.5 ± 2.32
2. Saline	17	27.2 ± .39	5,288 ± 294	20.8 ± 1.17	12.2 ± .68	65.5 ± 1.83
(D) Conventional Non-transfused						
1. ESF	18	27.0 ± .52	4,065 ± 277	14.6 ± .96	8.7 ± .48	47.7 ± 1.19
2. Saline	18	26.0 ± .43	3,582 ± 117	14.1 ± .78	8.6 ± .53	43.5 ± .98

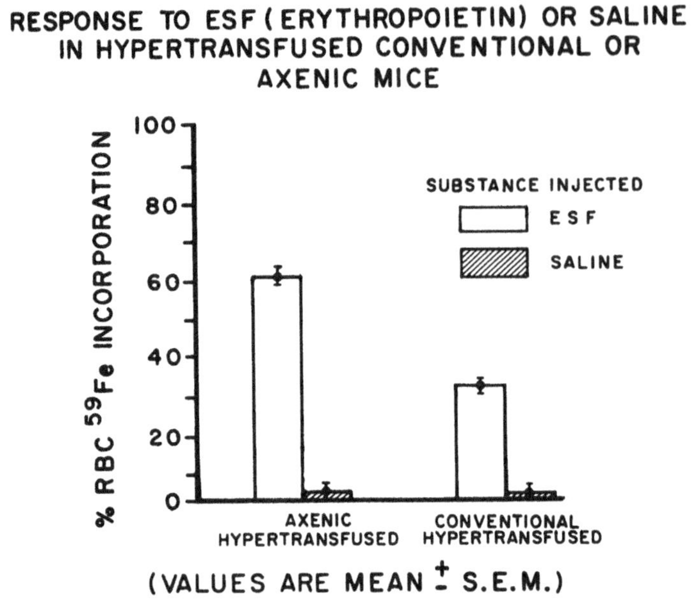
RESPONSE TO ESF (ERYTHROPOIETIN) OR SALINE IN HYPERTRANSFUSED CONVENTIONAL OR AXENIC MICE
% RBC 59Fe INCORPORATION
100
80
60
40
20
0
SUBSTANCE INJECTED
ESF
SALINE
AXENIC HYPERTRANSFUSED
CONVENTIONAL HYPERTRANSFUSED
(VALUES ARE MEAN ± S.E.M.)

Fig. 1

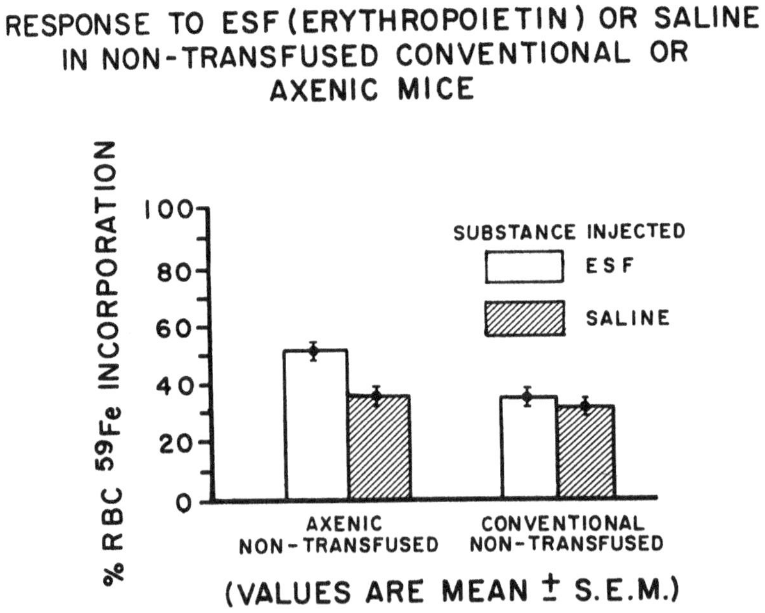
RESPONSE TO ESF (ERYTHROPOIETIN) OR SALINE IN NON-TRANSFUSED CONVENTIONAL OR AXENIC MICE
% RBC 59Fe INCORPORATION
100
80
60
40
20
0
SUBSTANCE INJECTED
ESF
SALINE
AXENIC NON-TRANSFUSED
CONVENTIONAL NON-TRANSFUSED
(VALUES ARE MEAN ± S.E.M.)

Fig. 2

DISCUSSION

The reason for the increased erythropoietic response to ESF in the germfree as compared to the conventional animal remains poorly understood; however, several possibilities to explain this difference are presented. One possibility is that greater numbers of responsive stem cells are present or that an increased sensitivity of these elements to ESF exists in the germfree animal. The axenic animal may also provide an environment more conducive to the augmented activity of the ESF. This could occur if the level of circulating erythropoiesis-inhibiting factor was lower in the germfree animal (7,8). Another possibility is that there is a greater renal elimination rate of ESF in the conventional than the germfree mouse. It is also conceivable that a decreased rate of hormone destruction occurs in the axenic mouse. In this regard, the liver has been implicated as a site of inactivation of the ESF (9,10). The finding that the liver is less developed, both morphologically and functionally, in germfree than in conventional animals (11,12) supports the possibility that a diminished rate of destruction of ESF occurs in the axenic animal. This is being investigated by us by determining the sojourn time of exogenously and endogenously-produced ESF in the circulation of germfree animals. Preliminary data in our laboratory demonstrate that the disappearance rate of plasma ESF in germfree mice is longer than in conventional mice. From these observations one can tentatively infer that the difference in erythropoietic response to ESF between germfree and conventional mice is due to the decreased rate of ESF destruction occurring in the germfree mouse.

SUMMARY

Germfree mice have an augmented response to erythropoietin (ESF) as compared to conventional mice. Several explanations for this have been discussed; however, it appears that the best explanation for the difference in the erythropoietic response is due to the germfree mouse having a longer disappearance rate of plasma ESF.

ACKNOWLEDGMENTS

This investigation was supported in part by research grants (Nos. CA-08847, CA-07745 and HE-03357-11) from the National Institutes of Health, Public Health Service, and the John A. Hartford Foundation, Inc. We extend thanks to Mary G. Walker, Joyce Jividen, Dolores Mickens, Raul Marti and Raul Baillo for technical assistance rendered.

REFERENCES

1. H. A. Gordon. Ann. N. Y. Acad. Sci. 78: 208, 1959.
2. B. P. Phillips, P. A. Wolfe, and H. A. Gordon. Ann. N. Y. Acad. Sci. 78: 183, 1959.
3. G. Keighley. In Erythropoiesis, p. 17. L. O. Jacobson and M. Doyle (Editors). Grune and Stratton, New York, 1962.
4. P. C. Trexler. Ann. N. Y. Acad. Sci. 78: 29, 1959.
5. E. A. Mirand and J. T. Grace, Jr. Nature 200: 92, 1963.
6. E. A. Mirand, A. H. Weintraub, A. S. Gordon, T. C. Prentice, and J. T. Grace, Jr. Proc. Soc. Exptl. Biol. Med. 118: 823, 1965.
7. T. Krzymowski and H. Krzymowska. Blood 19: 38, 1962.
8. W. H. Whitcomb and M. Moore. J. Lab. Clin. Med. 66: 641, 1965.
9. E. M. Jacobsen, A. K. Davis, and E. L. Alpen. Blood 11: 937, 1956.
10. T. C. Prentice and E. A. Mirand. Proc. Soc. Exptl. Biol. Med. 95: 231, 1957.
11. G. J. Thorbecke and B. Benacerraf. Ann. N. Y. Acad. Sci. 78: 247, 1959.
12. M. Miyakawa, Y. Uno and J. Asai. In Reticuloendothelial System, Morphology, Immunology and Regulation, p. 132. Nissha, Kyoto, Japan, 1965.

Viral Status of Germfree Rodents; Present and Future

Masahiro Kajima

Lobund Laboratory, Department of Microbiology

University of Notre Dame, Notre Dame, Indiana

The materials used in experimentation must be clearly defined as to their purity. Experimental animals are standardized by specifying species, strain, sex, age, diet, housing and microflora; for these factors are important for obtaining clear results and for correct interpretations of the results obtained. Considerable efforts have been directed toward producing and controlling animals as pure materials for experimentation. Germfree (GF) rodents that are caesarean-derived into a sterile isolator, fed a sterilized diet and maintained aseptically, are desirable subjects for some types of experimentation because they have proven to be free from bacteria, protozoa, fungi and mycoplasma (1,2,3,4). In the field of virology, possibly other fields (such as oncology, serology, and pathology) it is important to obtain "Virus-defined" or "virus-free" animals as reliable experimental tools. In this respect, viral flora in GF rodents, especially in mice has been explored extensively by various methods including: (a) Observation for symptoms suggestive of a viral disease; (b) histopathological examination for lesions in the tissues; (c) electron microscopy for virus detection in normal and neoplastic tissues of GF rodents; (d) the detection of a cytopathogenic agent in the tissue cultures derived from GF animals; (e) virus isolation by inoculation of tissue extract or excretions from GF rodents into tissue culture systems or into susceptible animals; (f) susceptibility tests of germfree rodents to standard viruses; and (g) serodiagnostic methods for virus infection in GF animals. The results obtained in each method have some limitations by sensitivity of the test or the observation itself and by the extent of the materials used. Nevertheless, the accumulation of results published indicates support of a theoretical assumption, drawn by analyzing technical details, that present germfree methods can not prevent the vertical transmission of virus but may be effective in preventing the horizontal transmission. This communication deals with the summarizing and analyzing of the results obtained by others and by ourselves on the search for viruses in GF rodents, to draw general conclusions on the viral status of GF rodents, and to suggest problems that should be overcome in the future.

Evidences and Comments

Germfree mice-On viruses such as Sendai (HVJ), mouse hepatitis (MHV), pneumonia virus of mice (PVM), mouse adenovirus, polyoma and K virus, there is no serodiagnostic evidence of their presence in germfree or in caesarean-derived (CD) mice in contrast with conventional mice which were positive (5,6,7,8). Theiler's encephalomyelitis virus infection was found by hemagglutination inhibition (HI) test with high incidence (33 and 42%) in conventional mice, but it was detected in a few CD mice and in none of GF mice (6,8). Mouse thymic agent was not isolated from specific pathogen-free (SPF) and GF mice (5). Vertically transmisible mouse salivary gland virus (9,10) was not isolated from SPF and GF mice (5) although very few were found in normal production mice. Lactic dehydrogenase (LDH) virus has also been known as an agent is possibly transmitted vertically to progeny (11); but there was no evidence of their presence in the GF mice examined (6). This may be partly because this virus is transmitted via the placenta in a mother infected during gestation but not in chronically infected ones (11). Lymphocytic choriomeningitis (LCM) virus and reovirus type 3 (reo-3) create a more complicated situation, because they were not detected in GF mice (6,7) as a rule, but in other instances reo-3 infection was recognized by HI test in SPF and GF mice (5) and CD mice (8) and also that LCM virus was present in GF mice derived by standard method from conventional mice with congenital LCM virus infection (from W.P. Rowe) (12). These facts are not surprising because LCM (13) and reo-virus (14) can be transmitted vertically into the progeny, yet a majority of conventional laboratory mice are free of them (5,6). If a source stock is clean, then the mouse colony derived from it may be free from the virus. As far as I know, it has not yet been determined as to whether or not the viruses such as ectromelia, epizootic diarrhea virus of suckling mice, lethal intestinal virus of infant mice, encephalomyocarditis virus and others, are present in GF mice. GF mice should be tested at least once for their presence or absence although it is assumed that they are not contaminated because these viruses are not ubiquitous, and the germfree method could prevent their infection. It has been proven that caesarean section and nursing on disease-free females produced a clean mouse stock for diarrhea virus of suckling mice (15); and there were no symptoms suggestive of ectromelia infection in

GF mice exposed by X-irradiation known as a stress which elicits disease in treated mice with latent ectromelia infection(7). Recently it was reported that mouse tumor tissues were widely contaminated with viruses such as LDH, polyoma, reo-3, Sendai, MHV, and Theiler's encephalomyelitis virus as tested by mouse antibody production (MAP) test (16). Similar study on tumors of GF mice has not yet been reported.

The early suggestions on the presence of leukemia virus in GF mice was made by the leukemogenic response of the mice to whole body X-irradiation (17) and by electron microscopy of normal thymic tissues of germfree mice (18). Subsequent studies indicated that lymphatic leukemia developed in all strains of GF mice exposed to X-irradiation (19,20) and spontaneously in GF AKR strain mice (21); and leukemia virus-like particles (tumor virus type C and its precursor)(Fig. 2)(22,23,24) were detected in their tissues (25). Similar particles were also noted in mammary tumor tissues (25) and in chemically induced fibrosarcoma tissues of GF mice; and it was suggested that they might be "passenger virus" in tumor tissues (26). By the detection of leukemia virus in prenatal mouse thymuses by electron microscopy it was strongly suggested that the virus could be transmitted vertically (27). Leukemia virus was found by electron microscopy in all 6 strains of normal GF mice examined (27). Therefore, it was attempted to find a virus-free mouse from conventional mouse strains; but the search has been unsuccessful (28). All strains of conventional laboratory mice examined were also carrying the virus. Cell-free filtrates prepared from radiation-induced leukemic tissues of GF mice were leukemogenic in test mice, although the incidence was very low (29).

The presence of mannary tumor virus (tumor virus Type B and its precursor) (Fig. 3 and 4)(22,23,24) was demonstrated by electron microscopy in spontaneous mammary carcinoma tissues of Lobund C3Hf mice (25). GF C3Hf mice produced in another laboratory also developed mammary tumors at a rate similar to that in their conventional counteparts and virus particles were found in the tissues by electron microscopy (30). From these results it can be assumed that the mammary tumor virus is capable of being transmitted vertically in addition to infection by means of milk. There is no report describing spontaneous mammary tumors in other strains of GF mice; but recently spontaneous mammary tumors have been observed in Lobund GF BALB/c mice (31). Purification and biological tests of mammary tumor virus from germfree mice have not been done yet. Another type of virus-like particles (cisternal type A tumor virus) (Fig. 1)(22,23,24) was recognized by electron microscopy in connection with the endoplasmic reticulum of the cells. They were detected in normal and tumor tissues of germfree mice (26). Significance of this particle in mice still

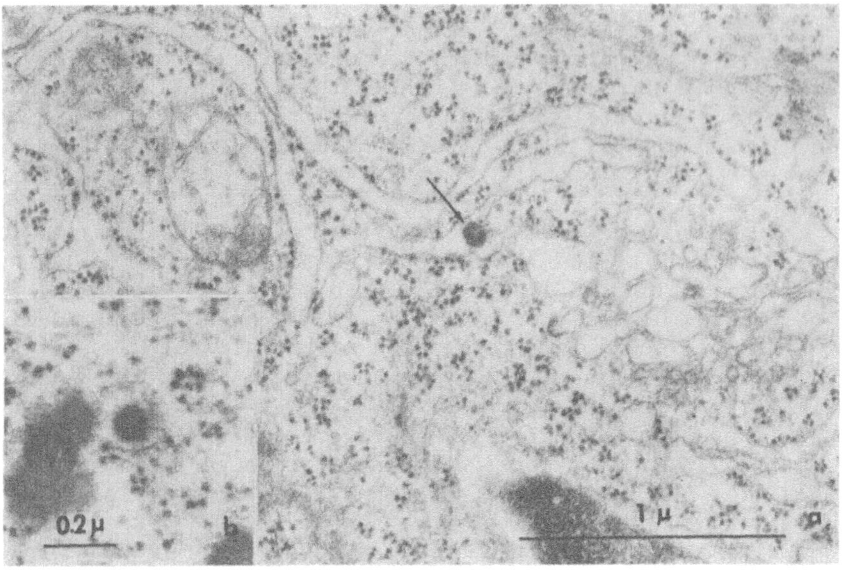

Fig. 1 a. Section of spontaneous skin carcinoma cell of GF mouse (Swiss-Webster). Formation of virus-like particles from the wall of cisternae of endoplasmic reticulum (arrow). Fig. 1 b. This represents virus-like particle (Type A) in cisternae of endoplasmic reticulum (Arrow).

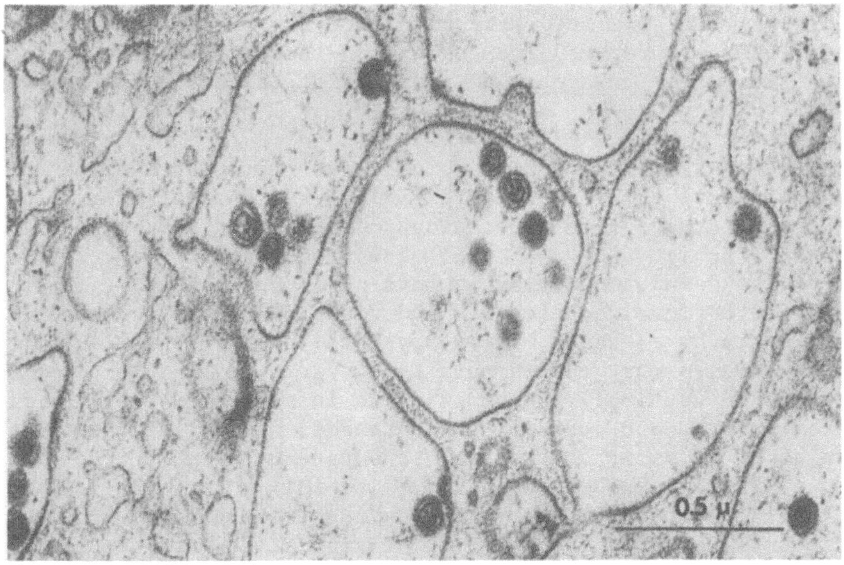

Fig. 2. Part of thymic epithelial cell of GF mouse (AKR). Tumor virus Type C and virus formation by budding can be observed.

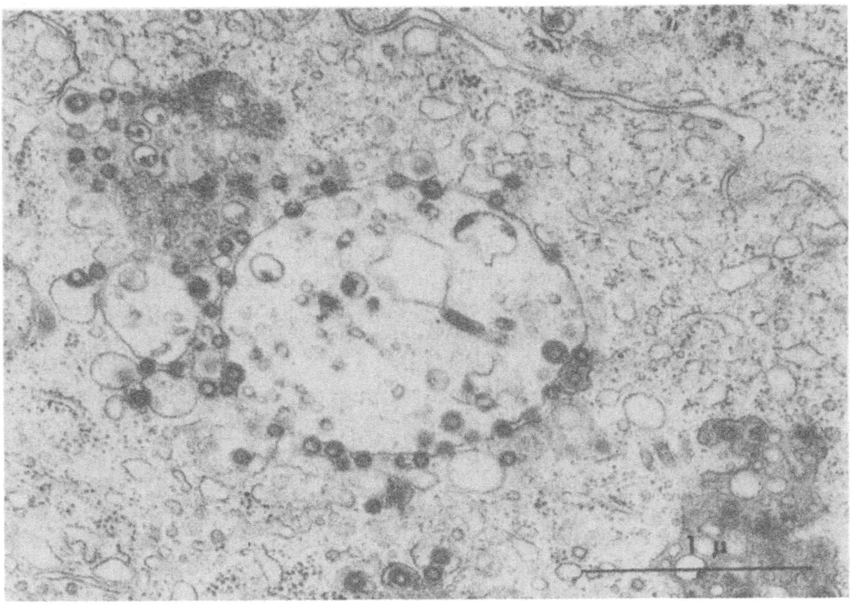

Fig. 3. Section of mammary adenocarcinoma cell of GF mouse (C3Hf). Doughnut-like particles (Type A) in cytoplasm, and the process of virla budding from the cell membrane is seen.

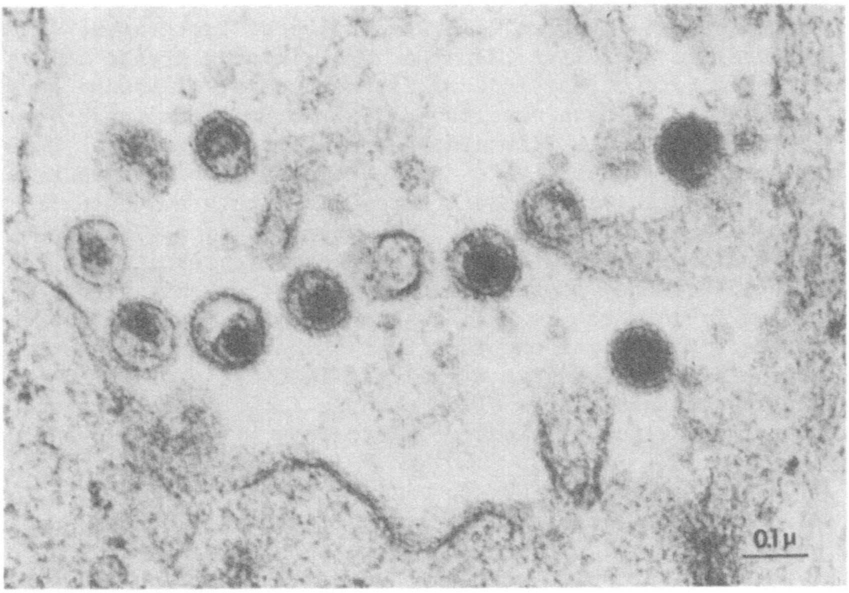

Fig. 4. Section of mammary adenocarcinoma cell of GF mouse (C3Hf). Tumor virus Type B and viral budding from microvilli can be observed.

obscure (22,24). The above studies done by electron microscopy indicate that the status of murine tumor virus in GF mice is quite similar to the results obtained by others in conventional mouse systems (22,23,24). Higher susceptibility of GF mice to leukemogenic treatments is reported (19,32).

Germfree rodents other than mice.-In addition to mice, other rodents such as rats, guinea pigs, and rabbits are now available as germfree animals. As far as I know, there has been no systematic search for virus infection in such animals, as described in germfree mice. Many known and unknown viruses may be able to infect rats, guinea pigs and rabbits naturally or experimentally; and they may manifest disease or stay in such animals as a latent infection. Some suggestive evidence of virus infection in GF rats has been given by the hemagglutinin test associated with the Kilham rat virus that was neutralized by the serum of the germfree rat (33), and that a new virus tentatively named rat submaxillary gland virus was isolated from GF rats as well as conventional ones (34). These results have not been confirmed in other laboratories as yet. It was reported that Lobund GF rats were free from antibody for Kilham's rat virus (7). Although virus-like particles resembling murine leukemia virus (22,23,24) has been noted by electron microscopy in chloroleukemia tissues of conventional rats (35) and in transplantable mammary tumor tissue (36) of conventional rats, we have failed to demonstrate any definite structure suggesting tumor virus in normal tissues and in various neoplastic tissues such as 7,12-Dimethylbenz(α)anthracene (DMBA)-induced myeloid leukemias, X-ray or DMBA-induced breast tumors and 3-Methylcholanthrene-induced fibrosarcomas, of Lobund germfree rats (26,37,38). These results are in contrast to those obtained in GF mice. Cell-free filtrates prepared from tissues of DMBA-induced myeloid leukemia were not leukemogenic in newborn rats and serodiagnostic methods failed to demonstrate evidence for the presence of Gross' leukemia virus in germfree rats with DMBA-induced myeloid leukemia (39). These negative results are not sufficient, but they do suggest that murine leukemia virus ubiquitous in germfree as well as conventional mice are not common in GF rats. Because of the fact that a leukemia virus was found in conventional guinea pigs with leukemia (40) it was planned to examine the tissues of the GF guinea pig for the presence or absence of the virus by electron microscopy. This is now being done.

In the future, the viral status of germfree rodents such as the rat, guinea pig, rabbit and others should be examined extensively as a base-line work, with special attention given to the viruses known or suspected to be transmitted vertically, for example Kilham rat virus (41), cytomegalovirus of the rat, reovirus and salivary gland virus of the guinea pig and guinea pig epizootics (LCM virus) etc. (42). It is quite possible to consider that other known or unknown viruses which are transmissible vertically will be detected in GF rodents with more extensive and advanced study in the future.

Summary

By the review of some previous studies in the search for virus in germfree rodents, the following conclusions can be drawn.
(a) Present germfree methods are very effective to minimize indigenous viruses in mice, presumably also in others, but not sufficient to exclude vertically transmitted viruses.
(b) Present germfree isolator system seems to prevent horizontal transmission of the virus because there is no evidence to suggest exogenous viral contamination through the barrier.
(c) Even so, the possibility of it can not be ruled out. Therefore it is desirable to monitor germfree animals periodically for viral contamination.
(d) At present, mouse leukemia virus is only known virus consistently detectable in all germfree mouse strains examined. In the future, it should be eliminated from germfree mice by more advanced methods or ideas.
(e) Viral status of germfree rats, guinea pigs and rabbits is relatively unknown. Therefore, it should be explored systematically as a baseline work in the future.

Acknowledgment

This paper was supported in part by a grant from the Leukemia Society of America, Inc. and the USPHS.

References

1. Reyniers, J.A.: Ann. N.Y. Acad. Sci., 78, 3 (1959).
2. Trexler, P.C.: Biomedical Purview, 1, 47 (1961).
3. Wagner, M.: Ann. N.Y. Acad. Sci., 78, 89 (1959).
4. Werderitsh, E.J., and Pollard, M.: Bacteriol. Proc., 66, (1963).
5. Rowe, W.P., Hartley, J.W. and Huebner, R.J.: Lab. Animal Care, 13, 166 (1963).
6. Parker, J.C., Tennant, R.W., Ward, T.G. and Rowe, W.P.: J. Nat. Cancer Inst., 34, 371 (1965).
7. Pollard, M.: Nat. Cancer Inst. Monogr., 20, 161 (1966).
8. Trentin, J.J., Van Hoosier, G.L. Jr., Shields, J., Stephens, K., Stenback, W.A., and Parker, J.C.: Ibid, 20, 147 (1966).
9. Weller, T.H., MacCauley, J.E., Craig, J.M. and Wirth, P.: Proc. Soc. Exp. Biol. Med., 94, 4 (1957).

10. Rowe, W.P., Hartley, J.W., Waterman, S., Turner, H.E. and Huebner, R.J.: Ibid, 92, 418 (1956).
11. Notkins, A.L.: Bacteriol. Rev., 29, 143 (1965).
12. Pollard, M., Kajima, M., and Sharon, N.: Perspectives in Virology, in press.
13. Traub, E.: J. Exp. Med., 69, 801 (1939).
14. Sabin, A.B.: Science, 130, 1387 (1957).
15. Cheever, F.S. and Mueller, J.H.: J. Exp. Med., 88, 309 (1948).
16. Cross, S.S. and Parker, J.C.: Bacteriol. Proc., 167 (1967).
17. Pollard, M., and Matsuzawa, T.: Proc. Soc. Exp. Biol. Med., 116, 967 (1964).
18. de Harven, E.: J. Exp. Med., 120 857 (1964).
19. Walburg, H.E. Jr., Uptom, A.C., Tyndal, R.L., Harris, W.W. and Cosgrore, G.E.: Proc. Soc. Exp. Biol. Med., 118, 11 (1965).
20. Pollard, M. and Kajima, M.: In the Proc. Intern. Conf. Radiation Biol. and Cancer, 175 (1967).
21. Pollard, M., Kajima, M. and Teah, B.A.: Proc. Soc. Exp. Biol. Med., 120, 72 (1965).
22. Bernhard, W.: Cancer Res., 20, 712 (1960).
23. de Harven, E.: In Tumor Induced by Viruses: Ultrastructural Studies, 183, (1962).
24. Dalton, A.J.: Federation Proc., 21, 936 (1962).
25. Kajima, M. and Pollard, M.: J. Bact., 90, 1448 (1965).
26. Kajima, M. and Pollard, M.: Cancer Res. 27, 980 (1967).
27. Kajima, M. and Pollard, M.: Electron Microscopy 1966, V.2, 215 (1966).
28. Kajima, M. and Pollard, M.: Nature 218, 188 (1968).
29. Pollard, M.: Unpublished observation.
30. Pilgrim, I.H.: (In form of discussion) Nat. Cancer Inst. Monogr., 20, 174 (1966).
31. Pollard, M.: Unpublished observation.
32. Mirand, E.A., and Grace, J.F.: Nature, 200, 92 (1963).
33. Kilham, L. and Olivier, L.J.: Virology 7, 428 (1959).
34. Ashe, W.K., Sherp, H.W. and Fitzgerald, R.J.: J. Bact. 90, 1719 (1965).
35. Weinstein, R.S. and Moloney, W.C.: Proc. Soc. Exp. Biol. Med., 118, 459 (1965).
36. Shirahama, S., Engle, C. and Dutcher, R.: Bacteriol. Proc., 174 (1968).
37. Pollard, M. and Kajima, M.: J. Nat. Cancer Inst., 39, 135 (1967).
38. Kajima, M. and Pollard, M.: Unpublished observation.
39. Sharon, N., Kajima, M. and Pollard, M.: Bacteriol. Proc., 175 (1968).
40. Opler, S.T.: Nature, 215, 184 (1967).
41. Kilham, L.: Nat. Cancer Inst. Monogr., 20, 117 (1966).
42. Andrews, C.: In Viruses of Vertebrates (1964).

Leukemia in Germfree AKR Mice

H. Ira Pilgrim, Richard C. Parks, and Lloyd W. Law

The Germfree Laboratory, Department of Surgery, University of Utah, Salt Lake City, Utah, and the National Cancer Institute, Bethesda, Maryland

Mice of the AKR strain have a very high incidence of lymphocytic neoplasms which appear to be caused by a vertically transmitted virus (see reviews by Kaplan, 1967; Moloney, 1962; Furth et al., 1964; Gross, 1958 and 1961). Since this agent is vertically transmitted, it is expected that germfree mice would also develop leukemia, unless the presence of a microbial flora, or other horizontally transmitted viruses, contribute to its development. That this is indeed the case has been demonstrated by Pollard and his co-workers (Pollard et al., 1966; Pollard, 1967) who showed that germfree AKR mice do develop leukemia in high incidence. The purpose of this paper is to compare the incidence of leukemia in germfree AKR mice with their conventional counterparts. We will also describe some of the characteristics of the germfree AKR mouse which make it somewhat different from other germfree inbred strains.

MATERIALS AND METHODS

AKR/Lw mice were foster-nursed on germfree Swiss mice. These animals were maintained by brother-sister mating in autoclavable aluminum isolators (Pilgrim and Thompson, 1962); were fed Wayne sterilizable Lab-Blox, and given canned water *ad libitum*. The animals used were breeding stock, and no attempt was made to distinguish animals on the basis of parity. The germfree animals were compared with AKR/Lw breeders kept at the National Cancer

---

Supported in part by Public Health Service Research Career Program Award CA-19,045; Grant #CA 05771 from the National Cancer Institute; and Contract #PH43-64-925 from the Cancer Chemotherapy National Service Center, National Institutes of Health.

Institute. These animals were fed a Morris diet (Derwood Farm Pellets), and given water ad libitum.

Autopsies were performed when animals died, or when they were sacrificed; which was done when they appeared to be debilitated. Sections were taken on sacrificed animals. Animals were considered to have leukemia when the thymus and/or lymphoid organs were grossly enlarged. Questionable diagnoses were confirmed histologically.

Some of the animals which are still alive at this writing are also included in Tables 1-3.

The data were evaluated by the method of Pilgrim and Dowd, (1963). This method makes a continuous correction for death due to extraneous causes. Animals dying of extraneous causes are utilized up to the time of death, and are then mathematically eliminated from subsequent consideration. The corrected data are then plotted by the standard actuarial technique of graphing the logarithm of the percent survivors against time. When plotted in this manner, the slope of any segment of the curve is the death rate; and the shape of the curve shows whether the mortality rate increases or decreases with age.

RESULTS

The raw data on causes of death at different ages in conventional, germfree, and monocontaminated AKR mice are presented in Tables 1, 2, and 3.

Figure 1 compares the occurrence of leukemia in germfree versus conventional mice. The first visible signs of disease appear somewhere between 4 and 6 months; following which tumor incidence increases rapidly. There is a difference of 1 to 2 months in the ages at which the male and female conventional mice develop leukemia. This is probably a real difference, as is evidenced from the fact that it has been consistently reported in the literature; and is a constant feature of the colony at the National Cancer Institute. This sex difference does not appear in the germfree animals. This may be because of the small size of the sample. A difference of 1 to 2 months would ordinarily not be considered as particularly significant unless it was a consistently repeatable occurrence. There is, however, no clear difference in the incidence of leukemia among germfree animals when compared to conventional mice. If a difference does exist, it is so small as to be of little or no biological significance.

The derivation of a truly germfree colony of AKR mice proved to be difficult. We made three separate derivations, by foster-nursing AKR mice on germfree Swiss mice; only to find that 2 of the 3 lines so derived were contaminated -- presumably via the transplacental route. One line was contaminated with Achromobacter delicatulus: This line was discarded. A second line was contaminated with Corynebacterium murisepticum. Due to the difficulty of isolating this organism, a large number of animals from this line were reared before the organism was detected. The third line was sterile by all methods used; this included culture on thioglycollate, blood, "L-form" media, and mycoplasma media. Figure 2 compares animals monocontaminated with C. murisepticum with the sterile line. There is no difference among the females, and little difference among the males.

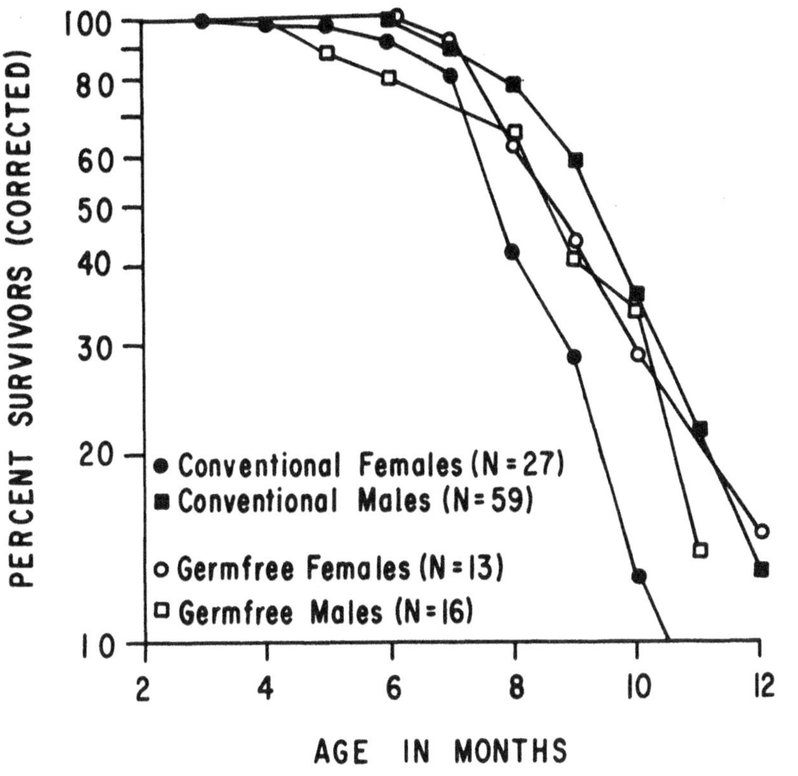

Figure 1

Plot of percent survivors against time, corrected to include only those mice exposed to the risk of developing leukemia (Pilgrim and Dowd, 1963). This figure compares germfree male and female AKR/Lw,Pi mice with conventional male and female AKR/Lw mice (see Tables 1 and 2 for raw data).

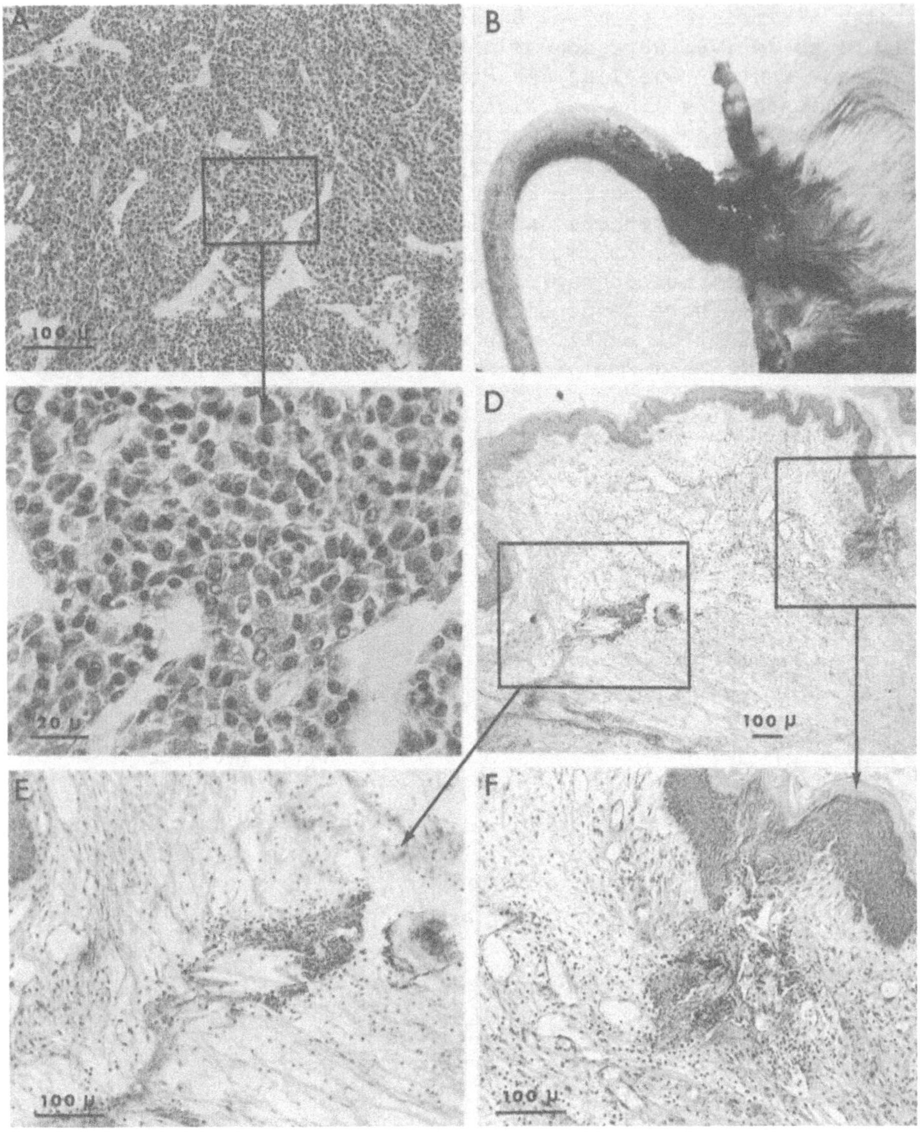

Plate 1

A and C: Inguinal node of mouse with proctitis. Note the large number of plasma cells (H&E).

B: AKR mouse with severe proctitis.

D, E, F: Sections through the skin of the anal area of a mouse with proctitis. Note the foreign bodies surrounded by neutrophils (H&E).

A syndrome involving inflammation of the anal area, similar to that described by Pollard (1967) was observed. The first noticeable sign was reddening around the anus. In animals containing *Corynebacterium murisepticum* this progressed to involve the entire base of the tail, resulting in extensive inflammatory changes and sloughing of the skin (Plate 1: B). The lymph nodes draining this area showed a typical chronic inflammatory response (Plate 1: A, C) with numerous plasma cells in the medullary cords. While this syndrome was severe in those animals contaminated with *C. murisepticum* it was also observed in animals that were sterile. *C. murisepticum* could be recovered from the lesions but we believe it to represent secondary contamination of the wound. Foreign bodies were observed embedded deeply in the anal area and surrounded by inflammation

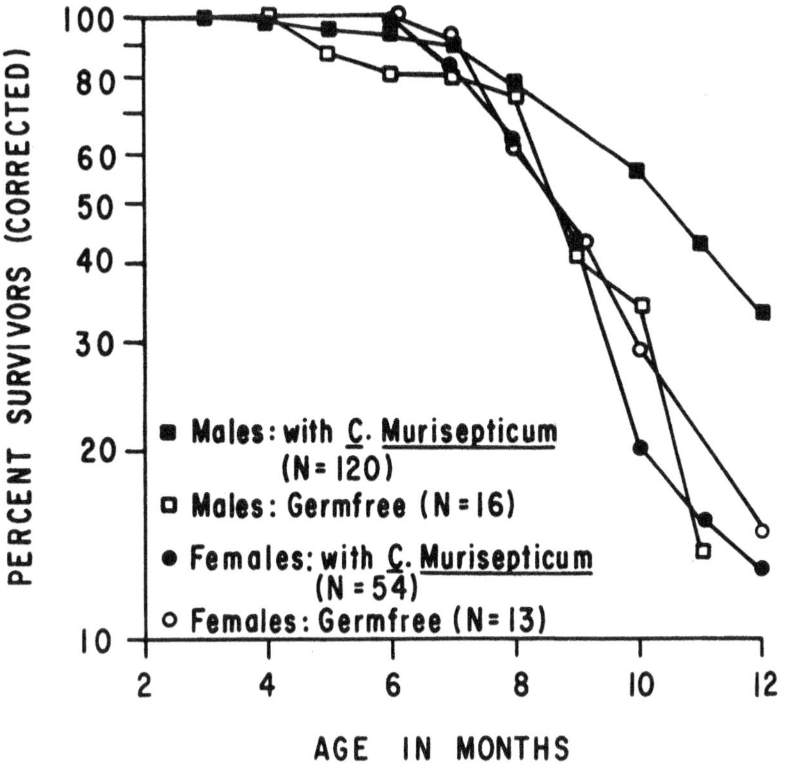

Figure 2

Plot of percent survivors against time, corrected to include only those mice exposed to the risk of developing leukemia (Pilgrim and Dowd, 1963). This figure compares male and female germfree AKR/Lw,Pi mice with similar mice found to be contaminated with *Corynebacterium murisepticum*. (See Tables 2 and 3 for raw data).

(Plate 1: D-F). Some of these were P.A.S. positive; others were not.

We can conclude that C. murisepticum is not necessary to the development of the initial lesion. We do not, however, know whether the foreign bodies observed in the lesions are the cause of the lesion or are secondary wound contaminants.

Figure 3 graphs the occurrence of leukemia in animals having proctitis against those not having this syndrome. One can conclude that if proctitis has any effect at all on the development of leukemia, it tends to delay it.

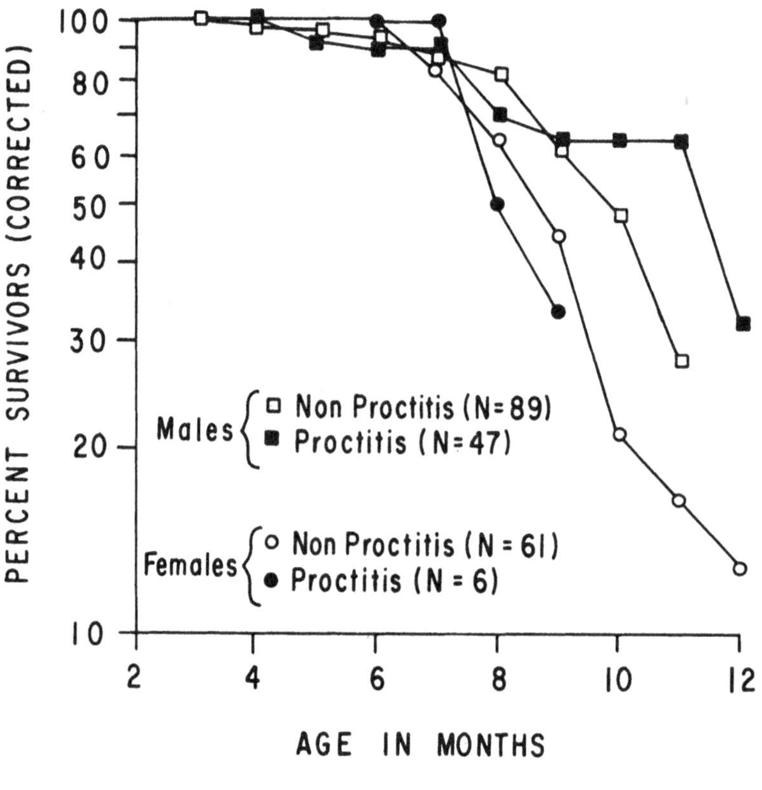

Figure 3

Plot of percent survivors against time, corrected to include only those mice exposed to the risk of developing leukemia (Pilgrim and Dowd, 1963). This figure compares male and female mice with and without proctitis. Most of the animals are also contaminated with C. murisepticum, although a few germfree animals also had mild proctitis.

DISCUSSION

The concave downward shape of the tumor decay curves in the germfree and conventional mice indicate that virtually every animal would develop leukemia if it lived long enough. In other words, the effective incidence of leukemia is 100%. Differences which exist between populations are not expressed as differences in incidence, but as differences in the age at which the leukemia develops. There is a difference of 1.6 months in the median age of tumor development between male and female conventional animals. This confirms the observations of others (see review by Gross, 1961). This difference is, however, so small that, were it not for numerous reports of such a difference in the literature, one might question its significance. It is difficult, therefore, to attach significance to differences between the germfree population and the conventional population. There is less than one month's difference between the median age of onset of leukemia between the conventional animals and their germfree counterparts. Conventional females have a median age of leukemia development of 7.8 months; conventional males have a median age of 9.3 months, while germfree males and germfree females have the same median age of 8.6 months. If there is any difference between germfree animals and conventional animals with regard to the development of leukemia, it is such a small one as to be of little biological significance.

Pollard (1967) has postulated that the "Puny Syndrome", which resembles the proctitis which we see in our animals, might possibly be a pre-leukemic syndrome. When we compare the occurrence of leukemia in animals with proctitis, with animals without it there is little or no difference. If any difference exists among the males, it is in favor of animals with proctitis developing less leukemia. Since animals which have proctitis are quite debilitated -- and therefore stressed -- it is not surprising that the occurrence of leukemia would be reduced since any stress will raise the age of occurrence of leukemia (see review by Gross, 1961).

CONCLUSIONS

We can conclude that the development of leukemia is not affected by the absence of a microbial flora. It also appears that the so-called "Puny Disease" described by Pollard is also not related to leukemia. Proctitis appeared in animals with and without detectable contaminants. It is not, therefore, related to detectable microorganisms; nor is it related to leukemia development. The fact that 2 of our 3 separately derived lines were contaminated, presumably via the transplacental route, indicates that this strain is different from most inbred strains in ways other than the development of leukemia.

		AGE IN MONTHS								
		4	5	6	7	8	9	10	11	12
Males	Died with Leukemia	0	0	0	3	3	5	6	3	2
	Died without Leukemia	0	0	0	0	0	1	1	0	1
	Number of Survivors	27	27	27	24	21	15	8	5	2
	Percent Survivors (Corrected)	100	100	100	89	79	59	34	22	12
Females	Died with Leukemia	1	0	3	7	23	8	9	2	1
	Died without Leukemia	0	0	0	0	0	1	2	1	0
	Number of Survivors	58	58	55	48	25	16	5	2	1
	Percent Survivors (Corrected)	98	98	93	81	42	29	12	6	3

TABLE 1

Distribution of causes of death at different ages in conventional AKR/Lw mice.

		AGE IN MONTHS								
		4	5	6	7	8	9	10	11	12
Females	Died with Leukemia	0	0	0	1	4	2	1	0	1
	Died without Leukemia	0	0	2	0	1	2	0	0	0
	Number of Survivors	15	15	13	12	7	3	2	2	1
	Percent Survivors (Corrected)	100	100	100	92	62	44	29	29	15
Males	Died with Leukemia	0	2	1	0	1	5	1	3	
	Died without Leukemia	1	0	1	0	0	0	0	2	
	Number of Survivors	16	14	12	12	11	6	5	0	
	Percent Survivors (Corrected)	100	100	88	81	75	41	34	14	

TABLE 2

Distribution of causes of death in germfree AKR/Lw,Pi mice.

		AGE IN MONTHS								
		4	5	6	7	8	9	10	11	12
Females	Died with Leukemia	0	0	0	9	11	10	10	2	1
	Died without Leukemia	17	2	8	0	2	3	1	0	0
	Number of Survivors	64	62	54	45	32	19	8	6	5
	Percent Survivors (Corrected)	100	100	100	83	63	43	20	15	13
Males	Died with Leukemia	1	3	2	3	8	7	5	6	4
	Died without Leukemia	15	12	13	8	10	6	1	2	1
	Number of Survivors	104	89	74	63	45	32	26	18	13
	Percent Survivors (Corrected)	99	96	94	90	79	67	56	43	34

TABLE 3

Distribution of causes of death in AKR/Lw,Pi mice found to be monocontaminated with *Corynebacterium murisepticum*.

Acknowlegments

The authors are inbedted to Dr. Gilbert Hill of the Department of Microbiology for the identification of the microorganisms discussed in this paper; to Mary Kelvington for her technical assistance; and to Dr. Barry C. Cole for culturing for mycoplasma.

REFERENCES

Furth, J.; Okano, H.; Kunii, A. 1964 Role of the thymus in Leukemogenesis. In: *The Thymus in Immunobiology: Structure, Function, and Role in Disease*. Harper & Row, Publishers, New York.

Gross, L. 1958 Viral etiology of "spontaneous" mouse leukemia: A review. Cancer Research, 18:371-381.

Gross, L. 1961 *Oncogenic Viruses*. Pergamon Press, Publishers, New York.

Kaplan, H. S. 1967 On the natural history of the murine leukemias: Presidential address. Cancer Research, 27:1325-1340.

McEndy, D. P.; Boon, M. C.; and Furth, J. 1944 The role of thymus, spleen, and gonads in the development of leukemia in a high-leukemia stock of mice. Cancer Research, 4:377-383.

Moloney, J. B. 1962 The murine leukemia. Fed. Proc., 21:19-31.

Pilgrim, H. I.; and Dowd, J. E. 1963 Correcting for extraneous death in the evaluation of morbidity or mortality from tumor. Cancer Research, 23:45-48.

Pilgrim, H. I.; and Thompson, D. B. 1962 An inexpensive, autoclavable germfree mouse isolator. Lab. Animal Care, 13:602-608.

Pollard, M. 1967 Preleukemic syndrome in germfree AK mice. Nature, 123:142-144.

Pollard, M.; Kajima, M.; and Teah, B. A. 1965 Spontaneous leukemia in germfree AK mice. Proc. Soc. Exper. Biol. & Med., 120:72-75.

RETICULAR NEOPLASMS IN IRRADIATED AND UNIRRADIATED GERMFREE MICE

H. E. Walburg, Jr. and G. E. Cosgrove

Biology Division, Oak Ridge National Laboratory

Oak Ridge, Tennessee

Although germfree (GF) animals were available in limited numbers in the early 1930's, studies on the pathogenesis of neoplasia in GF animals were not reported until the late 1950's, when tumors were induced in GF chickens by treatment with chemical carcinogens and oncogenic viruses (1). Spontaneous tumors in GF mice were reported by Ward in 1961 (2). Radiation-induced reticular neoplasms were first observed by Pollard and Matsuzawa (3) and subsequently by Walburg et al. (4). Spontaneous reticular neoplasms were first reported in germfree ICR mice by Walburg et al. (4) and subsequently in germfree Ak mice by Pollard et al. (5). None of these studies has adequately characterized the influence of microbial flora on development of the various types of reticular neoplasms which occur throughout life. This report summarizes the present status of our life-span studies on the latency and incidence of thymic and nonthymic lymphoma, myeloid leukemia and reticulum-cell sarcoma in irradiated and unirradiated germfree ICR, RFMf/Wg and AKRf/Wg mice.

MATERIALS AND METHODS

Mice used in these studies included: 1) germfree (GF) and conventional (CONV) noninbred ICR mice; 2) GF, CONV and conventionalized (CONVZ) inbred RFM mice and 3) GF and CONV inbred AKR mice. The derivation of GF lines and more detailed characteristics of the strains have been described elsewhere (4,6,7). Mice were irradiated with a GE Maxitron X-ray machine with the following factors: 300 kvp, 20 ma, hvl 0.98 mm Cu, and dose rate of approximately 100 R/min (6). The colonies were inspected twice daily, and dead or moribund mice were necropsied with histopathological examination whenever practical. With the exception of studies in AKR mice where the microscopic examinations are incomplete, all reticular neoplasms reported were identified histologically. The terminology used is similar to that in previous reports from this laboratory (8).

RESULTS

Preliminary experiments with ICR mice, which have been described in detail elsewhere (9), suggested that the microbial flora might influence the development of reticular neoplasms. Germfree male mice of this strain, whether irradiated or not, consistently had a higher incidence of lymphoma than their CONV counterparts and unirradiated GF mice of both sexes had a lower incidence of reticulum cell sarcoma than the CONV mice (Table 1). Additional studies were needed, however, since the numbers of animals involved were relatively small and the differences were not statistically significant ($P>0.01$).

Subsequent experiments with larger numbers of mice of a different strain, the RFM, supported these initial observations and more clearly demonstrated the influence of microbial flora on the development of reticular neoplasms. In these experiments the incidences of reticular neoplasms in GF mice reared and tested as previously described (4) were compared with the incidences of reticular neoplasms in CONV mice which were fed a standard diet (Purina Lab Chow) and supported an undefined but presumably typical microbial flora. Also compared were the incidences of

TABLE 1. Incidence of Reticular Neoplasms In Irradiated and Unirradiated Germfree and Conventional ICR Mice.

X-ray Dose(R)	Sex	Microbial Status	No. Mice	Incidence (%) Lymphoma	Incidence (%) Reticulum Cell Sarcoma
0	♀	GF	19	0	39
		CONV	20	0	45
	♂	GF	22	26	18
		CONV	24	16	33
600-900	♂	GF	31	32	6
		CONV	36	14	4

TABLE 2. Incidence of Reticular Neoplasms In Irradiated and Unirradiated Germfree, Conventional, and Conventionalized RFM Mice

X-ray Dose	Sex	Microb. Status	No. Mice	Incidence (%) Myel.	Thymic Lymph.	Nonthymic Lymphoma	Reticulum Cell Sarcoma
0	♀	GF	65	0	7	5	35
		CONV	113	0	4	3	59
	♂	GF	55	0	4	0	37
		CONV	99	1	2	3	55
30C	♀	GF	98	0	52	11	5
		CONV	91	7	41	17	16
		CONVZ	122	7	44	17	18
	♂	GF	82	1	37	4	21
		CONV	99	13	19	8	39
		CONVZ	117	20	15	10	24

such neoplasms in irradiated CONVZ mice whose ancestors had been removed from the GF colony and reared in an isolation facility, where they gradually accumulated a microbial flora consisting of a few bacterial species. At the time of irradiation, they were removed from this isolation facility and subjected to microbial shock by being placed in the same microbial environment as inhabited by the CONV mice, where they were exposed to a large number of microorganisms, some of them pathogens, with which they had no previous immunological experience.

The data are shown in Table 2. Radiation induced the development of myeloid leukemia in mice reared in the conventional microbial environment (7% in irradiated females and 13-20% in irradiated males) but failed to induce the disease in GF mice. A more detailed discussion of the reduced incidence of myeloid leukemia in germfree mice is presented elsewhere (10). As was the case in ICR mice, germfree RFM mice in this experiment had a consistently higher incidence of lymphoma than their conventional counterparts. The increased incidence was noted only with thymic lymphoma, since nonthymic lymphoma was as low or lower in GF as in CONV and CONVZ mice. Although the incidence of myeloid leukemia is lower in GF mice and the incidence of thymic lymphoma is higher, the total incidence of radiation-induced reticular neoplasms, i.e., lymphoma and myeloid leukemia, are not significantly different. For example, 42% of the GF male irradiated mice had such neoplasms, as compared with 40% and 45%, respectively, of the CONV and CONVZ irradiated males.

The incidence of reticulum-cell sarcoma, a late-occurring reticular neoplasm was reduced in all irradiated groups. This reduced incidence was primarily due to the increased mortality from early-occurring reticular neoplasms (thymic lymphoma and myeloid leukemia) in the irradiated mice, with a resultant decrease in the number of animals at risk to reticulum cell sarcoma. Despite a greater longevity (11) and thus more mice at risk to the neoplasm, GF mice in all groups, both irradiated and unirradiated, had a lower incidence of reticulum-cell sarcoma than their CONV or CONVZ counterparts.

The spontaneous development of reticular neoplasms is also being studied in germfree AKR mice. The total incidence of these neoplasms observed at necropsy is slightly but not significantly lower in GF than in CONV mice. CONV female AKR mice had a 97% incidence of such neoplasms, as compared with 94% in GF females and CONV males had a 92% incidence, as compared with 87% in GF males. Although histological examination of the neoplastic tissue is not complete at this time, certain observations have been made. In both GF and CONV mice most of the neoplasms occurring during the first 10 months are thymic lymphomas with generalized dissemination to many other organs. By the end of the first year, the lymphomas are often mixed with neoplastic reticular cells, and later neoplasms are more often exclusively reticulum cell sarcoma. No difference in the development of reticular neoplasms between GF and CONV females is noted until approximately 300 days of age (Fig. 1). In the last third of life, GF females show an accelerated death rate from these neoplasms when compared with their CONV counterparts. A similar difference is noted in a comparison of male AKR mice reared in GF and CONV environments, but the difference appears at the beginning of mortality and is maintained throughout life.

DISCUSSION

The role of virus-like agents in the pathogenesis of mouse reticular neoplasms is well established (12,13). The AKR viral agent was first described by Gross in 1951 (14), and viruses have been implicated in the development of radiation-induced myeloid leukemia (15) and thymic lymphoma (16) in RFM mice. Free virus has been demonstrated in untreated AKR mice (14), and radiation has been shown to release leukemogenic virus in other strains (17). Virus-like particles have been observed in many germfree mouse strains, including the Ak (5) and the RFM (4), and the development of reticular neoplasms in these mice suggests that at least some of these particles are leukemogenic viruses.

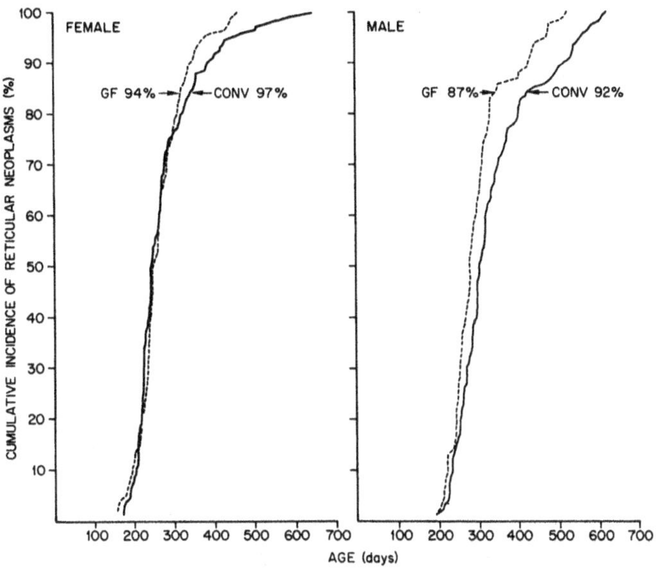

Figure 1. Influence of the germfree environment on cumulative incidence of reticular neoplasms observed at necropsy in male and female AKRf/Wg mice.

Conventional ———————— ♀ (156), ♂ (135)
Germfree - - - - - - - - - - ♀ (100), ♂ (99)

Numbers in parentheses represent number of animals entering experiment.

Possible mechanisms for the absence of radiation-induced myeloid leukemia in GF RFM mice have been discussed previously (10). The most attractive hypothesis at present is that absence of microbial flora and a decrease in inflammatory reactions influence the number and/or susceptibility of granulocyte precursor cells to neoplastic transformation by the leukemogenic virus. The total incidence of all types of radiation-induced reticular neoplasms in RFM mice is not affected by the microbial flora, but the decreased incidence of myeloid leukemia and increased incidence of thymic lymphoma in irradiated GF mice of this strain may reflect an alteration in the relative number and susceptibility of thymic and granulocytic precursor cells.

It appears evident that other factors are involved, since thymic lymphoma is also increased in unirradiated GF mice where the number and population kinetics of cells in thymus (18) and lymph nodes (19) are not increased or are reduced. The greater susceptibility to development of spontaneous thymic lymphoma in germfree mice may reflect a defect in immunological interference with virus infection or rejection of transformed cells. Our studies on the immunological response of germfree mice to particulate (20), soluble (21), and tissue antigens (22) does not support such an interpretation.

The increased incidence of reticulum cell sarcoma in CONV mice may be related to an increased antigenic stimulation of reticular cells in these mice. The most frequent site of origin of reticulum cell sarcoma in RFM mice is the mesenteric lymph nodes which drain the intestines, where the greatest amount of foreign antigenic material occurs. It has been demonstrated that antigen is trapped by the reticular cells which form the stroma of the germinal center (23,24). The decrease in antigenic stimulation of the mesenteric lymph nodes of GF mice is reflected in the decreased cellularity, DNA synthesis and number of germinal centers of that organ in GF mice (19).

REFERENCES

1. Reyniers, J. A. and Sacksteder, M. R. 1959. Ann. N. Y. Acad. Sci. 78: 328.
2. Ward, T. G. 1961. Fed. Proc. 20: 150.
3. Pollard, M. and Matsuzawa, T. 1964. Proc. Soc. Exp. Biol. Med. 116: 967.
4. Walburg, H. E., Jr., Upton, A. C., Tyndall, R. L., Harris, W. W., and Cosgrove, G. E. 1965. Proc. Soc. Exp. Biol. Med. 118: 11.
5. Pollard, M., Kajima, M. and Teah, B. A. 1965. Proc. Soc. Exp. Biol. Med. 120: 72.
6. Walburg, H. E., Jr., Mynatt, E. I. and Robie, D. M. 1966. Rad. Res. 27: 616.
7. Staats, J. 1967. Inbred Strains of Mice 5: 86.
8. Upton, A. C., Kimball, A. W., Furth, J., Christenberry, K. W. and Benedict, W. A. 1960. Cancer Research 20, No. 8 (Part 2): 1.
9. Walburg, H. E., Jr. and Cosgrove, G. E. 1967. Exp. Geront. 2: 143.
10. Walburg, H. E., Jr., Cosgrove, G. E. and Upton, A. C. 1968. Int. J. Cancer 3: 150.

11. Walburg, H. E., Jr., Cosgrove, G. E. and Upton, A. C. 1966. Radiation and Ageing. Proceedings of a Colloquium held in Semmering, Austria, June 23-24, 1966. Eds. P. J. Lindop and G. A. Sacher. pp. 361-365.

12. Gross, L. Oncogenic Viruses. International Series of Monographs on Pure and Applied Biology. Vol. 11. Pergamon Press 1961.

13. Moloney, J. B. 1962. Fed. Proc. 21: 19.

14. Gross, L. 1951. Proc. Soc. Exp. Biol. Med. 76: 27.

15. Jenkins, V. K. and Upton, A. C. 1963. Cancer Research 23: 1748.

16. Hiraki, K., Irino, S., Sota, S., and Ikejiri, K. 1962. Acta Haemat. Jap. 25: 816.

17. Haran-Ghera, N. 1966. Int. J. Cancer 1: 81.

18. Bealmear, M. and Wilson, R. 1966. Anat. Record 154: 261.

19. Olson, G. B. and Wostmann, B. S. 1966. J. Immunol. 97: 267.

20. Bosma, M. J., Makinodan, T. and Walburg, H. E., Jr. 1967. J. Immunol. 99: 420.

21. Hanna, M. G., Nettesheim, P., Walburg, H. E., Jr. 1968. Adv. Med. and Biol., this issue.

22. Walburg, H. E., Jr., and Cosgrove, G. E. 1968. Adv. in Exptl. Med. and Biol., this issue

23. Nossal, G. J. V., Abbot. J. M., and Lummus, Z. 1968. J. Exp. Med. 127: 277.

24. Szakal, A. K. and Hanna, M. G., Jr. 1968. J. Exp. Mol. Pathol. 8: 75.

(Research sponsored by the U. S. Atomic Energy Commission under contract with the Union Carbide Corporation.)

COMPARISON BETWEEN THE EFFECTS OF A SINGLE DOSE OF A CHEMICAL CARCINOGEN ON GNOTOBIOTIC AND CONVENTIONAL MICE

G. A. Grant & F. J. C. Roe

Chester Beatty Research Institute, Institute of Cancer Research, Fulham Road, London

In 1966 Miller and collaborators reported that neonatally thymectomised germ-free mice showed less immunological impairment, as measured by ability to reject foreign skin grafts, than their conventional counterparts. They concluded that the degree of immunological impairment caused by neonatal thymectomy is intensified by factors associated with conventional environmental conditions.

This finding suggested that under conventional conditions uncontrolled environmental factors may intensify slight immunological impairment and these should be taken into account in experiments where the results are known to be influenced by immunological competence.

Immunological impairment by neonatal thymectomy has been shown to facilitate the production and transplantation of polyoma virus induced tumours in hamsters and mice (Defendi and Roosa, 1965; Law, 1966) and to increase the incidence of tumours induced by some chemical carcinogens (Grant and Miller, 1965; Grant, Roe and Pike, 1966; Trainin, 1966; Johnson, 1968). It was thought, therefore, that the response to carcinogens might be different in gnotobiotic and conventional mice. In order to investigate this problem we tried to choose a carcinogenic stimulus such that only approximately half the animals exposed were likely to develop tumours in the course of 30 to 40 weeks. This, it was thought, would offer the best chance of detecting a small difference in response.

Two experiments have been completed both on C3H mice obtained "germ-free" from Gif-sur-Yvette, France and originating from the Roscoe B. Jackson Memorial Laboratory. Gnotobiotic mice were maintained in flexible film Trexler type isolators and conventional mice in a modern barrier-maintained "minimal disease" animal unit.

All mice were fed Allington Farm PRM diet sterilized by 5 Mrad gamma radiation at Wantage Research Laboratory.

Bacterial checks were carried out weekly on water, bedding and faeces from each isolator. Samples were incubated at 26°, 37° and 45°C on a range of media.

EXPERIMENT 1

Seven-day old mice were injected subcutaneously with 25 µg 9-10, dimethylbenzanthracene (DMBA) dissolved in arachis oil, the material being deposited between the scapulae. The carcinogen was sterilized by heating to 120°C on 3 separate occasions and material taken from the same batch was used for conventional and gnotobiotic mice. Animals were examined weekly for neoplasms at the injection site and those discovered measured and recorded when they had reached a size of 5 mm diameter. For the first 11 weeks of the experiment no bacterial or fungal contamination was detected but on week 12 two organizms, a Micrococcus sp. and Bacillus sp. were detected. The experiment was continued for a further 24 weeks with these two organisms but with no signs of further contamination.

TABLE 1. Development of Injection Site Tumours by C3H Mice given 25 µg DMBA subcutaneously at One Week of Age.

Status	Sex	Tumour-bearing mice/mice at Risk at Age designated (weeks)			
		18	24	30	36
Gnotobiotic	♂	0/16	0/16	2/16	2/16
Conventional	♂	1/28	5/28	5/27	7/24
Gnotobiotic	♀	1/15	1/15	3/13	5/13
Conventional	♀	4/25	9/25	11/24	11/24
Gnotobiotic	♂+♀	1/31(3%)	1/31(3%)	5/29(17%)	7/29(24%)
Conventional	♂+♀	5/53(9%)	14/53(26%)	16/51(31%)	18/48(37%)

TABLE 2. Comparison between Spleen and Liver Weights in Gnotobiotic and Conventional C3H Mice treated with DMBA.

Status	No. of Animals	Sex	Treatment	gms liver/100 gm Body wt (± S.D.)	Mg Spleen/100 gms Body wt (± S.D.)
Gnotobiotic	16	♂	No treatment	5.0 ± 1.0	199 ± 49
Conventional	23	♂	" "	6.4 ± 1.0	260 ± 84
Gnotobiotic	14	♂	DMBA (No tumour)	5.1 ± 0.8	233 ± 80
Conventional	17	♂	DMBA (No tumour)	7.0 ± 1.6	277 ± 86 *
Gnotobiotic	2	♂	DMBA (tumour+ve)	4.0	347 †
Conventional	7	♂	DMBA (tumour+ve)	6.3	557 †
Gnotobiotic	13	♀	No treatment	4.4 ± 0.4	264 ± 58
Conventional	23	♀	" "	5.5 ± 1.2	360 ± 113
Gnotobiotic	9	♀	DMBA (No tumour)	4.5 ± 0.4	292 ± 48
Conventional	13	♀	DMBA (No tumour)	5.9 ± 0.6	302 ± 29
Gnotobiotic	5	♀	DMBA (tumour+ve)	4.5	633 †
Conventional	10	♀	DMBA (tumour+ve)	6.6 ± 1.0	540 ± 216

* One mouse with necrotic areas in kidney and exceptionally large spleen omitted.

† Standard deviation omitted because too few animals in group.

Table 1 shows that injection site tumours arose more rapidly in conventional mice than in gnotobiotic mice and that by the time the experiment terminated at 36 weeks a higher proportion of conventional mice had developed tumours. Between the 24th and the 28th weeks, inclusively, the differences in tumour incidence were significant at the $P<0.01$ level.

Histologically the majority of the tumours were spindle cell sarcomas but there were some pleomorphic and some mixed spindle-cell and pleomorphic sarcomas. No difference could be detected between the sarcomas in gnotobiotic and conventional animals. Three mice, all of them conventional, developed papillomas at the site of injection.

Tumour-bearing animals were killed when tumours reached 15-20 mm diameter. All animals killed either during or at the end of the experiment were examined post mortem. Body weight and the weights of liver, spleen and kidneys were recorded.

All male mice, conventional or gnotobiotic, given DMBA and examined post mortem at 30+ weeks had one or more hepatomas. One female DMBA-treated gnotobiotic mouse had a hepatoma, so did one untreated conventional male.

Five mice (3 conventional females, 1 conventional male and 1 gnotobiotic female) developed pulmonary adenomas.

Table 2 shows spleen and liver weights relative to body weight for DMBA treated and untreated gnotobiotic and conventional animals. Treatment with DMBA had little effect on the mean size, or variation in size, of the spleen but the presence of a large sarcoma had a considerable effect in both groups.

EXPERIMENT 2

Comparable gnotobiotic and conventional five week old C3H mice were injected in the muscle of the right thigh with 50 µg 3-methylcholanthrene dissolved in 0.1 ml arachis oil. The solution was sterilized by heating to 140°C for 2½ hours. No bacterial or fungal contamination of the gnotobiotic mice was detected throughout the experiment.

Tumours began to arise after a latent period of 10 weeks and by 20 weeks most mice had developed injection site sarcomas (Table 3). The exceptions were one male out of 25 conventional mice and two females and one male out of 26 gnotobiotic mice. In these 4 mice the right lumbar node was enlarged but there was no evidence of an injection site tumour. In all the remaining mice tumours arose

TABLE 3. Development of Injection-site Sarcomas in C3H Mice following the Injection of 50 μg 3-Methylcholanthrene in 0.1 ml Arachis Oil when the Mice were Aged 5 weeks.

Status	Sex	No. of mice at risk	Tumour bearing mice (at time in weeks after injection)			
			13	15	17	19
Gnotobiotic	♂	14	2	5	9	11
Conventional	♂	11	2	5	10	10
Gnotobiotic	♀	12	2	4	7	9
Conventional	♀	14	2	8	12	14
Gnotobiotic	♂+♀	26	4	9	16	20
Conventional	♂+♀	25	4	13	22	24

after only a short latent interval and the difference between gnotobiotic and conventional mice although in the same direction as in the previous experiment, was not significant. Clearly in this experiment the dose of carcinogen exceeded the optimal.

Despite the results of the second experiment, the findings as a whole suggest that the absence of bacteria (and horizontally transmitted virus) may affect tumour incidence especially if a marginal dose of an oncogenic agent is used. Further experiments are planned by which it is hoped to define the limits within which the bacterial status of the experimental animal may influence response to carcinogenic stimuli.

TABLE 4. Comparison between Gnotobiotic and Conventional C3H Mice given 50 μg 3-Methylcholanthrene in 0.1 ml Arachis Oil.

Status	No. of Animals	Sex	Treatment	Mg Spleen/100 gms Body Wt.
Gnotobiotic	10	♂	Arachis oil	169 ± 14
Conventional	11	♂	" "	209 ± 45
Gnotobiotic	12	♂	MC (tumour+ve)	427 ± 159
Conventional	9	♂	MC (tumour+ve)	620 ± 252
Gnotobiotic	9	♀	Arachis oil	246 ± 36
Conventional	10	♀	" "	311 ± 54
Gnotobiotic	10	♀	MC (tumour+ve)	637 ± 264
Conventional	14	♀	MC (tumour+ve)	577 ± 281

REFERENCES

Defendi, V., Roosa, R. A. 1965. Cancer Res., 25, 300-306.
Grant, G. A.; Miller, J. F. A. P. 1965. Nature, 205, 1124.
Grant, G.A.; Roe, F. J. C.; Pike, M. C. 1966. Nature, 210, 603.
Johnson, S. 1968. Brit. J. Cancer, 22, 83-92.
Law, L. W. 1966. Cancer Res., 26, 551-574.
Miller, J. F. A. P.; Dukor, P.; Grant, G.; Sinclair, N., Sacquet, E. 1967. Clin. Exp. Immunol., 2, 531-542.
Trainin, N. 1966. 9th Int. Cancer Congress Abstracts 50183, 127.

ROUS SARCOMA VIRUS IN GERMFREE RATS*

Morris Pollard, Masahiro Kajima and Theodore P. Zacharia

Lobund Laboratory, University of Notre Dame

Notre Dame, Indiana

Seven years ago, there was general agreement among investigators that germfree animals were free of all microbial agents. Since then, extensive surveys, based on serological and tissue culture procedures have failed to reveal the presence of specific viral agents in them (1, 2, 3). Additional examinations, however, revealed virus-like particles in the thymic tissues of germfree mice, which were identified as leukemia virus when mice of the same strains were exposed to leukemogenic doses of x-rays (4, 5). Thus far, no mouse strain (germfree or conventional) has been found free of leukemia virus (6). It is very possible that those viruses which are transmitted to progeny by "vertical" or congenital route(s) would be found in germfree mice which had been derived from virus-contaminated stock. Thus far, three viral agents have been detected in germfree mice: leukemia virus, mammary tumor virus, and lymphocytic choriomeningitis virus (7, 8); and it is likely that others will be added to this list. For these reasons we feel that the "germfree" mouse cannot serve as an experimental subject to delineate the exclusive role of viruses in carcinogenesis, since a virus-free mouse is not yet available.

Viruses have not been detected in germfree rats of three strains (Wistar, Sprague-Dawley, and Fischer). The search for such agents has been extensive, and the results have been negative by those techniques with which viruses were disclosed in "germfree" mice. Germfree rats developed the same spectrum of neoplastic

* Supported by grants from The John A. Hartford Foundation and the U.S. Public Health Service (#CA 07271).

diseases following exposures to chemical and physical oncogenic stimuli as the conventional stock from which they had been derived; and viruses have not been detected in their tissues (9, 10). The germfree rats have provided unique experimental advantages for determination of host responses to neoplastic diseases: their lymphoreticular tissues were relatively inactive. They showed very few germinal zones, very few plasma cells, the splenic red pulps were relatively cell-free and inactive, and the serum globulin levels were uniformly about one-third that of conventional rats (11). In rats with tumors induced by subcutaneous injections of 3-methylcholanthrene in oil (fibrosarcomas) and by oral administration of 7,12 dimethylbenzanthracene in sesame oil (breast carcinomas), the lymph nodes were free of germinal zones, but had increased numbers of plasma cells. In such rats, the spleens were free of germinal zones, but the red pulps contained increased numbers of reticulum cells and small hyperchromatic lymphocytes (12). Tumor extracts did not contain demonstrable complement-fixing antigens with serums from the donor of the same tumor. This served to confirm the evidence accumulated by electron microscopy, tissue cultures, and serologies that the germfree rats and their tumors were virus-free. Based on studies with polyoma-, adenovirus 12-, and other induced tumors in mice, if viruses were involved in the oncogenic processes of rats, they should be detectable by serological procedures (13, 14). This was not demonstrable in the rats, bearing tumors which had been induced by chemical and physical agents.

As a further examination of the host response to neoplastic diseases, germfree and conventional Wistar and Sprague-Dawley rats were inoculated subcutaneously, at birth, with Schmidt-Ruppin strain Rous sarcoma virus (RSV) obtained from Dr. Ray Bryan, National Cancer Institute. They all developed large cysts in the axillary and inguinal regions and eventually subcutaneous fibrosarcomas. In some of the rats, tumors were located also in the lungs.

Rats which carried the RSV-induced primary tumors showed few unique tissue changes: the lymph nodes contained no germinal zones and the medullary areas had increased numbers of plasma cells. The spleens had no germinal zones and the red pulp areas contained increased numbers of reticulum cells and small lymphocytes. (It is of interest to note that one passage line of tumor cells in Wistar rats was associated with leukemoid changes in which circulating myelocyte levels above 100,000/cu ml were observed.

The tumors were transplanted by cell suspensions to newborn germfree and conventional rats of the same strain and to newly-hatched chicks; however, cell-free extracts of the rat tumors were not oncogenic in rats and chicks. Virus particles were not detected by electron microscopy in the tumor tissues of the rats, nor were complement fixing antibodies and antigens detected in the serums and tumors of the tumor-bearing rats. The original stock of Schmidt-Ruppin RSV induced large fibrosarcomas in inoculated chicks, in which numerous virus particles were observed by electron microscopy. Thus, RSV induced local and metastatic fibrosarcomas in neonatally-inoculated rats, but subsequently virus was not demonstrable in them. Possibly the virus was defective and required an associated virus for maturation, which was not available (15). In any case, the virus disappeared.

With many other investigators, we have conceded that viruses were oncogenic, but we have also contended that tumors could be induced by nonviral agents. The effects of RSV in germfree rats has served to modify this view. RSV induced neoplasms in germfree rats in which (a) changes in the lymph nodes and spleens resembled those in rats with MCA- and DMBA-induced tumors; (b) virus particles were not detected in the tumor cells; (c) cell-free extracts of the tumors were not oncogenic, and (d) in addition, no antigen-antibody relationship could be demonstrated in the tumor-bearing rats. The RSV-induced tumors differed little from the tumors which were induced by sterile chemical and physical agents in germfree rats. If virus-induced tumors resemble virus-free tumors, then are the latter actually free of virus?

We have maintained that viruses do not have an exclusive role in oncogenesis. However, in view of our observations on RSV-inoculated germfree rats, a reassessment of the problem is needed - especially as it may pertain to neoplasms in man from which viruses have not yet been detected. Some human tumors may actually be exhibiting a pattern of activity similar to that of RSV in rats.

REFERENCES

1. Pollard, M. 1963. The search for contaminants in gnotobiotic rats and mice. Proc. XVI Int. Congr. Zoology, Washington, D.C. (Abst.)

2. Parker, J. C., R. W. Tennant and T. G. Ward. 1965. Virus studies with germ-free mice. I. Preparation of serologic diagnostic reagents and survey of germ-free and monocontaminated mice for indigenous murine viruses. J. Nat. Cancer Inst., 34: 371.

3. Pollard, M. 1965. Viral status of "germ-free" mice. Nat. Cancer Inst. Monograph No. 20, p 167.

4. de Harven, E. 1964. Virus particles in the thymus of conventional and germ-free mice. J. Exp. Med., 120: 857.

5. Pollard, M. and T. Matsuzawa. 1964. Radiation induced leukemia in germfree mice. Proc. Soc. Exp. Biol. and Med., 16: 967.

6. Kajima, M. and M. Pollard. 1968. Wide distribution of leukemia virus in strains of laboratory mice. Nature, 218: 188.

7. Kajima, M. and M. Pollard. 1965. Detection of virus-like particles in germfree mice. J. Bacteriol., 90: 1448.

8. Pollard, M., N. Sharon and B. A. Teah. 1968. Congenital lymphocytic choriomeningitis virus infection in gnotobiotic mice. Proc. Soc. Exp. Biol. and Med., 127: 755.

9. Pollard, M. 1965. The use of germfree animals in virus research. Progr. Med. Virol., 7: 362.

10. Pollard, M. and M. Kajima. 1966. Leukemia in germfree rats. Proc. Soc. Exp. Biol. and Med., 121: 585.

11. Pollard, M. 1967. Germinal centers in germfree animals. In: "Germinal Centers in Immune Responses" (Eds.: H. Cottier and N. Odarchenko, R. Schindler and C. C. Congdon). Springer Verlag, N.Y. pp 343-346.

12. Pollard, M., M. Kajima and G. Lorans. 1968. Tissue changes in germfree rats with primary tumors. J. Reticuloendothelial Soc., 5: 147.

13. Habel, K. 1965. Specific complement-fixing antigens in polyoma tumors and transferred cells. Virology, 25: 55.

14. Huebner, R. J., W. P. Rowe, H. C. Turner and W. T. Lane. 1963. Specific adenovirus complement-fixing antigens in virus-free hamster and rat tumors. Proc. Nat. Acad. Sci. - USA, 50: 379.

15. Hanafusa, H., T. Hanafusa and H. Rubin. 1963. The defectiveness of Rous sarcoma virus. Proc. Nat. Acad. Sci. - USA, 49: 572.

7,12-DIMETHYLBENZ(α)ANTHRACENE-INDUCED MALIGNANT LYMPHOMA IN GERMFREE MICE

Masahiro Kajima

Lobund Laboratory, Department of Microbiology

University of Notre Dame, Notre Dame, Indiana

It has been well-established that conventional rodents that received a single subcutaneous injection of 7,12-Dimethylbenz(α)-anthracene (DMBA) develop local and/or distant neoplasms(1-10). The age of treated rodents, the dose of carcinogen, the difference in species of animals used and other factors (sex) seems to influence the incidence and type of DMBA-induced tumors(1,2,5,6,9). The incidence of malignant lymphoma in conventional mice is increased by subcutaneous injection of small dose of DMBA at birth (1-4). On the base of previous works by others, however, there is no such high incidence of lymphoma in other rodents, such as the rat(7,11), hamster(8,9) and rabbit(10) similarly treated with DMBA. At present, it is unknown what factor(s) makes this difference in response of mice distinct from other species to DMBA-injection at birth, although it can be assumed that leukemia virus widely present in "normal" laboratory mice may be responsible for this phenomenon.

High incidence of malignant lymphoma has been reported in special strains of mice (AKR, C58); and in mice treated with ionizing radiation, chemical or hormone and also in mice or rats inoculated with leukemia viruses(12,13). Thymic lymphoma is predominant in aged AKR strain mice(14), in C57Bl mice treated with whole body X-irradiations(15), and in Swiss mice injected with DMBA(16). In these, the first neoplastic cells seem to appear in the thymic tissue. In this regard, we have examined thymic tissues of "germfree" AKR strain mice and other Lobund "germfree" mice with radiation-induced lymphatic leukemia by electron microscopy for leukemia virus particles; and we have found virus particles in the tissues(17,18). As part of the continuing research of this type, the present study was undertaken to investigate(a) leukemogenic response of germfree mice to DMBA-injection, and (b) characteristics of DMBA-induced malignant lymphoma by electron microscopy.

Materials and Methods

Mice-Lobund germfree Swiss-Webster strain mice (both sexes) were used in this study. They were maintained under bacteria-free conditions for the duration of the experiment.

Chemicals-7,12-Dimethylbenz(α)anthracene (DMBA) was obtained from Eastman Kodak Co.

Preparation and administration of DMBA and control-DMBA was dissolved at the concentration of 100 μg in 0.05 ml of sesame oil, sterilized in ampoules, and introduced aseptically into the germfree isolator. Fourteen newborn germfree mice were inoculated into the subcutaneous tissue of the interscaplar area with 0.05 ml sesame oil containing 100μg of DMBA. Seven control newborn germfree mice were treated in a similar manner with 0.05 ml of sesame oil only.

Sample preparations and observations-The animals were observed daily for any sign of disease. Moribund mice were removed from the germfree isolator and examined macroscopically and examined by standard hematological and histological methods for tumors. The experiment was terminated by killing of survivors six months after the start of the experiment. The tissues (thymus, spleen, lymph nodes, liver, lung, and kidneys) from sacrificed animals were fixed in Bouins's solution and embedded in paraffin. Paraffin sections were stained with hematoxylin and eosin. In DMBA-treated mice with enlarged lymphoid organs and in five out of seven control mice, small parts of thymus, spleen, lymph nodes and bone marrow were taken as specimens for electron microscopy by the following procedures. Tissues samples were fixed in glutaraldehyde and sectioned by a razor blade into small pieces. Approximately ten selected pieces at random were post-fixed with osmic acid(19), dehydrated in ethanol solution and embedded in Epon-Araldite mixture(20). Thin-sectioned samples by a Porter-Blum microtome were mounted on bare grids, stained with uranyl acetate(21) and lead hydroxide(22)and examined by a Hitachi 7s electron microscope. In this experiment only thymic tissues were used for observations by electron microscopy. A local tumor was treated with similar manner.

Results

In this experiment all treated mice survived for the first 2 weeks. There after, six out of fourteen mice (42.9%) were moribund by DMBA toxicity within 6 weeks except one. After 11 weeks five mice (62.5% of survivors) developed malignant lymphoma. Mean latent period was about 13 weeks. Three mice inoculated with DMBA survived over 6 months of the experimental period without any sign of disease. Seven control mice inoculated with sesame oil remained disease-free for the duration of the experiment. Histological examination of their tissues showed no lesions at the termination of the experiment.

Macroscopically, there were some uniform alterations in the organs of moribund mice, which survived over the period of the toxic deaths. In all, the thymus was greatly enlarged into the mediastinal cavity. Enlargement of spleen and lymph nodes were also noted with some variations in four out of five mice. In one case a local solid tumor in the DMBA-inoculated area was observed.

Peripheral blood contained up to 38,700 white blood cells per mm^3 in three mice with lymphoma; in the other two diseased mice it was within normal limits. Giemsa-stained blood smears showed many mature lymphoid cells and their blast-type cells in the former three animals(Figs. 1 and 2).

Histological examination of diseased mice disclosed the presence of neoplastic cells having relatively large round nucleus containing distinct nucleolus, and a scanty cytoplasm in thymus, spleen, and lymph nodes which destroyed the normal histological structure of these organs (Fig. 3); and there were many cells in stages of mitosis(arrow in Fig. 3). Cells of similar type infiltrated into kidney, lung, and liver examined, at varying degrees (Figs. 4 and 5). Cytologically, neoplastic cells in diseased mice seemed to be diagnosed as either malignant lymphoma of the stem-cell type or of the lymphocytic type. A local tumor in the DMBA-inoculated area was a fibrosarcoma(Fig. 6).

Electron microscopy of normal and neoplastic thymus tissues indicated the presence of type C tumor virus which presumably are morphologically representative of murine leukemia virus(23,24,25). In normal thymus of control mice, the virus particles seemed to be associated mainly with thymic epithlial cells and very rarely present in intercellular space and in the formation from lymphoid cells. In thymoma tissues of DMBA-inoculated mice, the virus particles were also present in the vacuoles of thymic epithelial cells (Fig. 7); but in this case many phagocytic vacuole-like structures containing cell-debris were observed in the cytoplasm. There were a few type A tumor virus particles in the cisternae of the endoplasmic reticulum (arrow with A in Fig. 7). Leukemia virus was also observed in the stage of formation by budding from lymphoid cells and free in the intercellular spaces(Fig. 8). Because of some difficulty to distinguish immature lymphoid cells from malignant lymphoid cells by electron microscopy and because

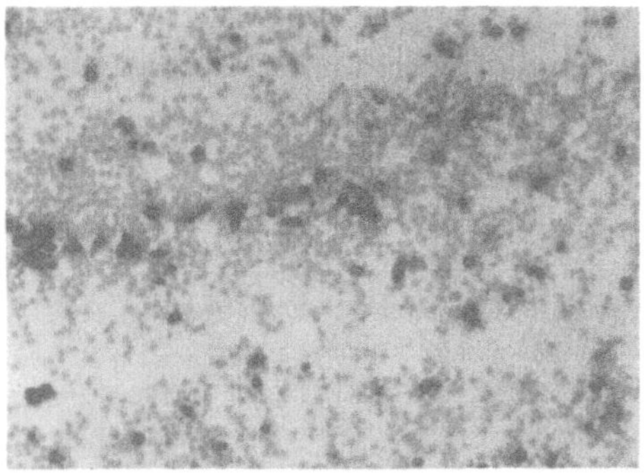

Fig. 1. Blood smear from GF mouse with DMBA-induced malignant lymphoma. Wright-Giemsa, X130.

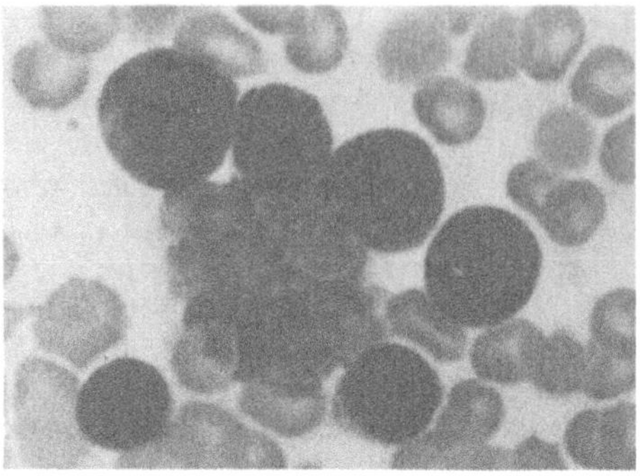

Fig. 2. Detail from Fig. 1. Showing blast cells, mature lymphocytes and red blood cells X1300.

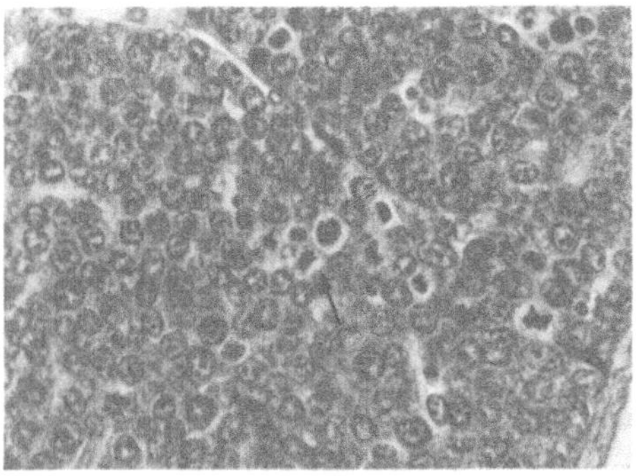

Fig. 3. Section of lymph node of GF mouse with DMBA-induced malignant lymphoma. H&E,X450.

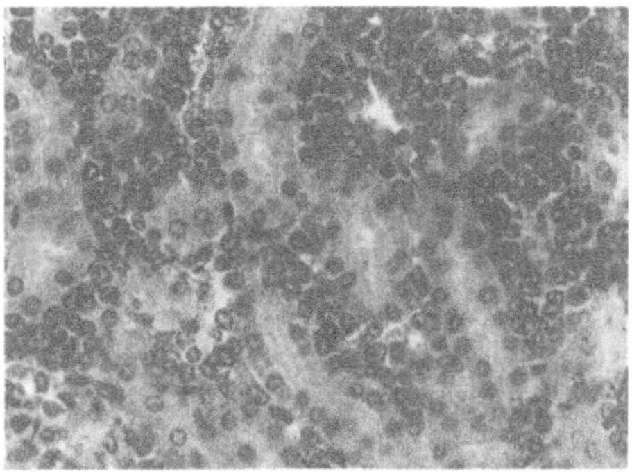

Fig. 4. Kidney section of GF mouse with DMBA-induced malignant lymphoma (Stem cell type). H&E, X 250.

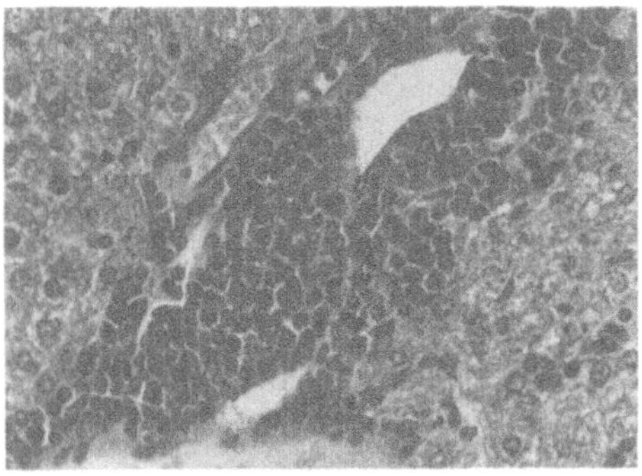

Fig. 5. Liver section of GF mouse with DMBA-induced malignant lymphoma (lymphocytic type). H&E, X 250.

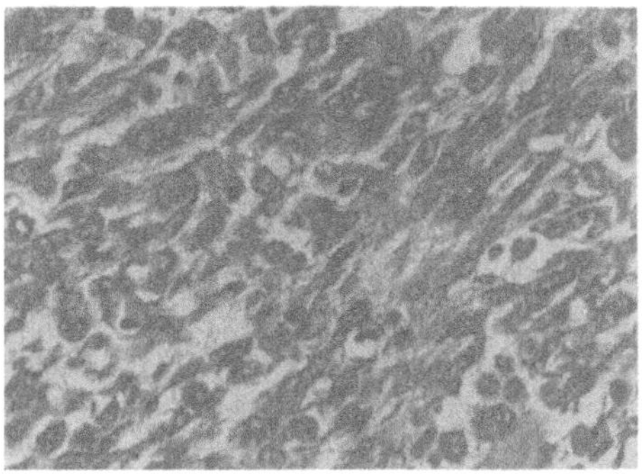

Fig. 6. Section of local tumor induced by DMBA-injection in GF mouse. H&E, X 300.

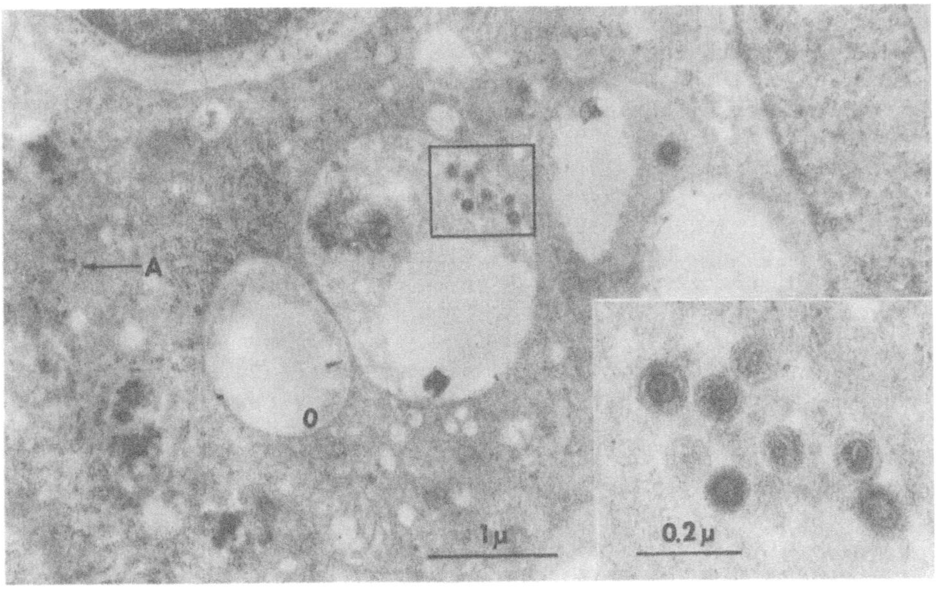

Fig. 7. Part of thymic epithelial cell which was observed in thymoma tissue of GF mouse with DMBA-induced malignant lymphoma. Unlike normal epithelial cell, cell debris can be observed in vacuoles of the cell. Cisternal Type A particles (arrow with A). Insert is high magnification of virus-like particles (type C tumor virus and precursor).

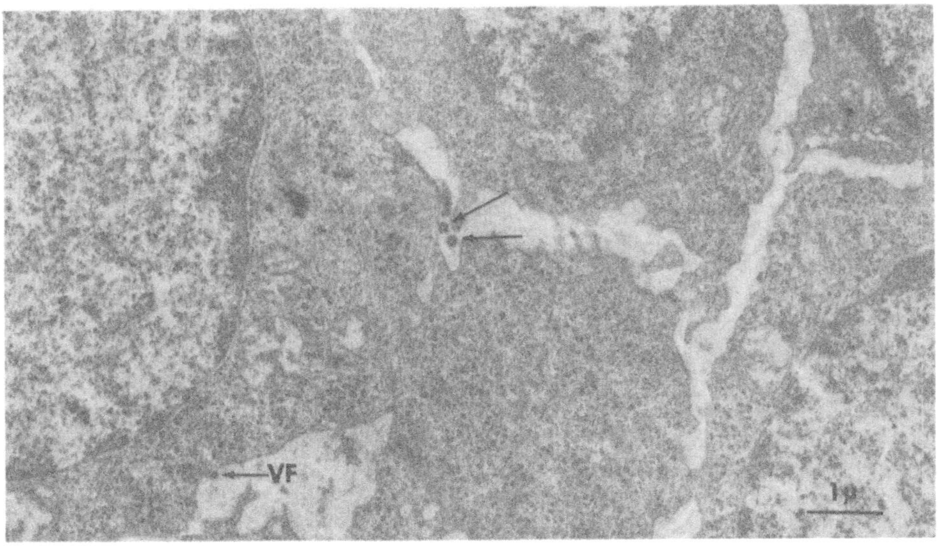

Fig. 8. Virus-like particles in intercellular space (arrows). Virus formation by budding from plasma membrane of lymphoid cell (arrow with VF).

normal lymphoid cells in normal thymic tissues rarely show virus formation by budding, we are unable to state that leukemia virus infection in lymphoid cells is an essential step in malignant transformation of the cells, which appeared after DMBA treatment, on the base of electron microscopy. However, it was of interest to note that frequency of virus detection in intercellular spaces and frequency of the observation of virus-infected lymphoid cells as judged by virus formation by budding were slightly higher in lymphoma tissues than in normal tissues. Details of quantitative analysis of the virus in the tissues will be presented in a separate paper.

Electron microscopy of the sarcoma tissue in DMBA-inoculated area disclosed the presence of type C tumor virus in intercellular space and its formation from tumor cells and also presence of type A tumor virus in association with endoplasmic reticulum in neoplastic cells (Figs. 9 and 10).

Discussion

As presented in this study, the responses (incidence, mean latent period, and type of tumor formation) of germfree mice to DMBA-treatment at birth resemble the results obtained by others in conventional mice similarly treated(1-4). Previous studies indicate that chromosome abnormality is observed in DMBA-induced malignant lymphoma cells taken from the mouse thymus(26) and that cell-free extracts prepared from DMBA-induced malignant lymphoma have leukemogenic activity in test animals(27,28). The results obtained by DMBA-injection in newborn mice is not incompatible with the immunological theory of carcinogenesis(3). This report points out the presence of leukemia virus (type C tumor virus) in DMBA-induced malignant lymphoma tissue and in normal tissues by electron microscopy. However, this is not surprising because leukemia virus is widely disseminated in normal laboratory mice (29). Because cell-free filtrates prepared from the tissue of spontaneous, x-ray-induced, or chemically-induced leukemic mice produced similar disease in inoculated mice(13,27) and because of the presence of leukemia virus-like particles in normal mice, it appears likely that DMBA-induced malignant lymphoma in mice may also be the result of activation of latent leukemia virus in mice. It is still not clear what is the mechanism of activation of the virus. In this regard, it should be emphasized from our results that virus particles in the intercellular spaces and in virus-infected lymphoid cells, as observed by electron microscopy, are much less in thymic tissues of mice with DMBA-induced malignant lymphoma than in those of AKR mice with spontaneous leukemia or in rats with Gross virus-induced leukemia. If leukemogenic action of DMBA in mice is considered in relation to the presence of leukemia virus, DMBA may act not through virus itself but through

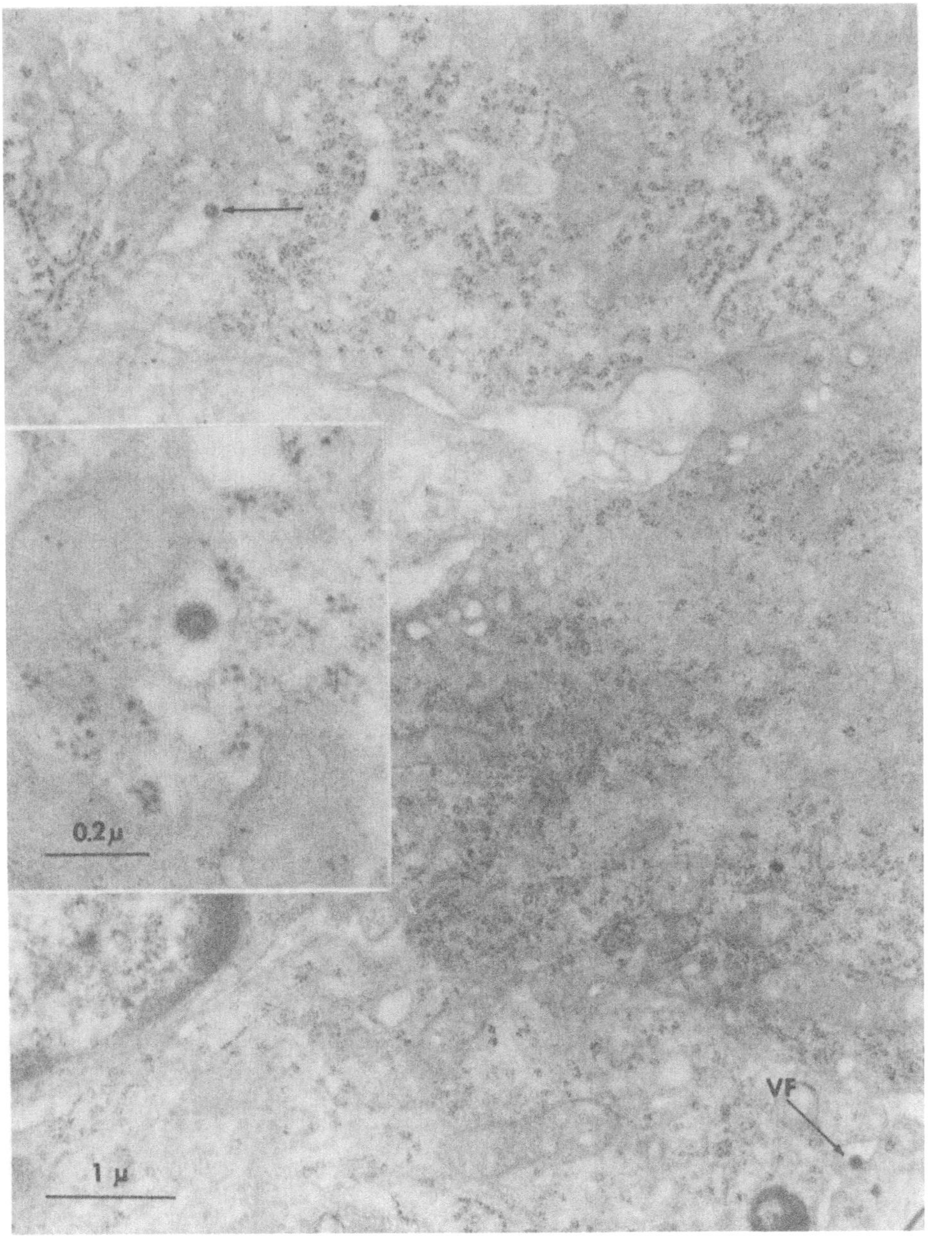

Fig. 9. Part of local sarcoma tissue of GF mouse, induced by DMBA injection. Virus-like particle in intercellular space (arrow) and virus formation by budding (arrow with VF).
Fig. 10. Type A particle in cisternae of endoplasmic reticulum of sarcoma cell.

enhanced host or host cell susceptibility or through increased probability that virus will meet with target cells responsible for malignant transformation. However, this assumption will require more supporting evidence.

The tumor virus which was found in sarcoma tissue may be interpreted as "passenger virus" on the basis of a previous study (30).

Summary

Response of germfree Swiss-Webster mice to subcutaneous injection of 7,12-Dimethylbenz(α)anthracene (DMBA) in sesame oil at neonatal age was investigated. The results can be summarized as follows: (a) Five out of eight mice which survived toxic death developed malignant lymphoma and in one case local sarcoma was observed. No neoplastic changes were found in control mice which were inoculated with sesame oil; (b) Neoplastic response of germfree mice to DMBA-inoculation were similar in many aspects (incidence, latent period, gross finding and histological examination) to that described by others in conventional mice similarly treated; (c) Tumor virus (presumably leukemia virus) was found by electron microscopy in neoplastic tissues induced by DMBA and in normal tissues of control mice; and it can be assumed that high incidence of malignant lymphoma which developed only in mice not in other rodents similarly treated with DMBA could be attributed to the presence of leukemia virus widely present in laboratory mice.

Acknowledgments

Acknowledgment is made to Miss B. J. Minczewski and Mrs. P. H. Luckert for their assistance.

This investigation was possible through the support of the Leukemia Society of America, Inc.

References

1. Pietra, G., Spencer, K., and Shubik, P.: Nature 183, 1689 (1959).
2. Pietra, G., Rappaport, H., and Shubik, P.: Cancer 14, 308 (1961).
3. Roe, F.J.C., Rowson, K.E.K. and Salaman, M.H.: Brit. J. Cancer 15, 515 (1961).
4. Kelly, M.G. and O'Gara, R.: J. Nat. Cancer Inst. 26, 651 (1961).
5. Toth, B., Rappaport, H. and Shubik, P.: ibid 30, 723 (1963).
6. Roe, F.J.C. and Walters, M.A.: Nature 214, 299 (1967).
7. Toth, B. and Shubik, P.: Brit. J. Cancer 17, 540 (1963).
8. Lee, K.Y., Toth, B. and Shubik, P.: Proc. Soc. Exp. Biol. Med. 114, 579 (1963).

9. Walters, M.A., Roe, F.J.C. and Levene, A.: Brit. J. Cancer 21, 184 (1967).
10. Roe, F.J.C., Carter, R.L. and Percival, W.H.: ibid 21, 815 (1967).
11. Pollard, M. and Kajima, M.: J. Nat. Cancer Inst. 39, 135 (1967).
12. Miller, J.F.A.P.: In Advances in Cancer Research Vol. 6, 291, Academic Press, New York (1961).
13. Gross, L.: ibid Vol. 6, 149 (1961).
14. Furth, J. and Boon, M.C.: In A.A.A.S Research Conference on Cancer, 1944. Washington A.A.A.S. p. 129 (1945).
15. Kaplan, H.S.: J. Nat. Cancer Inst. 8, 191 (1948).
16. Rapport, H. and Baroni, C.: Cancer Res. 22, 1067 (1962).
17. Kajima, M., and Pollard, M.: J. Bacteriol. 90, 1448 (1965).
18. Pollard, M. and Kajima, M.: In The Proc. Intern. Conf. Rad. Biol. and Cancer, Tokyo, 175 (1966).
19. Sabatini, D.D., Bensh, K. and Barnett, R.J.: J. Cell Biol. 17, 19 (1963).
20. Mollenhauer, H.H.: Stain Technol. 39, 111 (1964).
21. Watson, M.L.: J. Biophys. Biochem. Cytol. 4, 475 (1958).
22. Karnowsky, M.J.: ibid 11, 729 (1961).
23. Bernhard, W.: Cancer Res. 20, 712 (1960).
24. deHarven, E.: In Tumor Induced by Viruses; Ultrastructural studies, Academic Press, New York (1962).
25. Dalton, A.J.: Federation Proc. 21, 936 (1962).
26. Stich, H.F.: J. Nat. Cancer Inst. 25, 649 (1960).
27. Toth, B.: Proc. Soc. Exp. Biol. and Med. 112, 873 (1963).
28. Haran-Ghera, N.: ibid 124, 697 (1967).
29. Kajima, M. and Pollard, M.: Nature 218, 188 (1968).
30. Kajima, M. and Pollard, M.: Cancer Res. 27, 980 (1967).

GNOTOBIOTES AS USEFUL TOOLS IN STUDIES IN CARCINOGENESIS WITH CYCASIN, A NATURAL GLUCOSIDE

Gert L. Laqueur

National Institutes of Health, Bethesda, Maryland

Experimental studies in our laboratory have established marked oncogenic properties of the glucoside cycasin, β-D-glucosyloxyazoxymethane. The glucoside is extractable from seeds or tubers of ancient gymnosperms belonging to the family of Cycadacea. This plant was widely distributed during the mesozoic period, but todays members are found only in tropical and subtropical parts of the world. Starch prepared from seeds and tubers are used by many people as a source of food. It was apparently known to these populations for an unknown period of time that the crude starch contained a poison, and similar methods for removal of the poison have been practiced by them. The interested reader is referred to an excellent review of the toxicity of cycads by Marjorie Grant Whiting covering what was known about cycad toxicity up to 1963 (1).

Once the carcinogenic properties of the crude cycad starch were recognized (2) studies were undertaken to establish the carcinogenic agent in the crude material. The glucoside cycasin which had been isolated from cycad seeds in 1955 (3) was found to produce all the effects which had been observed with the crude material. Hence, it was thought that cycasin was the carcinogen (4). As our studies with cycasin progressed, it became apparent, however, that tumor incidence as well as the acute toxic manifestations varied among groups of experimental animals and from 15 to 25 percent of them essentially remained unaffected. This lack in uniformity of response required an explanation.

Information in the literature dating back to 1941 (5) and repeatedly confirmed since (6,7) indicated that acute toxic effects with cycad glucosides were obtained only in feeding experiments but not after parenteral injections. Thus the intestinal tract became implicated as contributing to the mechanisms by which cycasin exerted its toxicity. In 1963, Matsumoto reported on the toxic properties of the aglycone part of cycasin, methylazoxymethanol (MAM) and on its recovery following incubation of cycasin with β-glucosidases (7). Thus, the question could be restated by asking whether the lack of uniformity in response might be due to variations in enzyme activity in the intestinal tract by controlling the conversion of cycasin to MAM.

Among several experimental designs through which this problem could be approached, the one employing germfree animals was expected to give the most uncontestable information. If germfree rats hydrolyzed cycasin, the proper enzyme would have to be in the wall of the intestinal tract, particularly in view of the report by Dahlquist et al. (8) that the small intestine of conventional and germfree rats contained a β-glucosidase among other enzymes. Conversely, if enzymatic hydrolysis did not occur, the ingested cycasin should be excreted and toxic effects be absent. In that case the contribution of bacterial enzymes as providers of the enzyme would have to be explored by monocontaminating germfree rats with strains of organisms known to possess or to lack a β-glucosidase.

Experiments were started in late 1963 by feeding for 20 days an identical diet to which cycasin was added in a concentration of 200 mg percent to conventional and germfree Sprague-Dawley rats. Whereas the germfree rats more than doubled their body weight, over one-half of the conventional rats had died by the 20th day and those that survived had gained only 28 percent in weight. Histologic study of the tissues of 5 rats from the germfree group showed essentially normal structures, whereas severe, often hemorrhagic, centrilobular necrosis in the liver was found in all conventional rats. This then was the first indication that germfree rats apparently tolerated cycasin without showing ill effects (4). Long-term observations showed, furthermore, that germfree rats did not develop the kind of tumors commonly induced in conventional rats with cycasin although a limited number of "spontaneous tumors" were observed among them with advanced age (9).

In a subsequent experiment, the excretion of cycasin in urine and feces was compared between germfree and conventional rats. As in the previous experiment, cycasin was fed with the diet at a concentration of 200 mg percent but only for 3 days during which both groups of rats were kept in metabolic cages. Whereas germfree rats nearly quantitatively excreted the ingested cycasin, an average of 74 percent of the consumed cycasin was metabolized by the conventional rats (10). The results clearly indicated that, in contrast to conventional rats, germfree rats failed to hydrolyze cycasin to MAM. Closer inspection of the values obtained from conventional rats revealed considerable variations in the percentage of metabolized cycasin. If concentrations substantially smaller than the one used had been fed to animals, a situation could be imagined in which the total amount of hydrolyzed cycasin would become so small that toxic symptoms would not develop.

Having established that cycasin was not hydrolyzed in the intestinal tract under bacteria-free conditions, the introduction of microbes into the intestinal tract of germfree rats as monocontaminants was subsequently carried out. Although pure strains were not isolated from the feces of the conventional rats to perfrom the perfect experiment, strains of microorganisms were commercially obtained on which information on the presence or absence of a β-glucosidase was available. The strains were grown in our laboratory and assayed prior to use for the presence of the enzyme. P-nitrophenyl-β-D-glucoside, salicin and cycasin were used as substrates. Cell-free extracts of strains of Streptococcus fecalis and of Lactobacillus salivarius salicinius, both known salicin fermenters, and of Lactobacillus salivarius salivarius, known not to ferment salicin, were assayed for β-glucosidase activity. Monocontamination was performed on groups of germfree rats with each bacterial strain. After growth of the organism was well established, cycasin was administered by stomach tube in a dose of 750 mg/kg body weight. Urine and feces were collected on a 24-hour schedule for quantitative determinations of cycasin. The experiment was terminated on the 7th day after cycasin had been given and the livers of all rats including those that had died, were examined histologically. The results of this previously reported study (11) indicated that germfree rats monocontaminated with microorganisms with β-glucosidase activity hydrolyzed cycasin and that the extent of hydrolysis roughly paralleled enzyme activity of the strains used. On the basis of these collected findings, it was suggested that toxicity and carcinogenicity of cycasin were dependent on a bacterial intestinal flora containing organisms with β-glucosidase activity.

It remained then to be shown that germfree rats would develop neoplasms under conditions in which either the product of cycasin hydrolysis, namely MAM, was administered or in which a monocontaminant rich in β-glucosidase was successfully established in germfree rats with cycasin being administered a few days later. The first of these two experiments has been completed; the second one is still in progress.

Independent of the route of administration, the free MAM and the synthetic MAM acetate produced tumors in every germfree rat and at sites at which cycasin induced tumors in conventional rats (Tables I and II of reference 9). With parenteral administration, the majority of tumors occurred in the intestinal tract whereas inclusion of MAM acetate in the diet resulted in tumors of the liver and kidneys as well. The tumors in no way differed from those observed in conventional rats.

SUMMARY

The experiments cited in this report have clearly demonstrated the extraordinary usefulness of gnotobiotes in elucidating the mechanism by which cycasin is converted from a stabile nontoxic substance to a highly unstable, toxic and markedly carcinogenic compound in experimental animals. The studies have emphasized moreover the importance of the bacterial intestinal flora in accomplishing the conversion. Lastly, the results show that MAM, the aglycone of cycasin is the proximate carcinogen. Its effects are independent of the route of administration and conventional and germfree animals essentially respond in a similar way.

REFERENCES

1. Whiting, Marjorie G. Toxicity of cycads. Economic Botany 17: 269-302, 1963.
2. Laqueur, G. L., Mickelsen, O., Whiting, M. G., and Kurland, L. T. Carcinogenic properties of nuts from *Cycas circinalis* L. indigenous to Guam. J. Nat. Cancer Inst. 31: 919-951, 1963.
3. Nishida, K., Kobayashi, A., and Nagahama, T. Studies on cycasin, a new toxic glycoside, of *Cycas revoluta* Thunb. Part I. Isolation and the structure of cycasin. Bull. Agr. Chem. Soc. Japan 19: 77-83, 1955.
4. Laqueur, G. L. Carcinogenic effects of cycad meal and cycasin, methylazoxymethanol glycoside, in rats and effects of cycasin in germfree rats. Fed. Proc. 23: 1386-1387, 1964.
5. Cooper, J. M. Isolation of a toxic principle from the seeds of *Macrozamia spiralis*. Proc. Roy. Soc. New South Wales 74: 450-454, 1941.
6. Nishida, K., Kobayashi, A., Nagahama, T., Kojima, K., and Yamane, M. Studies on cycasin, a new toxic glycoside of *Cycas revoluta* Thunb. Part IV. Pharmacological studies of cycasin. Seikagaku 28: 218-223, 1956.
7. Kobayashi, A., and Matsumoto, H. Studies on methylazoxymethanol, the aglycone of cycasin: Isolation, biological, and chemical properties. Arch. Biochem. Biophys. 110: 373-380, 1965.
8. Dahlquist, A., Bull, B., and Gustafsson, B. E. Rat intestinal 6-bromo-2-naphthyl glycosidase and disaccharidase activities. I. Enzymic properties and distribution in the digestive tract of conventional and germfree animals. Arch. Biochem. Biophys. 109: 150-158, 1965.

9. Laqueur, G. L., McDaniel, E. G., and Matsumoto, H. Tumor induction in germfree rats with methylazoxymethanol (MAM) and synthetic MAM acetate. J. Nat. Cancer Inst. 39: 355-371, 1967.
10. Spatz, M., McDaniel, E. G., and Laqueur, G. L. Cycasin excretion in conventional and germfree rats. Proc. Soc. Exp. Biol. Med. 121: 417-422, 1966.
11. Spatz, M., Smith, D.W.E., McDaniel, E. G., and Laqueur, G. L. Role of intestinal microorganisms in determining cycasin toxicity. Proc. Soc. Exp. Biol. Med. 124: 691-697, 1967.

BIDIRECTIONAL PERMEATION OF WATER AND AMINO ACIDS IN THE CECUM OF GERMFREE RATS

Etiennette Combe* and Helmut A. Gordon

University of Kentucky, Lexington

The purpose of this work was to study transport across the cecal wall of germfree animals which has been found altered in previous experiments. Germfree (gf) and conventional control (conv) male Fisher (Ch.R., CD-F) rats, 8-10 months old, fed sterilized 5010C (water expt) or L-462 (amino acid expt) diets were used in urethane anesthesia (150 mg/100 g bodyweight). The labels employed were tritiated water, ^{14}C UL glycine and serine in doses of 10 μC/rat. Absorption was studied by injection of the label into the intact cecal sack which was ligated at the emergence of the ileum and of the colon, and by the successive recovery from its contents. Efflux into the cecum was observed after injection into the saphenous vein. The duration of the experiment averaged 20 minutes. The corrected body weight was approx. 300 g in the 24 gf and 26 conv rats used to date in this series. The ceca of the gf rats were considerably enlarged (avg. 11% of the body weight); their contents were rather liquid (avg. 18% dry matter) and their osmolality was similar to that of blood plasma (in mOsm, mean, blood plasma 335 ± 10, gf cecal contents 326 ± 24). The results indicated that the efflux of tritiated water from blood into the cecum was increased in the gf group in comparison to conv controls. However, when this value was expressed per unit mucosal surface area, the results were essentially the same in both groups. The absorption of labelled water from the gf cecum was found considerably reduced, particularly when this value was calculated per unit surface area (gf-conv ratio 1:7). The efflux of amino acids (glycine was studied to date most extensively) into the cecal lumen was very slight in both groups. The absorption of amino acids was almost undetectable in the gf group, while the disappearance of these substances from the cecum of conv rats was considerable, irrespective of the mode of expression (per unit surface area the gf-conv ratio was 1:130). These observations indicate that the increased water content of the gf cecum is based mainly on inhibition of water absorption, as suggested by the observations of Csaky (In: The Germfree Animal in Research, Academic Press, London; in press). The elevated levels of free amino acids reported in cecal contents of gf rats by Combe and Sacquet (C.R. Acad. Sc. Paris, 262: 685, 1966) may have originated from the diet and/or from secretions and express the greatly reduced absorption of these substances from the gf cecum, rather than an efflux from the tissues. The role of the intestinal flora in the diasppearance of the label from the conventional cecum is presently being evaluated. (Supported by grants from the U.S.P.H.S.) *On leave of absence from the National Institute for Research in Agronomy (INRA), Theix, France.

RECENT PROGRESS IN STUDIES ON A BIOACTIVE PIGMENT ("ALPHA PIGMENT") ISOLATED FROM CECAL CONTENTS OF GERMFREE RODENTS

Helmut A. Gordon, Eszter Kokas and Julius Magyary-Kossa*

University of Kentucky, Lexington and University of North Carolina, Chapel Hill

We have reported that cecal contents of germfree rats and mice contain an excess of various bioactive substances in comparison to conventional controls. One of these substances has recently been identified as a fecal kallikrein (Ann. N.Y. Acad. Sci., 147: 83, 1967). A pigment fraction isolated by centrifugation and column chromatography ("alpha pigment") caused vasodilation and a stimulatory effect on the spontaneous rhythmic contraction of intestinal villi in anesthesized dogs and occasionally exerted inhibitory effects to epinephrine on rat blood pressure. The same pigment fraction isolated from conventional control animals proved inactive (Fed. Proc., 26: 383, 1967). Germfree alpha pigment in the listed effects showed a close similarity to those of ferritin and apoferritin in their vasodepressor (VDM) form. In reduced form all three agents were active; in oxidized form all three were inert. Antiferritin serum neutralized these substances. The light absorption spectra of trypsinized ferritin and apoferritin indicated similarity to that of alpha pigment. The toxicity of germfree cecal contents previously described (Nature, 205:571, 1965) could be demonstrated in this pigment fraction. The concurrence of other autakoids or of other substances in the gut in these phenomena could be excluded either by the nature of their effects or by the use of blockers. It was indicated that alpha pigment is available for absorption from the intestine. It is theorized that alpha pigment is derived from apoferritin, the iron-carrier protein of the intestinal mucosa which is known to reach the gut lumen in considerable quantities on desquamation of the intestinal epithelium. Under conditions existing in the germfree gut, apoferritin or one of its derivatives assumes its bioactive form, while in conventional animals the intestinal flora is responsible for its (apparently irreversible) inactivation. Some anomalies observed in germfree rodents (refractoriness to epinephrine of mesenteric blood vessels, reduced metabolic rate and cardiac output) may be explained on this basis. (Supported by grants from U.S.P.H.S.). *On leave of absence from J.R. Geigy, S.A., Basel, Switzerland.

KININ-FORMING ACTIVITY IN CECAL CONTENTS OF GERM-FREE AND CONVENTIONAL MICE

N. Back, R. Steger, and E. A. Mirand

Department of Biochemical Pharmacology, School of Pharmacy
State University of New York at Buffalo and
Roswell Park Memorial Institute, Buffalo, New York

Recently, a substance was isolated from the cecal contents of both germ-free and conventional rodents that stimulated the smooth muscle of the perfused rat uterus.[1] Further studies identified this substance as cecal kallikrein possessing musculotropic activity both by direct and indirect kinin-forming action.[2]

This investigation was undertaken to study the nature of this kinin-forming agent in cecal contents of germ-free and conventional mice, and attempts to establish its relationship to components of the vasoactive polypeptide (kallikrein-kinin) system.

The vasoactive polypeptide system is represented schematically in Fig. 1. The kinin-forming enzyme kallikrein arises from prekallikrein, an inactive enzyme precursor present in both tissue fluids and plasma. Both physical[3] and enzymic mechanisms[4] are capable of effecting this activation. Activated Hageman factor also has been suggested as an activating mechanism.[5] Kallikrein acts on kininogen substrate to form several vasoactive kinins (bradykinin, a nonapeptide, and kallidin, a decapeptide). Other proteases, notably trypsin[6] and plasmin[6] have kinin-forming activity. A kininase plasma carboxypeptidase rapidly inactivates kinin.

Nanogram concentrations of kinin stimulate smooth muscle, increase capillary permeability, cause hypotension and pain. In view of some of the above actions, kinin was suggested as the possible mediator for the anomolous function and enlargement of cecum noted in the germ-free rodent. Cecal content extracts thus were analyzed for kinin-forming activity, kininogen and kinin levels, and kinin-destroying activity. Bioassay methods for these analyses have been published,[7] and are based on determining the

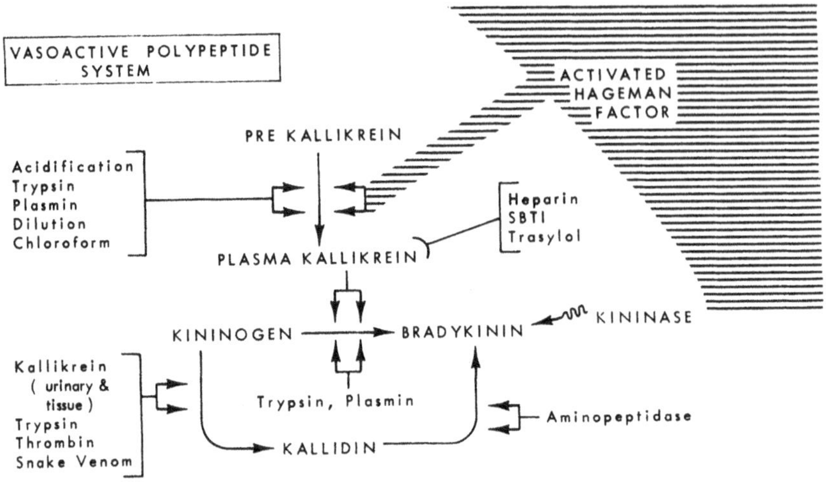

Fig. 1. Scheme of the vasoactive polypeptide (kallikrein-kinin) system.

amount of kinin present, formed, or destroyed when the extract is incubated with the appropriate substrate.

The cecal contents were collected from both germ-free and conventional ICR/Swiss mice killed by cervical fracture. Each cecum also was flushed with physiologic saline, and both contents and washings were homogenized, centrifuged at 3500 revolutions per minute for 20 minutes, and the supernatent lyophilized and stored at -20°C for further assay. No attempt was made to purify the extract since the overall activity of the extract was sought. Batches of cecal content extract were prepared from pools of 10 conventional mice or 5 germ-free mice. In some instances, cecal tissue also was homogenized and prepared for analysis.

On both a wet and dry basis, the average cecal content was greater in the germ-free animal compared to the conventional control mouse. The average cecal content in the germ-free mouse was 15 mls and 1.mls in the normal conventional animal. The average germ-free mouse cecum contained 148 mgms of contents on a dry basis compared to the 6.0 mgms in the average conventional mouse. Cecal tissue weights in the germ-free mouse on both a dry and wet basis were approximately twice in the conventional mouses 12.4 mgm compared to 5.8 mgm on a dry basis and 0.50 gms to 0.30 gms on a wet basis.

Neither the cecal contents nor cecal tissue had any kininase activity or kininogen. The cecal content extracts also had no prekallikrein activity. However, the cecal content extract did form kinin when incubated for one hour with kininogen substrate solution containing 1:10 phenanthroline to inhibit kininase activity. The incubation reaction was stopped with a protease inhibitor, Trasylol. A dose-response relationship obtained, and a 0.5 mgm amount of cecal content extract was capable of forming 0.2 micrograms of kinin per ml of 1% kininogen solution. Three batches of cecal content extract from germ-free mice and one batch from conventional mice all had similar activities.

Studies with a variety of natural mammalian, plant, and synthetic protease inhibitors strongly suggested that the kinin-forming activity in the cecal content extracts was trypsin or tryptic-like rather than kallikrein. The effect of protease inhibitors on the kinin-forming activity of germ-free and conventional cecal content extracts, trypsin, human plasmin, pancreatic kallikrein and mouse plasma kallikrein was compared.[a] The inhibitors used included a natural mammalian inhibitor from bovine lung known as Trasylol, soya bean and lima bean trypsin inhibitor (SBTI, LBTI), ovomucoid trypsin inhibitor from egg white, epsilon amino caproic acid (EACA) and its analogue 4-aminoethyl cyclohexane-1-carboxylic acid (AMCHA). The inhibitors were incubated in maximal concentrations for 5 minutes with the cecal content extracts or the enzyme and aliquots of the incubation mixture further incubated for one hour with a kininogen solution. Trypsin was the only enzyme that behaved toward inhibitor action in a manner similar to that of the cecal content extracts from both germ-free and conventional mice. All three activities were inhibited completely by Trasylol, LBTI, SBTI and OMTI, partially by EACA, and not at all by AMCHA. Kinin-forming activity of plasmin, pancreatic and mouse plasma kallikrein was not inhibited by OMTI. Furthermore, LBTI did not inhibit the kinin-forming activity of the kallikreins nor did Trasylol affect mouse plasma kallikrein.

Increased concentrations of cecal content extracts, trypsin, and plasma kallikrein (100-200 times that used to form kinin) stimulated the isolated rat uterus muscle directly. The direct smooth muscle stimulating activity of trypsin was inhibited by the protease inhibitors in the same manner as the cecal content extracts whereas the mouse plasma kallikrein was not.

Further comparisons of the kinin-forming activity of cecal content extract with trypsin established their close similarity. Dose-response curves of kinin-forming activities of trypsin and cecal content extracts were almost identical. One microgram of trypsin formed kinin equivalent to that amount formed by 100 micrograms of cecal content extract from germ-free mice. This amount of kinin caused a smooth muscle response equivalent to that obtained with 5×10^{-10} micrograms of standard synthetic bradykinin. The time course of kinin formation with trypsin and cecal content extract also

[a]Sources of inhibitors and enzymes have been published previously.[6]

was identical. At each time studied over a 90 minute time period, equivalent amounts of kinin were formed by each agent.

The pH profile associated with the kinin-forming activity of the cecal content extract compared most closely to that of trypsin. The kinin-forming activity of all the enzymes and cecal content extracts was studied over a pH range of 2-10. Optimum activity of cecal content extracts and trypsin was parallel at each pH point studied (2,4,6,8,10), both agents exhibiting a broad pH optimum from 4-8. The pH optimum for plasmin was steep and activity maximum at pH 6.0, whereas the pH optimum for pancreatic kallikrein was 6-7 and for plasma kallikrein 6.0. Neither of the two kallikreins ever yielded the high kinin levels achieved with cecal content or trypsin.

Thus, comparative kinetic enzyme studies, pH profiles, and studies with protease inhibitors, all indicate that the kinin-forming activity of mouse cecal contents is due to a trypsin or tryptic-like enzyme activity rather than kallikrein. The fact that the cecal contents and tissue contained no component of the vasoactive polypeptide system would suggest strongly that kinins are not involved in the enlarged cecums characteristic in the germ-free mouse.

Summary

Cecal content extracts from both germ-free and conventional mice contained kinin-forming activity, and, in very high concentrations, directly caused contraction of smooth muscle. Comparative kinetic enzyme studies, pH profiles and studies with protease inhibitors suggest that trypsin or a trypsin-like enzyme is responsible for the kinin-forming activity, and not kallikrein. The absence of kininogen, prekallikrein, as well as kininase in both cecal contents and tissue support the view that kinins are not involved in the anomalous function and enlargement of the cecum characteristically noted in germ-free mice.

---

[1] Supported in part by Public Health Service Grants #HE-03786, National Heart Institute and #HE-11492, National Cancer Institute.

Bibliography

1. Wiseman, R.F., Gordon, H.A.: A bioactive substance with caecum of germ-free animals. Nature, 205: 571, 1965.

2. Gordon, H.A.: A substance acting on smooth muscle in intestinal contents of germ-free animals. Ann. N.Y. Acad. Sci. 147: 83, 1967.

3. Armstrong, D., Jepson, J.B., Keele, C.A., Stewart, J.W.: Pain-producing substance in human inflammatory exudates and plasma. J. Physiol. London, 135: 350, 1957.

4. Back, N., Guth, P.S., Munson, A.E.: On the relationship between plasmin and kinin. Ann. N.Y. Acad. Sci. 104: 53, 1963.

5. Margolis, J.: Activation of plasma by contact with glass: evidence for a common reaction which releases plasma kinin and initiates coagulation. J. Physiol. London 144: 1, 1958.

6. Back,N., Steger, R.: Effect of inhibitors on kinin-releasing activity of proteases. Fed. Proc. 27: 96, 1968.

7. Back, N., Wilkens, H., Steger, R.: Fibrinolysis and vasoactive peptides in anaphylaxis. In Hypotensive Peptides. Eds. E.G. Erdos, N. Back, and F. Sicuteri, Springer-Verlag, pg. 485, 1966.

EFFECTS OF CLOSTRIDIUM DIFFICILE, LACTOBACILLUS CASEI, BACILLUS SUBTILIS AND LACTOBACILLUS SP. AS MONO- AND DICONTAMINANTS ON THE CECUM OF GERMFREE MICE

Laszlo Pesti*, Eszter Kokas and Helmut A. Gordon

University of Kentucky, Lexington and University of North Carolina, Chapel Hill

The purpose of this study was to test the effects of various enterally seeded microorganisms on the cecal enlargement and on other parameters in germfree mice. Experimental animals were HaM/ICR (Ch.R.CD-1) mice fed sterilized 5010C diet, aged 11-16 months with approx. the same sex distribution. The microbial status of various groups was: (1) Germfree (gf), (2) Conventional (conv). Ex-germfree monocontaminated with (3) *Cl. difficile* or (4) *L. casei*. Ex-germfree dicontaminated with (5) Cl. difficile and *L. casei* or with (6) *B. subtilis* and Lactobacillus sp. The mice were sacrificed 1 and 3 weeks after insemination. In the cecal contents the following were recorded (in parenthesis the characteristics of gf rodents are given in comparison to conv controls): (a) weight (elevated); (b) dry per cent (reduced); (c) pH (elevated); (d) Eh (more positive); (e) kallikrein content (elevated) and (f) "alpha pigment" (an epinephrine inhibitory chromatographic fraction which displays bioactivity in gf and not in conv cecal contents). Observations (a)-(d) were made with commonly used methods; (e) was assayed on rat uterus *in vitro*; (f) was tested by its effect on the spontaneous motility of intestinal villi of dogs *in vivo*. The results showed the following trends: (A) *B. subtilis* and Lactobacillus sp. dicontamination caused the most widespread, yet incomplete shift in the ex-germfree mice from gf to conv characteristics. (B) *L. casei* as a monocontaminant or in association with *Cl. difficile* changed Eh to almost conv-like values and neutralized the epinephrine inhibitory effect of alpha pigment, but did not affect the other gf characteristics of the animals. (C) *Cl. difficile* as a monocontaminant neutralized alpha pigment and left most other characteristics at gf levels. (D) In contaminations (3), (4) and (5), after the lapse of 1 week, the ceca were found reduced to approx. 50% of the original value. At 3 weeks in all three instances the ceca returned to gf levels. These observations failed to identify a microbial species which as a single contaminant is capable of eliminating all tested anomalies of gf animals. They have suggested, however, that various anomalies may be selectively redressed by various species of the commensal flora. (Supported by grants from U.S.P.H.S.).

*On leave of absence from the Veterinary Medical Research Institute of the Hungarian Academy of Sciences, Budapest.

SIMILARITIES BETWEEN GERMFREE MICE AND MICE WITH AN ANTIBIOTIC DECONTAMINATED DIGESTIVE TRACT

D. van der Waaij

Radiobiological Institute TNO

Rijswijk Z.H., The Netherlands

To prevent that mice die from infections caused by "conditioned pathogenic micro-organisms" after lethal irradiation, a technique has been described elsewhere (1), by which it is possible to decontaminate the digestive tract of conventional mice with the help of oral non-absorbable antibiotics. In the present study the effect of antibiotic decontamination on the morphologic appearance and the weight of the various parts of the digestive tract, the liver, the kidneys and the lymphatic organs, was investigated. In addition, the susceptibility of these animals to colonization of the digestive tract by certain antibiotic resistant bacteria following oral contamination was studied. An investigation of the resistance to oral contamination with non-pathogenic bacteria in antibiotic decontaminated mice seemed indicated since a number of investigators have described an enhanced susceptibility to infections with pathogenic bacteria after treatment with high doses of oral antibiotics (2, 3, 4).

MATERIALS AND METHODS

Animals. Non-inbred female mice of the ND2 strain, both conventional and germfree, were used. The animals varied in age between 9 and 10 weeks and weighed 20-30 grams. Housing and diet as described in a previous paper (1).

Antibiotics were selected as described elsewhere (1). Streptomycin was given in combination with bacitracin both in a concentration of 4 mg/ml of drinking water. Pimaricin in amounts of 0.1 mg/ml was added to eliminate possible yeasts.

The bacterial concentration of the cecum and in some cases

of the duodenum was determined daily after the onset of treatment in groups of 10 mice. Tenfold serial dilutions of the intestinal contents were made in broth and Brewer's semi-solid thioglycolate. In the experimentally contaminated animals sub-inoculations were made on Endo agar with 25 mg streptomycin per ml to evaluate the concentration of the "contaminating strain" of bacteria in the duodenum and the cecum. In addition, in this experiment the cecum was weighed full and empty. The cecum weights of 20 germfree mice of the same strain, sex and age, were measured for the sake of comparison.

Oral contamination was performed with an overnight culture of S(treptomycin) R(esistant) strain of *Escherichia coli*, isolated from the heart blood of a lethally irradiated, streptomycin treated mouse. The various concentrations used for oral contamination were checked afterwards by performing tenfold serial dilutions in broth. Small doses ($< 10^2$ cells) were determined by making counting plates. The suspension was administered orally with a syringe in amounts of 0.1 ml per mouse. This volume was usually readily ingested when the drinking water had been withdrawn for 6-8 hours.

Organ cultures of the spleen and the mesenteric lymph nodes were made by removing these organs and cutting them into small pieces under aseptical conditions. Then the organs were cultured in broth. Positive cultures were sub-inoculated on Endo agar with 25 mg streptomycin per ml to identify the original strain of E. coli used for contamination. The cultures were recorded negative when no growth was observed after 4 days of incubation.

Weight and morphology were determined of the various parts of the digestive tract and of the mesenteric lymph nodes, the spleen, the liver, the thymus and the kidneys. To determine changes in weight of the intestinal wall, pieces with standard length of 2.5 cm of various parts of the digestive tract were washed and carefully dried with filterpaper before weighing. Each group of mice, sacrificed after 1, 2, 4, 7 and 14 days of antibiotic treatment, consisted of 20 animals.

EXPERIMENTS AND RESULTS

As is shown in figure 1, the bacterial count of the cecum decreases rapidly to zero within 4 days after the onset of antibiotic treatment. During this time the wall of the cecum and the wall of the other more densely populated parts of the intestines, like the ileum and the colon, increase in weight (fig. 2). The cecum, however, continued to increase in size after the first 4 days and finally equalled the weight of the cecum of germfree mice of the same age in 7 to 9 days. During the first 4 days the cecum is filled with a watery suspension. In mice sacrificed after 1 week of antibiotic treatment the contents of the cecum

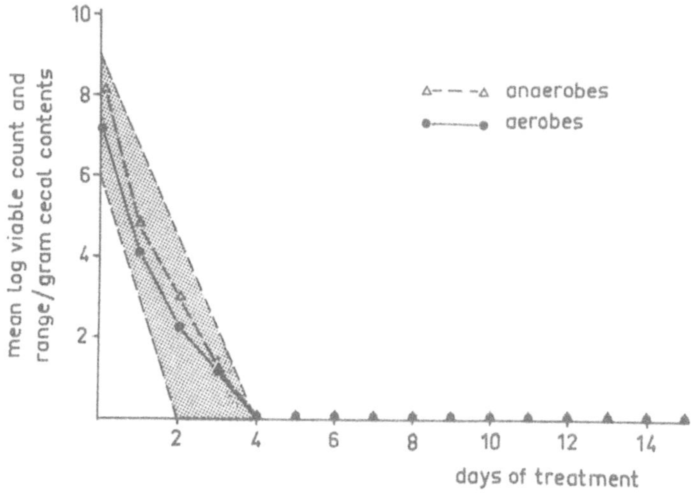

Figure 1: Bacterial concentration of the cecum during antibiotic treatment.

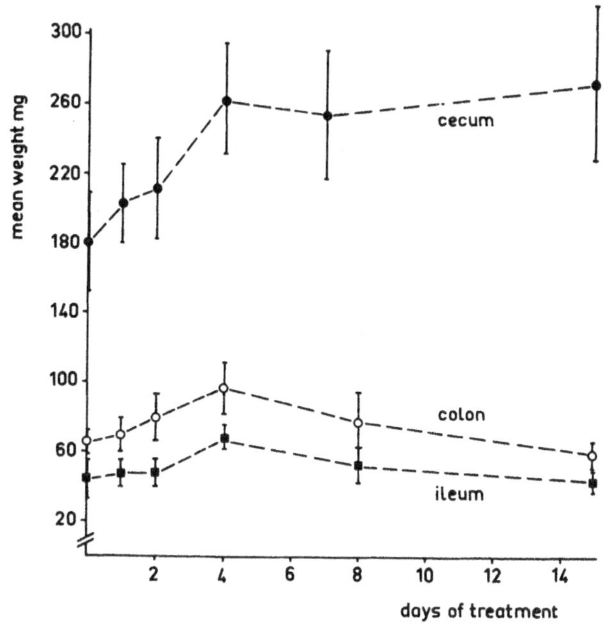

Figure 2: Mean weight* of the densely bacterial populated part of the intestines.

* The vertical lines represent the standard deviations.

were more solid, while in the 10 to 14 days treated animals, there was no difference in appearance with those of germfree mice. The more sparsely bacterial populated parts of the intestines like the duodenum and the jejunum, also showed a slight increase in weight (fig. 3). However, more data are necessary to investigate whether this is significant. The decreased daily water intake during the first 2 days of antibiotic treatment, reflected in the temporary decrease of the kidney weight and also evident in the weight of the spleen (fig. 4), may have an influence on the intestinal wall. This includes that the weight increase of the intestines during the first days of treatment could also be influenced negatively by the temporary dehydration, so that their increase in weight may be even stronger than was found. The thymus showed a slight but significant decrease in mean weight during the first 2 weeks from 51 mg (SD = 2.7) to 40 mg (SD = 6.8), while the mean spleen weight decreased from 110 mg (SD = 16) to 78 mg (SD = 9.9) during this interval (fig. 4).

The histological appearance of the thymus revealed no changes. In the spleen and in the mesenteric lymph nodes, however, the reaction centers showed more activity than the normal controls. In some mice massive accumulation of lymphocytes were seen in the submucosa of the colon during the first week of treatment while in others no differences could be observed with the untreated control animals. Increased lymphocyte infiltrations were also occasionally seen in the wall of the ileum. A microphoto of a "lymphocyte invaded" wall of the colon of a mice sacrificed on day 4 is given in figure 5. Apart from these extreme cases seen in the first week, the histological appearance of the intestinal wall was more of the conventional type than of germfree type even after 2 weeks of antibiotic treatment. In the liver and in the

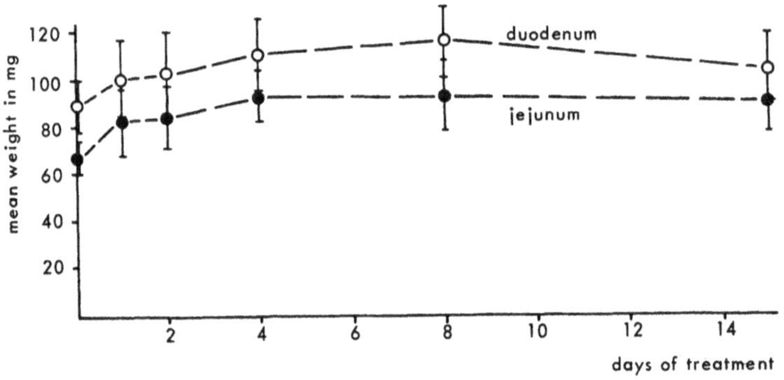

Figure 3: Mean weight* of the sparsely bacterial populated part of the intestines.

* The vertical lines represent the standard deviations.

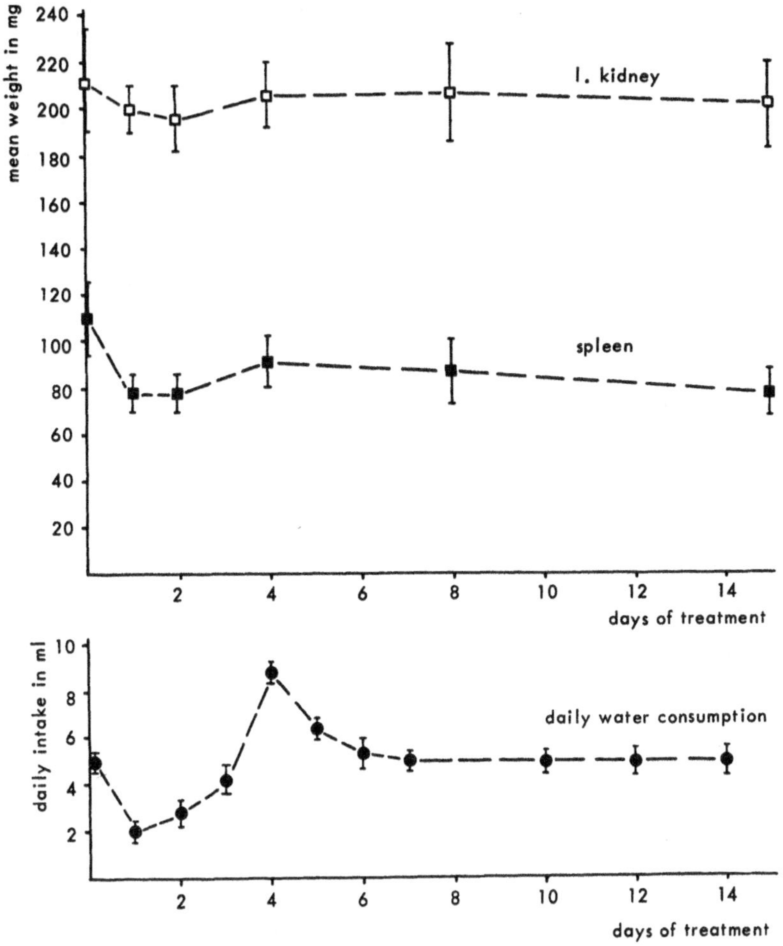

Figure 4: Mean organ weights* after various intervals of antibiotic treatment.

*The vertical lines represent the standard deviations.

kidneys no pathologic changes or signs of intoxication were observed. In mice that had been supplied with antibiotics for 1 week, the susceptibility to colonization of the digestive tract following oral contamination with a S(treptomycin) R(esistant) strain of E. coli, was similar to that seen in germfree mice (5). In both types of mice the susceptibility was found to be very high (1). This was further investigated in groups of 10 to 20 mice sacrificed at various intervals after contamination. Doses varying between 8 and 20 bacteria were sufficient to cause long lasting colonization of the digestive tract in all animals. During this experiment the mice were maintained individually in "isolation cages" (6), after contamination. This "decrease in resistance" was also found for antibiotic resistant strains of klebsiella, pseudomonas, proteus and enterococci. In normal conventional mice colonization of the digestive tract occurred only after the administration of very large doses of bacteria (10^8 and more) and persisted frequently for a few weeks at most. In 6 out of 20 animals sacrificed 48 hrs after contamination, colonization of the digestive tract was correlated with positive cultures of the mesenteric lymph nodes. When the mice were sacrificed 3 days after contamination, the S.R. strain of E.coli could be isolated in low concentrations only from the large intestines and the ileum. Finally, in normal conventional mice, sacrificed 2 weeks after contamination, the contaminating strain was recovered only in 9 out of 20 mice.

To investigate how soon following the onset of antibiotic supply the resistance to colonization decreases, mice, drinking antibiotic water, were orally contaminated 1, 6 and 12 days after the beginning of treatment. The dose given, 10^5 S.R.-E.coli cells, had been found marginally effective in causing a temporary "take" in the digestive tract of most animals. This made it possible to determine slight increases in the "susceptibility". The mice were sacrificed 48 hours after contamination because in previous experiments a "peak" of invasion and an increased intestinal bacterial concentration was found after this interval

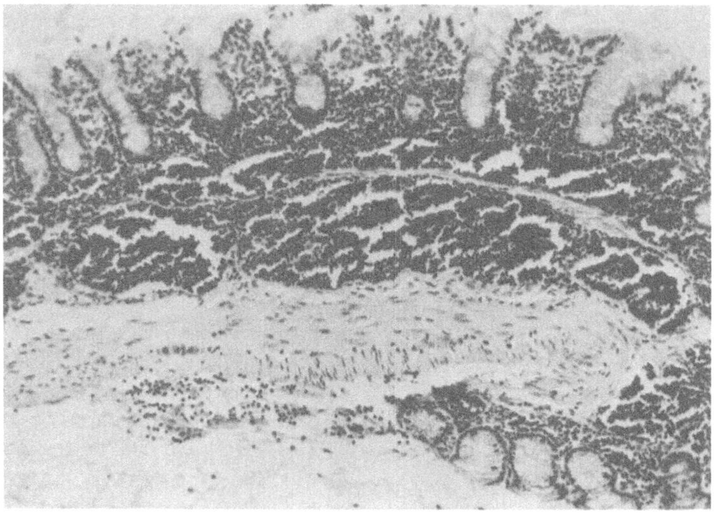

Figure 5: The wall of the colon after 4 days of antibiotic treatment in an extreme case.

* Large accumulations of lymphocytes in the submucosa.

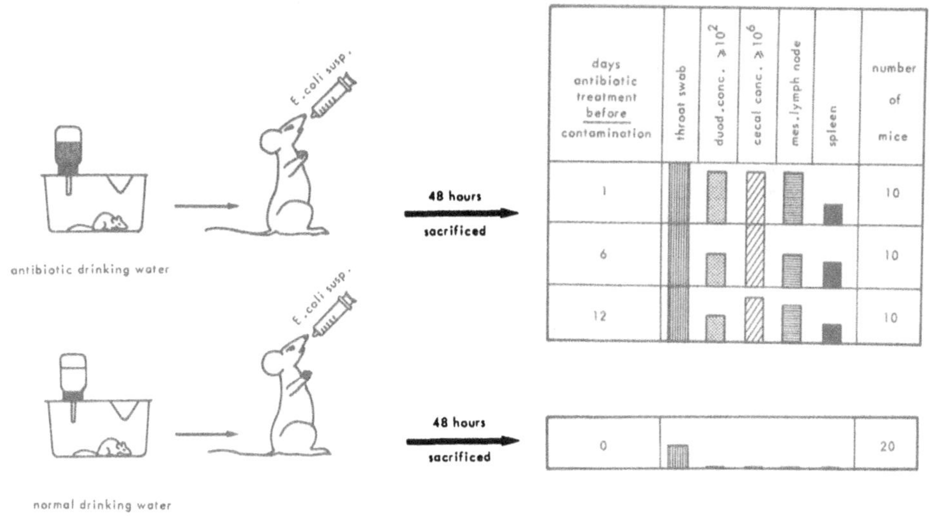

Figure 6: Percentage of mice with increased intestinal concentrations and positive organ cultures after oral contamination with antibiotic resistant E. coli (10^5).

following an oral dose with 10^8 E. coli cells. The results of this experiment, depicted in fig. 6, show that already after 24 hours of antibiotic treatment, the resistance to oral challenge is strongly decreased. After this relatively short interval, the effect of 10^5 S.R.-E. coli cells was comparable to that found in the mice that had been supplied with antibiotics for 6 and 12 days.

DISCUSSION

Although no overt signs of intoxication or illness otherwise were observed in mice treated with concentrations of non-absorbable antibiotics, changes were found in the lymphatic organs and in the intestinal wall. A reduction of the mean weight of the spleen of about 30 per cent accompanies an increase in the weight of the intestinal wall. In some animals large accumulations of lymphocytes were seen in the submucosa of the colon. In other parts of the intestines this phenomenon was less evident, but consisted merely of increased numbers of lymphocytes. These findings could tentatively be explained by the invasion of bacteria (still present during the first day(s) of treatment) through the intestinal mucosa as a result of the rapidly decreased resistance to colonization of the digestive tract (within 24 hours after the onset of treatment). A bacterial invasion could also explain the active "reaction centers" found in the mesenteric lymph nodes and in the spleen. The decrease of the spleen weight cannot so easily be interpreted as the increase of the weight of the intestinal wall, but may result from lymphocyte migration from the spleen to the intestines.

The decrease in "resistance" is "non-selective", since all bacterial species tested, resulted in colonization of the digestive tract and frequently in invasion reaching the lymph nodes and the spleen, after very low oral doses without causing illness or signs of inflammation. This is essentially different from what is seen after lethal irradiation. In this case after 4-7 days the resistance to oral challenge is only strongly decreased for some conditioned pathogenic bacterial species like Pseudomonas aeruginosa. In these cases bacterial invasion is always followed by death.

SUMMARY

Antibiotic decontamination of the digestive tract causes a strong and rapid (within 24 hours) decrease of the normal resistance to colonization of the digestive tract following low doses of oral contamination. The elimination of bacteria from the intestines is accompanied by an increase of weight of the intestinal wall. This is most evident in the more densely populated large intestines, which could be explained by the lymphocyte accumulations seen in the submucosa. Both this lymphocyte infiltration and the increased number of active reaction centers in the mesenteric lymph nodes and in the spleen are tentatively explained by invasion of bacteria that are still present after 24 hours of treatment when the "resistance" has already substantially decreased.

ACKNOWLEDGEMENTS

The advice of Prof. Dr. M.J. de Vries in the evaluation of the histological data is gratefully acknowledged. Most of the weighing and culturing of organs reported, was accurately performed by Mrs. J.E.C. Lekkerkerk and Miss J.M. de Vries.

REFERENCES

1. Waaij, D. van der, and Sturm, C.A. (1968). Antibiotic decontamination of the digestive tract of mice. Technical procedures. Lab. Animal Care, 18, 1-10.
2. Freter, R. (1955). The fatal enteric cholera infection in the guinea pig, achieved by inhibition of normal enteric flora. J. infect. Dis. 97, 57-65.

3. Bohnhoff, M., Drake, B.L. and Miller, C.P. (1954). Effect of streptomycin on susceptibility of intestinal tract to experimental salmonella infection. Proc.Soc.exp.Biol. (N.Y.) 86, 132-137.
4. Cooper, G.N. (1959). Experimental shigellosis in mice. I. Chronic infection with shigella dysenteriae type 2. Aust.J.exp. Biol.med.Sci. 37, 193-200.
5. Schaedler, R.W., Dubos, R. and Costello, R. (1965). Association of germfree mice with bacteria isolated from normal mice. J.exp.Med. 122, 77-82.
6. Waaij, D. van der. (1968). The persistent absence of Enterobacteriaceae from the intestinal flora of mice following antibiotic treatment. J.infect.Dis. 118, 32-38.

THE EFFECT OF INTESTINAL FLORA ON TOXICITY OF NITROGEN MUSTARD

Judith A. O'Malley, Joyce Jividen, Edwin A. Mirand, James F. Holland

Roswell Park Memorial Institute, Buffalo, New York

The clinical use of germ-free isolators in cancer medicine is based on numerous investigations which have demonstrated reduced toxicity from chemotherapeutic agents or radiation in axenic animals (1-3). Earlier workers have attempted to show a similar decrease in toxicity in antibiotic-sterilized animals (1,4). The clinical approach tacitly assumes that results of experiments with axenic animals can be approximated by results in antibiotic-sterilized man. Our experiments with mice were designed to test the validity of this assumption by comparing nitrogen mustard (HN_2) toxicity in axenic, conventional and antibiotic-sterilized animals.

Six-week old Ha/ICR male mice weighing 30 grams were allocated randomly to groups as shown in Figure 1. Each subgroup contained 12 mice. The mice were purchased from Charles River Breeding Laboratories and axenic mice were studied in the germ-free state or after conventionalization to a normal gut flora by feeding a suspension of normal mouse feces eleven days prior to HN_2 injection. Antibiotics were incorporated into the drinking water of appropriate groups. A triple-antibiotic mixture which we have used in clinical barrier-isolation patients was used. The mixture, PPV, contained Polymyxin, 1.5 mg/day, Paromomycin, 12 mg/day, and Vancomycin, 6 mg/day, the same dose levels per square meter of body surface as were used in the patients. A group with Polymyxin alone, 0.75 mg/day, and another with Tetracycline alone, 60 mg/day, were also included. Three days after the antibiotic regimens were begun, the conventional and antibiotic-treated mice were transferred to sterilized plastic isolators. After four days in germ-free isolation, all animals were given a single 1.0 ml intraperitoneal injection of 0, 6 or 9 mg/kg of HN_2 diluted in 0.01 N HCl. This acid

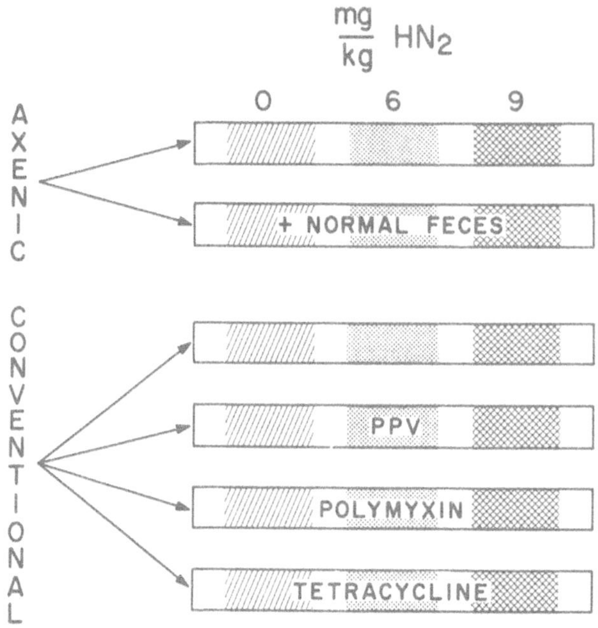

Figure 1. Randomization of axenic and conventional Ha/ICR male mice. PPV represents Polymyxin, 1.5 mg/day, Paromomycin 12 mg/day, and Vancomycin, 6 mg/day. Dosage for the group receiving Polymyxin alone was 0.75 mg/day. The Tetracycline group received 60 mg/day.

pH was found by White and Claflin to enhance the therapeutic index of HN_2 in axenic mice (1).

Representative fecal cultures from each group were done at various intervals throughout the experiment. Fecal homogenates were overlayed with a salt-free agar in order to inhibit the spreading of Proteus (5), thereby facilitating quantitation. A liver broth was used to support anaerobic growth.

At the 6 mg/kg dose level of HN_2, axenic mice did not die. Mortality was greatest in the three groups receiving antibiotics, (Figure 2). At 9 mg/kg HN_2, the mortality of conventional mice was similar to that of antibiotic-treated mice.

The pre-treatment intestinal flora of conventional, and conventionalized mice included Proteus, E. coli, Staphylococci and enterococci of the order of magnitude of 10^5-10^6 bacteria per gram of feces. Although mice receiving Polymyxin by mouth had no decrease in flora, those which were kept axenic or had received

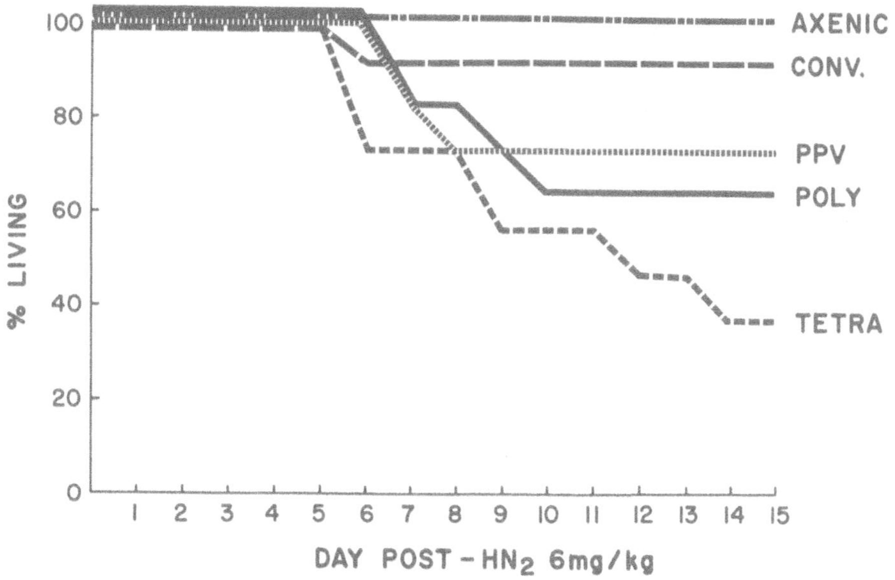

Figure 2. Effect of three oral antibiotic regimens on lethal toxicity of HN_2.

triple antibiotics for three days were sterile. After HN_2 at 6 mg/kg or 9 mg/kg, the mixed bacterial population increased fivefold in animals which had had mixed gut floras, reaching a maximum on day 3 or day 6. Axenic mice remained sterile, but the triple-antibiotic group gained a flora by day 6, almost exclusively enterococci.

The increase in gut bacteria after HN_2 was further studied in a new experiment. Conventional, Polymyxin and triple-antibiotic treated mice were studied in isolators and injected with 0, 3, 6 or 9 mg/kg of HN_2. All four animals in each group were cultured. No deaths occurred in the 0 or 3 mg/kg dose level of HN_2. At 6 mg/kg, the mortality of the triple-antibiotic group was greater than the conventional group (Figure 3). Mortality of conventional mice was similar to that of antibiotic treated mice at the 9 mg/kg HN_2 concentration.

Cultures of conventional mice seven days after 6 mg/kg HN_2 injection showed a fivefold increase in the number of the usual types of bacteria. The Polymyxin-treated group injected with 6 mg/kg HN_2 also showed substantial increase in the number of bacteria by day 7. The flora in the triple-antibiotic group after HN_2 in this experiment was entirely yeast, shown on day 3 with an X representing 10 yeast per gram of feces (Figure 3). All groups injected with 9 mg/kg HN_2 also showed increases in the number of organisms by day 7.

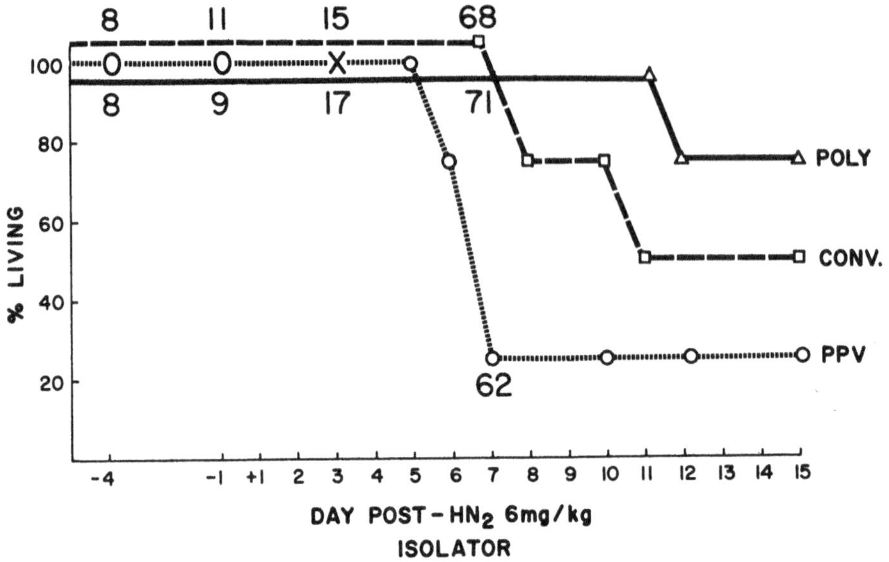

Figure 3. Effect of oral antibiotics on survival and number of organisms after HN_2 injection. Numbers on the curves represent the median number of organisms/10^5 per gram of feces. The experiment was conducted in isolators.

This experiment was then repeated out of isolators. Again, the mortality in both the antibiotic-treated groups was greater than in the conventional group which received no antibiotics (Figure 4). Cultures of conventional animals after 6 mg/kg HN_2 showed a doubling in the number of bacteria per gram of feces on day 4. The phenomenon of increased flora was also seen in the group receiving Polymyxin alone. By the seventh day, gut flora was decreasing toward normal. The triple-antibiotic treated animals retained sterile guts but died sooner nonetheless.

Changes in flora of mice who received 9 mg/kg HN_2 were similar, although no quantitation was possible on day 7 because all animals in these groups had died.

Other data with DBA-2 mice, collected in these laboratories, has shown that HN_2 induces leukopenia with a nadir at day 4 which subsides by day 7 after injection. This may explain the overgrowth of bacteria during this same post-injection period and the eventual decrease of bacteria to pre-injection levels. Contemporaneously with the leukopenia, it is also possible that HN_2 damages gut mucosal structure or function. Experiments are under way to clarify the effect of cytotoxic drugs on gut flora.

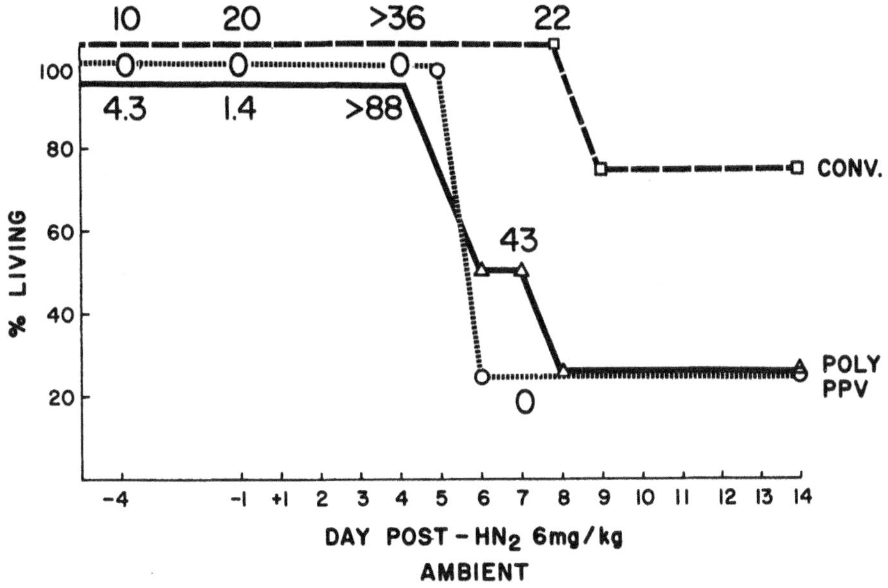

Figure 4. Effect of oral antibiotics on survival and number of organisms after HN_2 injection. Numbers on curves represent the median number of organisms/10^5 per gram of feces. The experiment was done under ambient conditions.

In all experiments, the cultural patterns of the ceca were closely paralleled by the results of feces cultures. As has been noted previously, the ceca of axenic mice were larger than those of conventional animals. Also of interest is the fact that the ceca of triple-antibiotic treated animals had enlarged ceca (Figure 5), and that in previous experiments with formerly-axenic, or conventionalized mice, these animals showed a decrease in cecal size that approached the size of ceca of conventional animals.

In summary, we have shown that use of three oral antibiotics sterilizes the gut of mice, but does not provide the same protection against death from toxic doses of HN_2 that the axenic state affords. In fact, greater deaths than in the conventional animals may represent cumulative toxicity of HN_2 and antibiotics. Possibly because of this, our observations afford little support to the hypothesis that intestinal bacteria play an important role in the systemic toxicity of HN_2. In mice whose gut flora was not eliminated, an increase of bacteria per gram of feces occurred after HN_2 administration but such increase in flora did not necessarily cause disease.

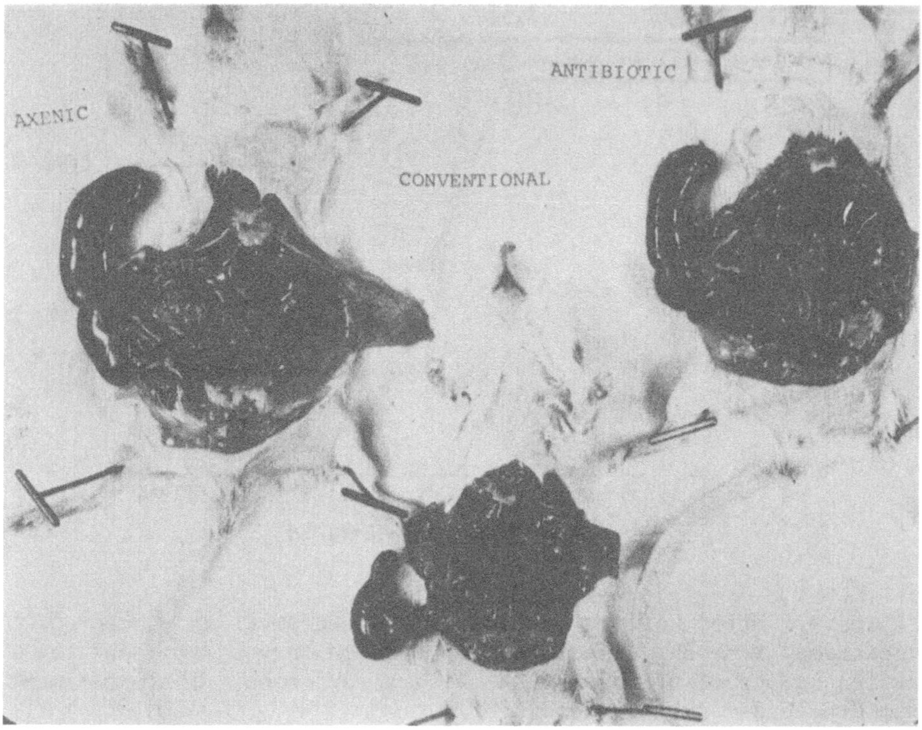

Figure 5. Effect of triple-antibiotic mixture (Polymyxin, Paromomycin and Vancomycin) on cecal size.

REFERENCES

1. White, L.P., and Claflin, E.F. "Nitrogen Mustard: Diminution of Toxicity in Axenic Mice." Science 140, 1400-1401, June 28, (1963).

2. McLaughlin, M.M., Dacquisto, M.P., Jacobus, D.P., and Horowitz, R.E. "Effects of the Germ-free State on Responses of Mice to Whole-Body Irradiation." Radiation Research, 23, 333-349, (1964).

3. Wilson, B.R. "Survival Studies of Whole-Body X-Irradiated Germ-free (axenic) Mice." Radiation Research, 20, 477-483, (1963).

4. Rosoff, C.B. "The Role of Intestinal Bacteria in the Recovery from Whole Body Radiation." J. Exp. Med., 118, 935-943, Dec., (1963).

5. Schneirson, S. "Production of Discrete, Nonswarming Colonies by Proteus on Medium Deficient in Sodium Chloride and other Salts. J. Bact., 82, 621-622, (1961).

EFFECTS OF THE NORMAL FLORA ON HOST DEFENSES AGAINST MICROBIAL INVASION *

Gerald D. Abrams, M.D.

Department of Pathology, The University of Michigan

Ann Arbor, Michigan

INTRODUCTION

It is well recognized that the normal microbial flora contributes to the defense of the host by virtue of its competition with or its suppression of invading microorganisms, particularly within the gastrointestinal tract (1,2). What is less well recognized is that the resident flora may also contribute to the defense of the host by virtue of its direct impact on host tissues, apart from any ecological effect on other microorganisms. This sort of direct effect on the host is the topic of the present communication, which traces the background of some of the work in this area, and presents examples of this direct effect on host defenses both within the habitat of the flora, as in the gastrointestinal tract, and beyond.

GASTROINTESTINAL DEFENSES

In recent years the small intestine has been recognized as an organ which reflects the tremendous impact of the normal flora upon the host. Comparison of the small intestine of the germfree animal with that of the conventional reveals the details of this impact: The "normal" lamina propria of the conventional animal manifests to a significant degree, what might be termed a physiologic inflammatory reaction to the presence of the intestinal flora, this layer of the bowel wall being poorly developed in the germfree animal (3). The surface area of the small intestine is significantly smaller in the germfree animal (4). The architecture of the mucosa is different in the two groups of animals, the crypts being relatively larger in the

*Supported by NIH grant AI-05827-05

presence of the flora than in the germfree animal. This difference reflects the fact that the life cycle of mucosal cells is significantly affected by the flora, the entire process of epithelial cell renewal proceeding approximately twice as rapidly in the conventional animal as in the germfree. The net result of this difference in cell renewal is that the intestinal lining of the germfree animal consists of cells of greater average age than those in the conventional (3).

Given these differences, it seemed that the mucosa of the germfree animal might be a deficient barrier to microbial invasion as compared to the mucosa of the conventional animal, apart from the ecologic situation obtaining in the bowel lumen. In order to test the validity of this notion, infection with *Salmonella typhimurium* was studied in the two groups of animals (5).

Germfree and conventional mice were challenged by peroral delivery of Salmonellae into the lumen of the gastrointestinal tract. At the end of various experimental intervals following challenge, the animals were sacrificed and the bacterial populations of the intestines and mesenteric lymph nodes were estimated by routine quantitative microbiological techniques. The number of organisms reaching the mesenteric lymph nodes in each case was taken as an index of translocation of organisms across the mucosal barrier.

In animals sacrificed 24 hours after intragastric challenge (Table 1) more organisms were indeed found to be present in the nodes of germfree mice than in their conventional counterparts. However, this could not be accepted as evidence of a difference in the barrier properties of the mucosa in the two groups of animals, since in germfree mice, in the absence of a flora, bacterial counts within the lumen were strikingly higher than in the conventional mice. Under these circumstances, the difference in the number of organisms reaching the mesenteric nodes could simply have been a reflection of the difference in intraluminal population.

Therefore, the attempt was made to create a situation in which the number of organisms in the bowel of germfree and conventional mice would be held at equivalent levels during the penetration period. Intraduodenal challenge and the utilization of a shorter test interval failed to produce the desired situation (Table 1). The results in this latter instance were parallel to those in the first, i.e., more Salmonellae once again had reached the mesenteric nodes of the germfree mice but again this was seen to be associated with a larger number of organisms within the bowel, impinging upon the mucosal barrier.

Finally, the ligation of the terminal ileum was used in order to create an infection in a closed tube in the two groups of animals. By means of this maneuver (Table 1) the intestinal

Table 1. Geometric mean counts of *Salmonella typhimurium* in small intestine and mesenteric lymph nodes of germfree and conventional mice.

Conditions of Challenge	Status of Mice	Organisms in Small Intestine	Organisms in Mesenteric Node
10^9 organisms. Intragastric. 24 hrs. before sacrifice	Germfree	8.6×10^9	8.9×10^3
	Conventional	3.0×10^3	1.9×10^2
10^7 organisms. Intraduodenal. 16 hrs. before sacrifice	Germfree	6.1×10^8	8.1×10^2
	Conventional	4.1×10^4	2.0×10^1
10^7 organisms. Intraduodenal Ileum ligated. 16 hrs before sacrifice.	Germfree	9.6×10^{10}	2.0×10^2
	Conventional	1.1×10^{11}	4.0×10^2

populations of Salmonallae in the two groups of animals were kept equal. Under these conditions, the number of Salmonellae reaching the lymph nodes in germfree animals was almost identical to that in the conventionals.

These results demonstrated that the original hypothesis was incorrect, i.e., given equivalent loads of organisms with which to deal, there is no difference in the barrier function of the intestinal mucosa of germfree and conventional animals despite anatomic differences. However, these results led to the elucidation of a most interesting effect of the flora. The fact that the growth of Salmonella was several logs less in the conventional than in the germfree (Table 1) when the ileum was not ligated, certainly indicated some protective effect of the flora. The fact that this effect disappeared with ileal ligation, despite the presence of the flora in the ligated conventional bowel, suggested that the protective effect of the flora was mediated not via suppression of the invader in the small intestine, but by enhancement of intestinal emptying in the presence of the flora.

In order to test this possibility a suspension of carmine dye was next fed to germfree and conventional mice, and the progress of the dye marker through the gastrointestinal tract was observed in the two groups of animals by serial sacrifice of members of both groups. This experiment indicated that although visible dye had disappeared from the small intestine of conventional mice within six hours after intragastric feeding, abundant dye remained in the small intestine of germfree mice at the end of the same interval, indicating that the normal flora does indeed act to accelerate intestinal emptying as compared to the situation in the germfree.

These dye studies were then refined by the use of radioactive yttrium-91 in place of carmine dye. In this latter experiment germfree and conventional mice were fed a radioactive meal by peroral intubation of the stomach, and animals of both groups were sacrificed at intervals to follow the passage of radioactivity through various parts of the gastrointestinal tract (6).

These studies revealed that in the absence of the normal microbial flora, propulsion along the entire gastrointestinal tract was slower than in the conventional situation. In Table 2 the results with respect to the small intestine are shown. It can be seen that at each interval tested, the percent radioactivity remaining in the small intestine of the germfree animal was greater than that in the conventional.

Other studies in the literature (7) indicate that the main defense of the small intestine against the build-up of bacterial populations is the rapid propulsion of invading organisms into the cecum and colon (where other defense mechanisms can take over). The flora, then, by accelerating intestinal propulsion makes a direct contribution to host defense against infection within the small intestine of the conventional animal. The small intestine of the germfree animal, in contrast, having a significant residual even at 24 hours after entry of material into the gastrointestinal tract, affords the opportunity to invading organisms of significant multiplication during this interval.

PARENTERAL DEFENSES

Contributions of the normal flora to the defense of the host are not limited to the immediate habitat of the flora, although the evidence in support of this contention is not as clear as in the

Table 2. Percent of radioactive meal remaining in small intestine of germfree and conventional mice after intragastric feeding at various intervals.

Status of Mice	2 hrs.	4 hrs.	6 hrs.	16 hrs.	24 hrs.
Germfree	23.5	14.1	16.2	9.8	8.7
Conventional	16.2	12.2	4.2	1.0	0.1

case of the gastrointestinal tract. It has been known for some time that germfree animals are unusually susceptible to infection by a variety of routes, not just the gastrointestinal route. In addition, it has been shown that mice can be rendered hypersusceptible to parenteral infection by depressing the enteric flora with antibiotics (8).

Mediation of such distant effects of the flora remains largely unknown. However, because of the obvious importance of leukocyte function in the defense of the host against infections, it seemed reasonable first to examine and compare the ability of germfree and conventional mice to mobilize leukocytes from the blood stream into an area of injury, particularly an area not inhabited normally by the flora.

In order to effect this comparison, starch-induced peritonitis was studied in germfree and conventional mice. The animals were compared with respect to their ability to pour leukocytes into the irritated but sterile peritoneal cavity in response to the same irritant (9).

Our initial studies indicated that the peritoneal inflammatory response was qualitatively similar in germfree and conventional mice. In both groups of animals there was a rapid outpouring of predominantly polymorphonuclear leukocytes into the irritated peritoneal cavity (Table 3). Whereas in the non-irritated peritoneal cavity there was no statistically significant difference in the number of leukocytes in the two groups of animals, at three and six hours the conventional animals were found to be able to pour many more leukocytes into the peritoneal cavity than their germfree counterparts. This difference, highly significant statistically at the six hour peak, was maintained to a lesser degree throughout the entire period of observation.

Simply stated, these data indicate that the presence of the normal flora elsewhere in the host significantly enhances the process of leukocytic exudation, a process of great defensive importance to the host.

Table 3. Millions of leukocytes in starch-induced peritoneal exudates of germfree and conventional mice at various intervals after injection.

Status of Mice	0 hrs.	3 hrs.	6 hrs.	10 hrs.	16 hrs.	24 hrs
Germfree	6.1	11.2	18.3	6.3	7.8	10.1
Conventional	4.3	14.0	33.4	9.4	11.6	12.1

Given the difference in the ability of the two groups of animals to pour leukocytes out of the vascular system and into the peritoneal cavity in response to an irritant, the question logically arises as to whether this difference might be related to some difference in the availability of leukocytes from the blood stream. This question is particularly important in the light of the fact that the resting germfree mouse has a lesser number of neutrophils in circulation than does its conventional counterpart. Recent experiments have been directed at demonstrating any possible influence of the flora on the ability of animal to mobilize leukocytes into the circulating blood and at learning whether or not conventional animals would still exhibit enhanced extravascular mobilization of leukocytes as compared to germfree animals given conditions of equal numbers of blood leukocytes at the time of induction of peritonitis in the two groups.

In order to answer these questions, we took advantage of the fact that intravenous injection of killed *Bordetella pertussis* in mice produces a striking hyperleukocytosis in circulating blood (10). In an initial experiment, 0.1 cc of a commercial pertussis vaccine was administered to each of a number of germfree and conventional mice. The response to this injection was compared in the two groups of animals with respect to the mobilization of polymorphonuclear neutrophils into the blood stream.

As seen in Table 4, the resting (uninjected) conventional mouse has more polymorphs in circulation than does the germfree, this difference being statistically significant. In both groups there is a rapid and massive mobilization of polymorphs into the peripheral blood in response to administration of the vaccine. The magnitude is enormous, particularly at days 3 and 7. The data indicate that the two groups of animals are essentially equal in their response, there being no statistically significant difference in leukocyte counts at any of the postinjection intervals tested. These data indicate that the normal flora or the absence thereof has no significant effect on the ability of the host to mobilize leukocytes into the blood from the bone marrow.

Table 4. Number of polymorphonuclear leukocytes per cubic mm. of blood of germfree and conventional mice after intravenous injection of *Bordetella pertussis*.

Status of Mice	Uninjected	1 day	2 days	3 days	7 days
Germfree	910	5,380	5,790	19,800	13,600
Conventional	1,580	2,840	9,400	13,100	8,530

Given this means of elevating both germfree and conventional blood leukocyte counts to equivalent levels, it became possible to compare the ability of the two groups of mice to mobilize leukocytes into the irritated peritoneal cavity from an equivalent intravascular pool. For this experiment, germfree and conventional mice were pretreated by prior intravenous injection of pertussis vaccine and at a time when peripheral leukocyte counts were known to be extremely high (days 3 and 7) events within the irritated peritoneal cavity were evaluated, using the six hour exudative peak as an endpoint.

As seen in Table 5, in the absence of pretreatment, the conventional animal is able to mobilize a far greater number of polymorphs into the peritoneal cavity than the germfree. Interestingly, when peritonitis was induced three days following treatment with pertussis vaccine, at a time when both groups of animals had high blood leukocyte counts, both germfree and conventional animals showed no increase in the ability to mobilize cells into the peritoneal cavity. In fact, the figures, although not statistically significant, even indicate the possibility of a decrease in mobilization at this interval. When peritonitis was induced seven days after injection of pertussis vaccine, there was a significant augmentation of leukocytic exudation into the peritoneal cavity in both groups of mice. However, even though at this interval germfree animals had certainly no fewer (and possibly even more) circulating blood leukocytes, the augmentation was greater in the conventional group. Thus even in the presence of equal numbers of circulating leukocytes to draw from, the animal having a flora is able to mobilize leukocytes more readily than the animal raised under germfree conditions. Apparently the flora exerts some influence over whatever mechanisms control the exudative process *per se*.

GENERAL DISCUSSION

The mechanisms by which the normal flora exerts its influence on these aspects of host resistance are almost completely obscure.

Table 5. Millions of polymorphonuclear leukocytes in starch-induced 6 hr. peritoneal exudates of germfree and conventional mice pretreated at various intervals with intravenous *Bordetella pertussis*

Status of Mice	Non-pre-treated	3 days	7 days
Germfree	14.8	8.4	19.7
Conventional	22.4	17.5	34.3

In the case of the gastrointestinal tract there are some hints that the effects of the flora on motility might be related to such things as the metabolism of neurons in Auerbach's plexus (11), the intestinal complement of pharmacologically active amines (12), the luminal content of musculo-active substance (13), or perhaps the threshold of neuromuscular response of the bowel to effector agents (14).

In the case of leukocyte mobilization, the possible mechanisms are more obscure because of our basic ignorance of the means of control of the process of leukocytic exudation. Conceivably, the complement system may be involved (15). There is also evidence that certain plasma substrates are activated during the acute inflammatory process (16). Whether these systems are affected by the presence or absence of the flora, or whether the leukocytes *per se* are affected are matters of current interest in our laboratory.

Not only do we not know the identity of the members of the flora responsible for these defense-enhancing effects, but we are quite ignorant of the locus of significant host-flora interaction, i.e., whether this locus is the lumen of the bowel, the lamina propria, the mesenteric lymph nodes, or some other parenteral site.

Whatever the means by which the flora exerts its influence on the host, there is no doubt that such an influence does exist; the flora enhances intestinal motility, thus promoting the protection of the small intestine against infection. The flora enhances exudation of polymorphonuclear neutrophils, with obvious potential importance for the host. Other effects of the flora could be added to this list from reports in the literature. For instance, there is evidence that mobilization of histiomonocytic cells is enhanced by the presence of the flora (17) and there are data to indicate that intracellular digestion in macrophages is enhanced by the presence of the flora (18). These all constitute significant direct contributions of the flora to the defense and the well-being of the host.

SUMMARY

The normal microbial flora contributes to the defense of the host not only by virtue of its ecological effects but also by virtue of its direct impact on host tissues. This impact is evident beyond as well as within the habitat of the flora. *Salmonella typhimurium* has been found to proliferate to a significant greater extent in the small intestine of germfree as compared to conventional mice following peroral challenge. However, proliferation of the Salmonellae was identical in the two groups when intestinal emptying was prevented by ileal ligation, indicating that the conventional flora does not act in this respect by suppression of the invader but by enhancement of intestinal motility. Subsequent studies have confirmed that intestinal emptying is significantly more rapid in animals harboring a conventional flora than in germ-free animals. Beyond the habitat of the flora, studies of aseptic peritonitis in the two groups of animals have shown that extravascular leukocyte mobilization is significantly enhanced when a flora is present elsewhere in the body. This difference has been found to be independent of any effect of the flora on the number of circulating leukocytes, indicating that the flora has an impact on the control mechanisms of leukocytic emigration per se. These effects, as well as others reported in the literature, undoubtedly constitute important direct contributions of the normal flora to host defense.

ACKNOWLEDGMENTS

The invaluable help of Miss Jane E. Bishop and Mr. Eddie J. Burks is gratefully acknowledged.

REFERENCES

1. R. Freter, "In vivo and in vitro antagonism of intestinal bacteria against Shigella flexneri. II. The inhibitory mechanism," J. Infect. Dis., 110:38, 1962.
2. M. Bohnhoff, C.P. Miller, and W.R. Martin, "Resistance of the mouse's intestinal tract to experimental Salmonella infection. I. Factors which interfere with initiation of infection by oral inoculation," J. Exptl. Med., 120:805, 1964.
3. G.D. Abrams, H. Baver, and H. Sprinz, "Influence of the normal flora on mucosal morphology and cellular renewal in the ileum," Lab. Invest., 12:355, 1963.
4. H.A. Gordon, and E. Bruckner-Kardoss, "Effect of normal microbial flora on intestinal surface area," Amer. J. Physiol., 201:175, 1961.
5. G.D. Abrams, and J.E. Bishop, "Effect of the normal microbial flora on the resistance of the small intestine to infection," J. Bact., 92:1604, 1966.
6. G.D. Abrams, and J.E. Bishop, "Effect of the normal microbial flora on gastrointestinal motility," Proc. Soc. Exptl. Biol. Med., 126:301, 1967.
7. J.M.S. Dixon, "The fate of bacteria in the small intestine," J. Path. Bact., 79:131, 1960.
8. P. Dineen, "Effect of alterations in intestinal flora on host resistance to systemic bacterial infection," J. Infect. Dis., 109:280, 1961.
9. G.D. Abrams, and J.E. Bishop, "Normal flora and leukocyte mobilization," Arch. Path., 79:213, 1965.

10. S.I. Morse, and S.K. Riester, "Studies on the leukocytosis and lymphocytosis induced by Bordetella pertussis. I. Radioautographic analysis of the circulating cells in mice undergoing pertussis-induced hyperleukocytosis," J. Exptl. Med., 125:401, 1967.
11. J.R. Dupont, H.R. Jervis, and H. Sprinz, "Auerbach's plexus of the rat cecum in relation to the germfree state," J. Comp. Neurol., 125:11, 1965.
12. M.H. Beaver, and B.S. Wostmann, "Histamine and 5-hydroxytryptamine in the intestinal tract of germfree animals, animals harboring one microbial species and conventional animals," Brit. J. Pharm. Chem., 19:385, 1962.
13. H.A. Gordon, "Demonstration of a bioactive substance in caecal contents of germfree animals," Nature, 205:571, 1965.
14. K. Strandberg, G. Sedvall, T. Midvedt, and B. Gustafsson, "Effect of some biologically active amines on the cecum wall of germfree rats," Proc. Soc. Exptl. Biol. Med., 121:699, 1966.
15. P.A. Ward, C.G. Cochrane, H.J. Muller-Eberhard, "The role of serum complement in chemotaxis of leukocytes in vitro," J. Exptl. Med., 122:327, 1965.
16. J.V. Hurley, "Substances promoting leukocyte emigration," Ann. N. Y. Acad. Sci., 116:918, 1964.
17. E.H. Perkins, P. Nettesheim, T. Morita, and H.E. Walburg, Jr., "The engulfing potential of peritoneal phagocytes of conventional and germfree mice," Advances in Experimental Medicine and Biology, 1:175, 1967.
18. H. Bauer, F. Paronetto, W.A. Burns, and A. Einheber, "The enhancing effect of the microbial flora on macrophage function and the immune response," J. Exptl. Med., 123:1013, 1966.

Ageing and Resistance to Infection in Germfree C3Hf Mice

Henry C. Outzen

Department of Anatomy and The Germfree Laboratory of the Department of Surgery, University of Utah College of Medicine, Salt Lake City, Utah

Germfree C3Hf mice were observed to have a high mortality rate which appears to be dependent on both the age and the sex of the mice at the time of their exposure to our conventional C3H mouse colony (Outzen, and Pilgrim, 1967). When germfree male mice were introduced into our conventional colony before two months of age, approximately 73% survived six weeks' exposure. However, in animals 2-6 months old at the time of exposure, only 34% survived, and if the germfree mice were more than six months of age at the time of exposure, less than 10% survived.

The purpose of this report is to ascertain whether an effect similar to that seen previously would be observed and to document the effect of age and sex on the ability of germfree C3H mice to survive challenge by selected live enteric pathogens within a defined environment.

MATERIALS AND METHODS

Male and female germfree C3Hf mice were maintained in aluminum autoclavable isolators (Pilgrim and Thompson, 1963). Autoclavable Wayne Lab-Blox and water were supplied ad libitum. Males and females were divided into four groups by age; under 2 months, 2-6 months, 6-12 months, and more than 12 months of age. Male mice were caged together at weaning to minimize fighting.

This work was supported by Contract No. PH43-64-925 from the Cancer Chemotherapy National Service Center, and Grant No. CA05771-08 from the National Cancer Institute, National Institutes of Health.

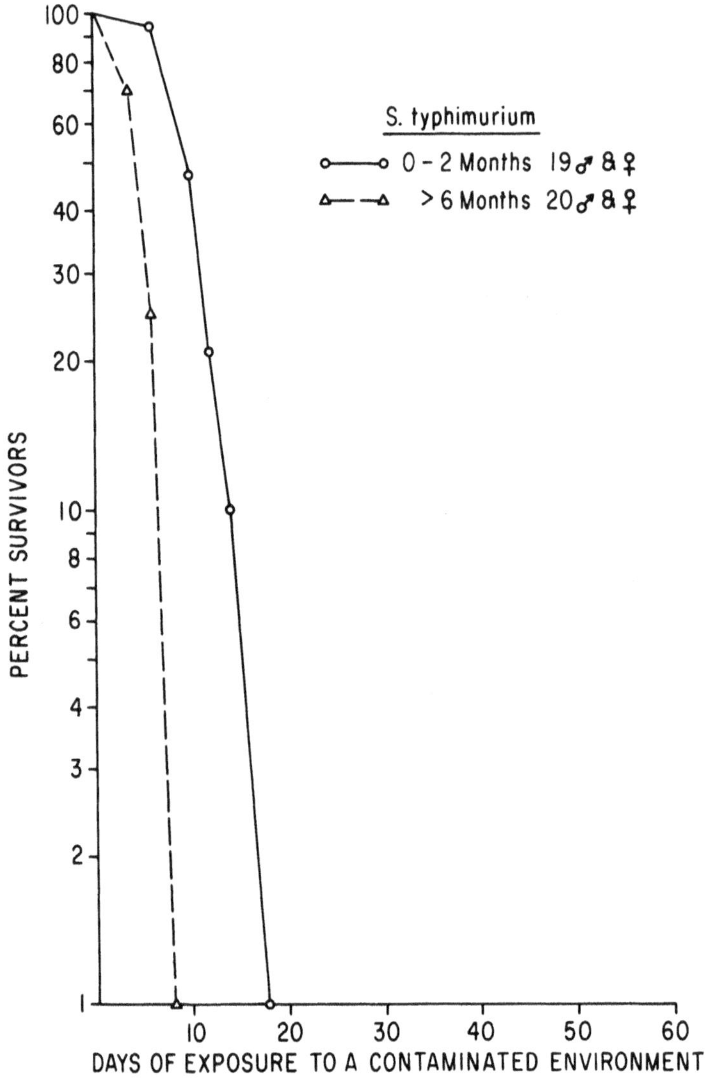

Figure 1. Survival of two different age groups of germfree C3Hf/Pi mice challenged with *S. typhimurium*.

The mice were caged in small aluminum cages with approximately 3-5 mice per cage. Sterility checks, using fluid thioglycollate, were performed on each isolator prior to their being contaminated, and further contamination checks were performed throughout the course of the experiment. Contamination of the mice with the bacteria being studied was achieved by either spraying the organism onto the mice or by administering approximately 10^7 organisms orally, using a blunt needle on a syringe. Bacterial cultures isolated from conventional mice were kindly furnished by Dr. Russell Schaedler of Rockefeller University; they consisted of two strains of lactobacilli, an anerobic (group N) streptococcus, a bacteroides, an enterococcus, and a slow-lactose-fermenting Escherichia coli (Schaedler, et al., 1965).

The data were evaluated by the statistical method described by Pilgrim and Dowd (1964). This method makes a continuous correction for death due to extraneous causes (e.g. sacrifice for histological study). Animals dying of extraneous causes may be utilized up to the time of death, and are then mathematically eliminated from subsequent consideration. The corrected data are then plotted using the standard actuarial technique of plotting the logarithm of the percent survivors against time. When plotted in this manner, the slope of any segment of the curve is the death rate and the shape of the curve shows whether the mortality rate increases or decreases with age.

RESULTS

The age of germfree mice at the time of their exposure to <u>Salmonella typhimurium</u>[1] has an effect on the mortality rate. Younger mice have a slightly longer latent period before succumbing to the pathogen (Table I, figure 1). All of the mice died within 20 days after exposure to <u>S. typhimurium</u>. The mortality rate during this period is very high and quite constant as shown by the slopes of these curves. It is apparent that there is an age-related difference; with the median age of death occurring 4-5 days earlier in the older mice. The extremely high mortality rate suggested the use of a less pathogenic species of Salmonella. Studies by Schaedler, et al. (1965) also suggested that an attempt at partially protecting the mice with part of the normal flora might be advisable. In subsequent experiments we used Schaedler's flora and <u>Salmonella newport</u>[2]. Previous contamination with Schaedler's

---

1 The <u>S. typhimurium</u> was a human-isolated strain from the Utah State Department of Health.

2 The <u>S. newport</u> was obtained through the courtesy of Dr. Warren Hoag of the Jackson Memorial Laboratory.

TABLE I. Raw data on survival of germfree C3Hf/Pi mice exposed to *S. typhimurium*.

Age prior to challenge with *S. typhimurium*		Days of Exposure 0	2	4	6	8	10	12	14	16	18	20
2-8 weeks old	Alive at start of interval	19	19	19	19	18	18	9	4	2	2	0
	Died during interval	0	0	0	1	0	9	5	2	0	2	0
More than 6 months old	Alive at start of interval	20	20	20	14	5	0	0	0	0	0	0
	Died during interval	0	0	6	9	5	0	0	0	0	0	0

flora made little or no difference in the mortality rate after challenge with *S. newport* (Table II and figure 2). No apparent difference was noticed in either spraying the *S. newport* directly onto the mice or by administering approximately 10^7 organisms orally.

The age of the mice at the time of challenge with *S. newport* exerts a profound effect on their subsequent survival. By the 15th day after *S. newport* challenge the difference in the mortality rates of the four groups of male mice are readily apparent (figure 2). Mice that are 12 months or older dying the most rapidly. After the 15th day the mortality rates began to decrease and following three weeks exposure the mortality rates approached zero. Mice less than two months of age challenged with only *S. newport* appeared to have a slightly lower mortality rate than those previously contaminated with Schaedler's flora (figure 2). However, at the end of the 60-day observation period there was essentially no difference, with approximately 80% of the mice in both groups surviving. Mice that survived past 60 days were then resistant to subsequent challenges with either *S. typhimurium* or *S. newport*.

In the two groups of male mice more than two months old which were previously contaminated with Schaedler's flora, there were 42% survivors in the 2-6 month-old mice and 3.5% survivors in the 6-12 month-old mice. In male mice more than 2 months of age exposed only to *S. newport* there were 55% survivors in the group 2-6 months

TABLE II. Raw data on survival of germfree C3Hf/Pi mice exposed to either S. newport or Schaedler's flora and S. newport.

Age prior to S. newport challenge		Days of Exposure to S. newport 0	2	4	6	8	10	12	14	16	18	20	22	24-60	Total No. of Mice
2-8 weeks of age	G 1	0(0)	0(2)	1(3)	0(1)	1(0)	0(0)	0(0)	0(1)	0(0)	0(0)	1(0)	0(0)	2(4)	29
	S 2	0(0)	1(3)	0(2)	0(1)	6(0)	2(0)	0(0)	0(0)	0(0)	0(0)	0(0)	0(0)	0(0)	47
	S 3	0(0)	0(1)	0(2)	0(1)	4(0)	4(0)	0(0)	0(0)	0(0)	0(0)	0(0)	0(0)	0(9)	34
2-6 months of age	G 1	0(0)	0(0)	2(0)	1(5)	0(0)	0(0)	1(0)	3(0)	1(0)	0(0)	0(0)	0(0)	0(0)	22
	S 2	0(0)	0(2)	0(3)	0(2)	3(0)	2(0)	2(0)	0(2)	0(0)	0(0)	0(0)	0(0)	0(2)	19
	S 3	0(0)	0(2)	0(3)	0(2)	1(0)	0(0)	1(0)	0(2)	0(0)	0(0)	0(0)	0(0)	0(2)	13
6-12 months of age	G 1	0(0)	0(0)	35(0)	4(0)	4(0)	0(0)	2(0)	2(0)	1(0)	1(0)	0(0)	0(0)	0(0)	49
	S 2	0(0)	0(2)	1(1)	3(0)	16(0)	4(0)	2(0)	0(0)	1(0)	0(0)	0(0)	0(0)	0(0)	31
	S 3	0(0)	0(2)	0(1)	0(0)	8(0)	2(0)	1(0)	1(0)	0(0)	0(0)	0(0)	0(0)	0(3)	20
12-17 months of age	G 1	0(0)	7(0)	10(0)	0(0)	0(0)	0(0)	0(0)	0(0)	0(0)	0(0)	0(0)	0(0)	0(0)	17
	S 3	0(0)	0(0)	3(0)	0(0)	12(0)	5(0)	2(0)	1(0)	0(0)	0(0)	0(0)	1(0)	0(0)	26

G-1 = Germfree males exposed only to S. newport.

S-2 = Germfree males exposed to Schaedler's normal mouse flora and two weeks later challenged with S. newport.

S-3 = Germfree females exposed to Schaedler's normal mouse flora and two weeks later challenged with S. newport.

number in parenthesis indicates number of mice sacrificed during the interval; number out of parenthesis indicates animals that died during the interval, eg. 5(2) means 5 mice died and two mice were sacrificed, a total of 7 mice in that interval.

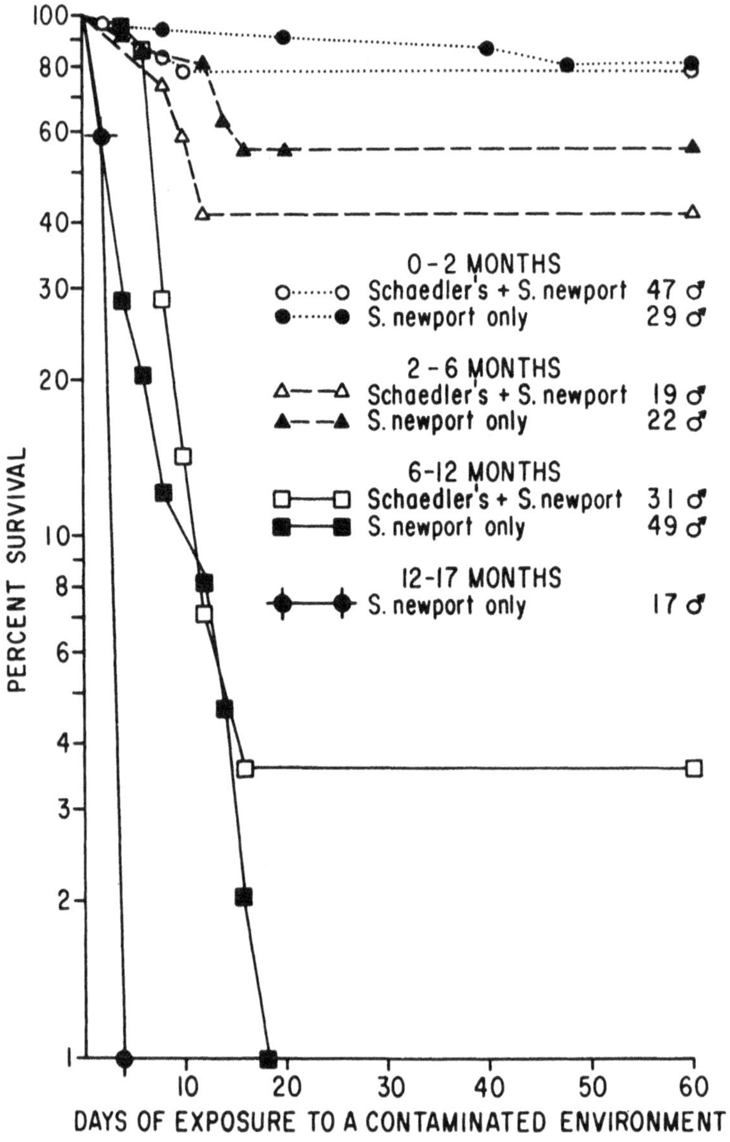

Figure 2. Survival of 4 different age groups of male gnotobiotic C3Hf/Pi mice challenged with *S. newport*. (Each group was composed of both germfree mice and mice contaminated with Schaedler's flora, excepting the oldest group).

of age. In the group 6-12 months of age there was 100% mortality by day 18, and in the group more than 12 months of age there was 100% mortality by the 4th day.

The female mice, which were always contaminated with Schaedler's flora, were not as susceptible to the effects of challenge with S. newport (Table II, figure 3). However, an age response is still apparent by the 15th day of exposure in the 4 groups of female mice. Female mice less than 2 months of age and from 2-6 months of age have similar mortality rates during the first 10 days after S. newport challenge, with 73% survivors at this time. The mortality rates in the female mice 6-12 months and more than 12 months of age were much higher with 41% and 23% respectively surviving on the 10th day of exposure. Following the 10th day of exposure the mortality rates decrease and approached zero in all groups of female mice by the end of one month's exposure. At the end of 60 days exposure there were 73% survivors in the group under 2 months of age, 67% survivors in the 2-6 month group, 29% in the 6-12 month group, and about 8% in the group of female mice greater than 12 months of age.

It has not been possible to establish a relationship between the normal mouse flora used and resistance to Salmonella infection. This may possibly be due to the fact that we are contaminating the mice with a large number of organisms and are overpowering the protective or inhibitory effect thought to be provided by the normal flora in the mouse.

Histologically the picture seen in the gastrointestinal tract is that of a severe enteritis. The most dramatic change is seen in the cecum and colon within 24 hours after S. newport challenge (plate B). Dilation of the lymphatic vessels and much submucosal edema occur at this time. This picture of inflammation gradually subsides, and by the 13th day a definite cellular infiltration is noted (plates C and D). The small intestine shows only slight edema in the villi, and no submucosal involvement. The relationship of the pathology that is seen when germfree mice are challenged with S. newport and the mortality rates in the various age groups is not understood at this time.

DISCUSSION

The age-dependent immunological response has been described by Makinodan and Peterson (1966). They demonstrated that the peak primary response to rat erythrocytes in the murine spleen occurs between one and 4 months of age. It then slowly declines with increasing age. A decrease in the number of progenitor cells is thought to be one of the causes of the decreased immune response

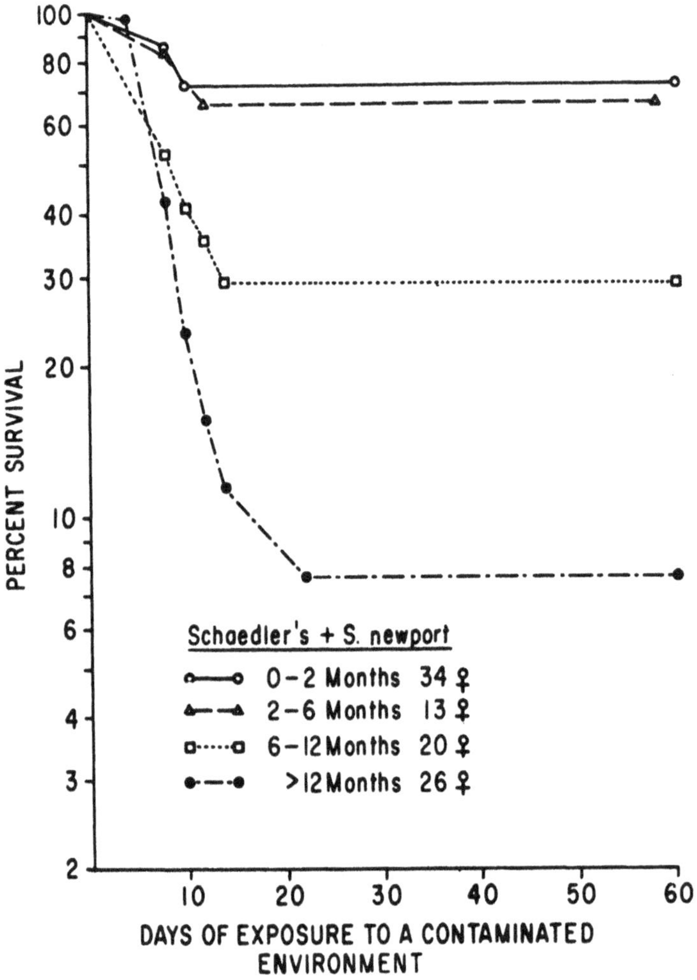

Figure 3. Survival of 4 different age groups of female gnotobiotic C3Hf/Pi mice contaminated with Schaedler's flora 2-3 weeks prior to their being challenged with *S. newport*.

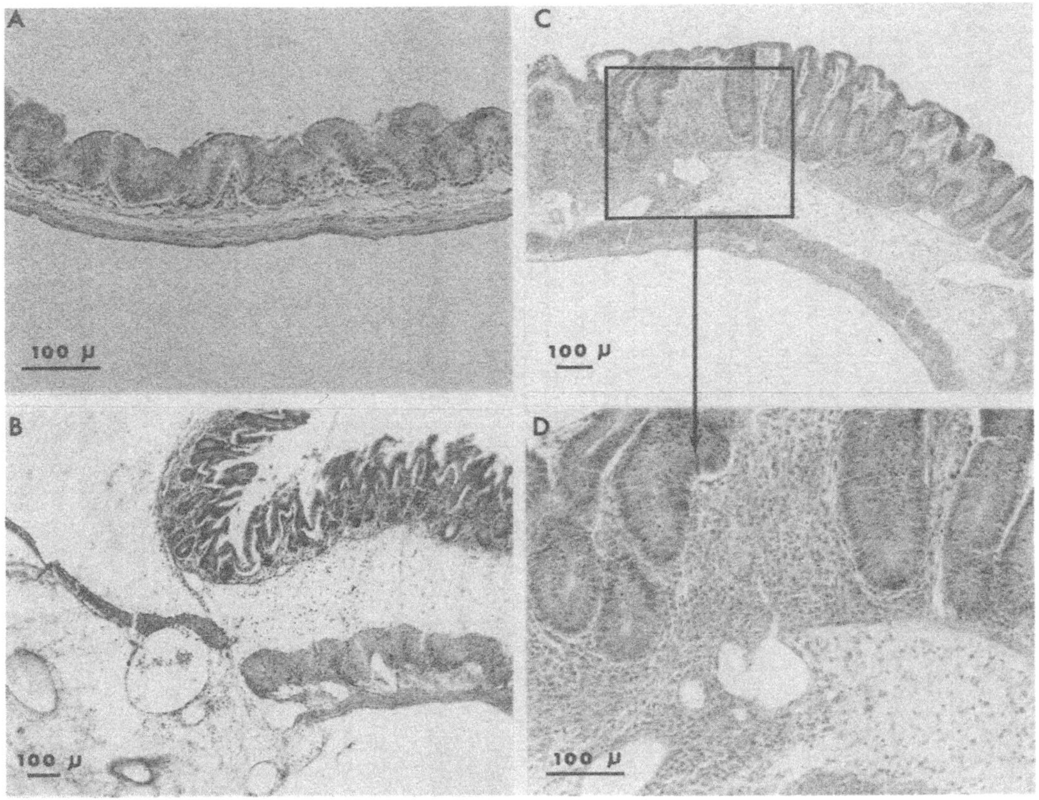

<u>Plate A.</u> Cecum from germfree C3Hf male mouse.
<u>Plate B.</u> Cecum from a germfree C3Hf/Pi male mouse which has been exposed to *S. newport* for one day. Note the extensive lymphatic dilation and submucosal edema.
<u>Plate C and D.</u> Cecum from a germfree C3Hf/Pi male mouse which has been exposed to *S. newport* for 13 days. There is an apparent increase in cellularity in the lamina propria and submucosal tissue. The amount of submucosal edema is less than seen earlier.

which occurs with advancing age, according to Albright and Makinodan (1966). Three-year-old mice were found by Hanna, et al. (1967) to retain only 10% of the hemagglutinin responsiveness to sheep erythrocytes as compared with 12 to 14 week-old mice. They also observed that splenectomy appreciably reduced the immunological response. Olson and Wostman (1966) have shown that germfree mice exhibit peak plasmacytic responses to killed S. typhimurium in the mesenteric and cervical nodes 6 to 10 days after intraperitoneal stimulation. Maximum lymphoid hyperplasia in the spleen occurs about the same time.

Bosma, et al. (1967) documented the development of immunological competence (using sheep erythrocytes) in germfree and conventional mice. They demonstrated that the primary antibody-forming potential in the spleens of germfree and conventional mice follow typical growth curves; developing exponentially during the first five weeks after birth, and reaching a maximum plateau level about 9 weeks of age. Administration of estrogens to conventional animals produces increased phagocytosis, decreased blood lymphocyte counts, and increased resistance to specific bacterial infections (Von Haam & Rosenfield, 1942; Ferrel, 1947; Nichol, et al., 1964). Female C3H mice have been shown to produce more antibodies more rapidly and over a longer period of time than males (Chouroulinkov, et al., 1964). Terres, et al. (1968) have shown that there is a quantitative difference between male and female mice in their immune response to Bovine serum albumin. The females developed stronger, and longer-lasting immunity; and were more responsive to smaller doses. We have demonstrated previously that C3Hf/Pi germfree male mice have a higher mortality rate than females when transferred to a conventional colony (Outzen and Pilgrim, 1967). The causes of this difference in susceptibility which seems to be dependent upon both age and sex is unknown.

SUMMARY

Resistance to microbial infection in germfree C3Hf mice appears to be age-dependent. The death rate remains high for approximately 20 days after contamination of the mice with S. newport. Following this period the death rate stabilizes and approaches zero. The death rate appears to be slightly higher in males than in females and much higher in older animals than in younger ones.

Acknowledgments

The author gratefully expresses his appreciation for the guidance provided by Dr. H. I. Pilgrim. He would also like to acknowledge the assistance of Richard Parks, and Carlos Smith.

REFERENCES

Albright, J. F., and Makinodan, T. 1966 Growth and senescence of antibody-forming cells. J. Cell Physiol., 67:185.

Bosma, M. J., Makinodan, T., and Walburg, H. E. Jr. 1968 Development of immunologic competence in germfree and conventional mice. J. Immunol., 99:420.

Chouroulinkov, I., Old, L. J., and Boyse, E. A. 1964 Facteurs qui jouent un role dans production des isoanticorps chez la souris sexe, voie d'immunisation, splenectomy et choc operatoire. J. Reticuloendothelial Soc., 1:315.

Ferrel, P. 1947 The role of hormones in immunity. Master's Thesis. Department of Microbiology, University of Utah.

Hanna, M. G. Jr., Nettesheim, P., Ogden, L.; and Makinodan, T. 1967 Reduced immune potential of aged mice: Significance of morphologic changes in lymphatic tissue. Proc. Soc. Exper. Biol. and Med., 125:882.

Makinodan, T., and Peterson, W. J. 1966 A secondary antibody-forming potential of mice in relation to age -- its significance to senescence. Develop. Biol., 14:96.

Nicol, T., Bilbey, D. J., Charles, L. M., Cordingsley, J. L., and Vernon-Roberts, B. 1964 Oestrogen: The natural stimulant of body defense. J. Endocrin., 30:277.

Olson, G. B., and Wostmann, B. S. 1966 Cellular and humoral immune response of germfree mice stimulated with 7S or HGG or Salmonella typhimurium. J. Immunol., 97:275.

Outzen, H. C., and Pilgrim, H. I. 1967 Differential mortality of male and female germfree C3H mice introduced into a conventional colony. Proc. Soc. Exper. Biol. & Med., 124:52.

Pilgrim, H. I., and Dowd, J. E. 1963 Correcting for extraneous death in the evaluation of morbidity or mortality from tumor. Cancer Res., 23:45.

Pilgrim, H. I., and Thompson, D. B. 1963 An inexpensive autoclavable germfree mouse isolator. Lab. Animal Care., 13:602.

Schaedler, R. W., Dubos, R., and Costello, R. 1965 Association of germfree mice with bacteria isolated from normal mice. J. Exper. Med., 122:77.

Terres, G., Morrison, S. L., and Habicht, G. S. 1968 A quantitative difference in the immune response between male and female mice. Proc. Soc. Exper. Biol. and Med., 127:664.

Von Haam, E., and Rosenfield, I. J. 1942 The effect of various sex hormones upon experimental Pneumococcus infection in mice. J. Infect. Dis., 70:243.

STUDIES OF THE EFFECTS OF IMMUNE SERUM ON THE COURSE OF *E. coli* INFECTIONS IN GNOTOBIOTIC PIGS

E. M. Kohler

Department of Veterinary Science
Ohio Agricultural Research and Development Center
Wooster, Ohio

In previous studies (1) it was shown that clinical signs of *E. coli* diarrhea could be observed in gnotobiotic pigs 12 to 24 hours following oral administration of 10^7 viable enteropathogenic *Escherichia coli* (O8:K·:H21-Arnold). No diarrhea or other clinical illness was detected when a nonenteropathogenic strain of *E. coli* (O8:K·:H16 CDC1466-56) was administered to gnotobiotic pigs in the same manner (1,2). In addition immune serum was produced by vaccinating some of the survivors of these tests (3).

In studies to attempt to evaluate the effects of immune serum on *E. coli* caused diarrhea, it was considered desirable to simulate as nearly as possible the natural conditions in which a conventional pig acquires antibodies from the colostrum and milk of the sow. Conversely it was desired to evaluate the effect of the antibody in the lumen of the intestinal tract in the absence of parenteral antibody. This was readily accomplished by conducting the tests in pigs 4 to 6 days old when they no longer absorbed detectable levels of antibody from the intestinal tract.

Since pigs normally nurse every 45 minutes to 1½ hours, it was considered necessary to feed 7 ml of serum in 7 ml sterilized diet every 1½ hours. The first serum was fed 1½ hours prior to infection at 4 to 6 days of age. Within this general framework, experiments were conducted with gnotobiotic pigs.

The effect of the serum was evaluated on the basis of the occurrence, or lack of occurrence, of greenish-yellow liquid feces frequently containing clumps of undigested diet and gas bubbles (4). Diarrhea was observed 12 to 24 hours after oral infection of 10 pigs fed no serum, 4 pigs fed nonimmune serum from nonvaccinated germfree pigs, and 5 fed serum with low titers. No diarrhea was observed in 12 pigs fed immune serum until 12 to 24 hours after the cessation of serum feeding. This was 40 to 55 hours after oral administration of *E. coli*.

TABLE I. Summary of Results

Number Pigs	Serum Administered	Number of Pigs	Onset of Diarrhea
14	No serum	12 2	12-24 hours PI 24-36 hours PI
4	Nonimmune serum (7 ml/1½ hr)	4	12-24 hours PI
5	Low titer (Bactericidal 2-64) (7 ml/1½ hr)	5	12-24 hours PI
12[a]	Immune serum (Bactericidal titer 500-1000)	12	40-55 hours PI (12-24 hours after last serum was fed)
7[b]	Immune serum (7 ml/1½ hr)	7	"
3[b]	Heated immune serum[c] (7 ml/1½ hr)	3	"
2[b]	Immune serum (3½ ml/1½ hr)	2	"

[a] The pigs reported in this group are further divided into three subgroups b.

[b] Subdivisions of group a.

[c] Heated (56°C for 30 minutes) immune serum.

In this study it was shown that the bactericidal effect of the serum was probably not responsible for the protection observed. We endeavored to consider possibilities for the mechanism of protection of the immune serums and thought, at that time, that perhaps the antibody was inactivating the endotoxin. On the basis of the results of current studies with enteropathogenic *E. coli* (5), it is apparent that the endotoxin is not the primary factor in *E. coli* diarrhea of young pigs and thus the possible association of the antibody with endotoxin is probably not pertinent.

In a subsequent series of experiments (6) we endeavored to attempt to evaluate the effects of oral and parenteral administration of immune serum in experimentally infected gnotobiotic pigs. The serums used were produced in conventional swine. Both O8 and O8:K· antiserums were used. For studying the effects of parenterally administered antibodies, 14 ml of immune serum were injected intraperitoneally into 4- to 6-day-old gnotobiotic pigs 16 to 20 hours prior to infection.

Five groups of gnotobiotic pigs were studied. The 18 pigs in the first group were infected and given no serum. The second group of 5 pigs was infected following the oral administration of O8:K· antiserum as previously described. The third group was composed of 7 pigs infected following the intraperitoneal administration of O8:K· antiserum. Three pigs were infected following intraperitoneal and oral administration of O8:K· antiserum. The three pigs in the fifth group were infected following intraperitoneal and oral administration of O8 antiserum. Four pigs were used as germfree controls. The pigs were euthanitized at previously determined time intervals between 6 and 48 hours postinfection.

The results were evaluated on the basis of 1) clinical observations, 2) necropsy observations, 3) counts of viable E. coli in segments of the small intestine, 4) attempts to isolate E. coli from the heart, liver, and bile, 5) microscopic examination of fixed intestinal sections to determine the location of E. coli and morphologic evidence of host response, and 6) determination of the pH of the various portions of the gastrointestinal tract.

These studies confirmed the protective value of the orally administered immune serums. In addition the counts of viable bacteria indicated that the numbers of viable bacteria recovered per gm of the segments of the intestinal tract were not markedly different whether serum was administered or not.

TABLE II. Means of counts of viable E. coli recovered from intestinal segments 6-12 hours postinfection

Treatment		No Serum	Oral O8:K· Antiserum	IP O8:K· Antiserum	Oral + IP O8:K· Antiserum	Oral + IP O8 Antiserum
Serum O8 Titer		<1	<1	128-256	1024	512
Serum K Titer		<1	<1	8-16	128	<1
Log 10 of range and means of viable E. coli/gm.						
Fresh Tissue	Duo.	6.6	6.7	NT	2.8	6.2
	Jej.	5.8	5.0	NT	5.6	5.6
	Il.	7.9	8.1	NT	6.6	7.3
Frozen Tissue	Duo.	5.9	5.0	4.7	3.0	6.6
	Jej.	5.8	4.5	5.8	NT	5.3
	Il.	6.9	6.2	6.7	5.5	7.0

TABLE III. Means of counts of viable E. coli recovered from intestinal segments 15-48 hours postinfection

Treatment		No Serum	Oral 08:K· Antiserum	IP 08:K· Antiserum	Oral + IP 08:K· Antiserum	Oral + IP 08 Antiserum
Diarrhea present		7/10	0/3	0/4	0/2	0/2
Serum 08 Titer		<1	<1	64-256	1024	1024
Serum K Titer		<1	<1	4-8	128	<1
Log 10 of range and means of viable E. coli/gm.						
Fresh Tissue	Duo.	7.4	6.4	NT	6.5	7.0
	Jej.	6.8	6.0	NT	6.0	6.7
	Il.	8.2	7.7	NT	7.2	8.3
Frozen Tissue	Duo.	6.5	5.4	5.5	5.6	5.8
	Jej.	6.9	5.2	6.4	4.8	5.6
	Il.	7.8	5.9	7.5	5.4	6.3

Antibodies to the 08, or 08 and K antigens were detected in the serums of those pigs injected with antiserum but not in the serums of those fed antiserum at 4 to 6 days of age. The orally or parenterally administered antiserums did limit the occurrence of bacteremia.

TABLE IV. Bacteriological Findings 6-12 Hours Postinfection (Prior to Onset of Diarrhea)

Treatment		No Serum	Oral 08:K· Antiserum	IP 08:K· Antiserum	Oral + IP 08:K· Antiserum	Oral + IP 08 Antiserum
Serum 08 Titer		<1	<1	128-256	1024	512
Serum K Titer		<1	<1	8-16	128	<1
E. coli cultured from	Heart	1/8	0/2	0/3	0/1	0/1
	Liver	1/8	0/2	0/3	0/1	0/1
	Bile	0/8	0/2	0/3	0/1	0/1
E. coli detected microscopically	Villi	4/8	0/2	0/3	0/1	0/1
	Submuc.	1/8	0/2	0/3	0/1	0/1
	Mesentery	1/8	0/2	0/3	0/1	0/1

TABLE V. Bacteriological Findings 15-48 Hours Postinfection

Treatment		No Serum	Oral 08:K· Antiserum	IP 08:K· Antiserum	Oral + IP 08:K· Antiserum	Oral + IP 08 Antiserum
Diarrhea present		7/10	0/3	0/4	0/2	0/2
Serum 08 Titer		<1	<1	64-256	1024	1024
Serum K Titer		<1	<1	4-8	128	<1
E. coli cultured from	Heart	10/10	1/3	0/4	0/2	0/2
	Liver	10/10	1/3	0/4	0/2	0/2
	Bile	3/10	0/3	0/4	0/2	0/2
E. coli detected microscopically	Villi	7/10	1/3	0/4	0/2	2/2
	Submuc.	5/10	1/3	0/4	0/2	1/2
	Mesentery	7/10	1/3	0/4	0/2	1/2

This alone does not, however, explain the enteric protective effects of immune serum for three reasons. Disintention of the intestine with fluid was noted in the pigs infected with 08:K· antiserums even though diarrhea did not occur and no bacteria were isolated parenterally. Secondly, fluid had accumulated in the intestinal lumen of serum-deprived pigs by 12 hours postinfection, but _E. coli_ was isolated from the heart and liver of only 1 of 8 of these pigs. Finally, _E. coli_ was isolated from the heart and liver of a pig fed 08:K· antiserum and killed at 48 hours postinfection, but no intestinal disorder was detected.

The results with pigs that had been fed and injected with 08 antiserum were conducted with only 3 pigs and could not be evaluated with certainty. It appeared that the 08 antiserum was less effective than the 08:K· antiserum but it also appeared that there was some degree of protection.

There was also correlation with degrees of protection and the pH of the intestinal contents. (See Table VI) The foregoing has answered some questions at least in part and has been useful in our general understanding of _E. coli_ infections in young conventional pigs. The information concerning the incubation time and the need for continuity in receiving antibodies have practical implications. These implications are apparently valid since it is usually difficult to reproduce _E. coli_ diarrhea experimentally in pigs after they have started nursing (apparently all sows have a moderate level of antibodies against a variety of O antigens) (7), but a high percentage develop diarrhea if infected immediately prior to initial nursing. This situation apparently occurs frequently in natural outbreaks.

TABLE VI. Means of pH of Gastrointestinal Contents 15-48 Hours Postinfection

Treatment	No Serum	Oral O8:K· Antiserum	IP O8:K· Antiserum	Oral + IP O8:K· Antiserum	Oral + IP O8 Antiserum	Germfree
Number Pigs	10	3	4	2	2	4
Stomach	4.3	3.5	3.0	3.2	4.5	3.5
Duodenum	7.5	6.5	6.6	6.5	7.0	6.8
Jejunum	8.1	7.5	7.8	7.4	8.0	7.2
Ileum	8.3	7.6	8.0	7.6	8.0	7.6
Cecum	8.4	7.3	8.1	7.5	7.5	7.1
Colon	8.0	7.1	7.8	7.0	7.0	7.0
Bile	7.4	7.0	7.3	7.1	7.1	7.1

As is often the case, we have a surplus of unanswered questions which are more complex, and perhaps more important, than those for which we have acquired answers.

This brings us back to the question -- how does orally administered OK antiserum protect against the enteric effects of enteropathogenic *E. coli*. I do not know. In studies with conventional pigs (5) we have found that we can produce enterotoxic filtrates from 3 strains of enteropathogenic *E. coli*. These filtrates cause diarrhea in young pigs when administered intragastrically even though the pigs have received colostrum and are nursing their dams. This is true even if the dams have been vaccinated with the homologous filtrate and have high serum anti O titers. Thus, it appears that the protective value of the antiserums referred to in these experiments is not related to direct action against the enterotoxic portion but must in some way interfere with the production or release of this factor. We have not answered the question regarding the value of vaccination of sows with O and K antigens - primarily because the *in vivo* titration of orally administered antiserum would be quite involved. Further the numerous possible serotypes which can be involved in *E. coli* diarrhea of baby pigs creates vast problems in development of practical vaccines with a sufficiently broad spectrum of antigenicity.

It would appear that a more useful immunizing agent would be the material responsible for the enterotoxic activity, but the results of our recent immunization attempts as well as some of the characteristics of this material indicate it may be difficult or impossible to immunize against the enterotoxin.

Gnotobiotic pigs would be very useful tools to study the enterotoxic activity of filtrates or extracts, particularly if protective antibodies could be produced and evaluated.

REFERENCES

1. Kohler, E. M. and E. H. Bohl. Studies of Escherichia coli in Gnotobiotic Pigs. I. Experimental Reproduction of Colibacillosis. Can. J. Comp. Med. & Vet. Sci., 30 (1966) 199-203.

2. Kohler, E. M. Studies of Escherichia coli in Gnotobiotic Pigs. IV. Comparison of Enteropathogenic and Nonenteropathogenic Strains. Can. J. Comp. Med. & Vet. Sci., 31 (1967) 277-282.

3. Kohler, E. M. and E. H. Bohl. Studies of Escherichia coli in Gnotobiotic Pigs. II. The Immune Response. Can. J. Comp. Med. & Vet. Sci., 30 (1966) 169-174.

4. Kohler, E. M. and E. H. Bohl. Studies of Escherichia coli in Gnotobiotic Pigs. III. Evaluation of Orally Administered Specific Antisera. Can. J. Comp. Med. & Vet. Sci., 30 (1966) 233-237.

5. Kohler, E. M. The Enterotoxic Activity of Filtrates of Escherichia coli in Young Pigs. Am. J. Vet. Res. Accepted for publication.

6. Kohler, E. M. Studies of Escherichia coli in Gnotobiotic Pigs. V. Evaluation of the Effects of Oral and Parenteral Administration of Immune Serum. Can. J. Caomp. Med. & Vet. Sci., 31 (1967) 283-289.

7. Kohler, E. M., Ruth B. Moore and Sharon Smith. Serologic Response of Swine to Escherichia coli Antigens. Am. J. Vet. Res. In Press.

DIFFERENTIATION OF POTENTIALLY IMMUNOCOMPETENT CELLS IN THE MOUSE

G. Cudkowicz and G.M. Shearer

Department of Experimental Biology

Roswell Park Memorial Institute, Buffalo, New York

The administration of antigen to immunologically competent animals initiates a series of complex biochemical and cellular events culminating in the synthesis of several molecular species of specific antibody (1). The lymphoid cells that react with immunogens do not synthesize antibody themselves, but they interact with cells of bone marrow origin so as to induce the latter to generate antibody-forming progeny (2-4). The lymphoid cells reacting with immunogen have been called "antigen sensitive" or "target" cells (5,6). We prefer to designate the cells involved in these processes, e.g., in reactions with antigen, in specific instruction of marrow or "antibody factory" cells, and in production of immunocytes, as antigen-sensitive units (ASU), because the complementary cell types are integrated in functional units in organs such as spleen, lymph nodes, and peripheral blood.

There is good evidence for at least three discrete sequential events between antigen administration and antibody synthesis. Each of these events is mediated by a specialized cell type recognizable by a unique function. Antigen is processed by macrophages (7,8) and rendered highly immunogenic by complexing it with RNA (9,10); immunogen interacts with lymphoid cells that do not become capable of synthesizing antibody, but rather of instructing or cooperating in a yet unknown way with marrow-derived cells in the production of specific immunoglobulin (2-4). Some of the cell types mediating these events are naturally separated in organs such as thymus and bone marrow which, unlike spleen and lymph nodes, do not contain complete antigen-sensitive units (11). However, ASU can be reconstituted by preparing artificial mixtures of cells from these and other organs (11,12).

We shall analyze some of the properties of splenic (complete) and thymus-bone marrow (reconstituted) ASU. Our attention will be focused on the question of whether ASU produce only one or more than one type of antibody-forming cells. The latter are specialized for the exclusive synthesis of a given molecular class of antibody (13).

SPLENIC ANTIGEN-SENSITIVE UNITS

Antigen-sensitive units can be detected by transplantation of spleen cells and antigen (SRBC) into heavily irradiated syngeneic recipient mice (14). Due to radiation injury, cells of the host animal are incapable of responding to antigen by antibody formation. However, the donor cells become established in the hemopoietic organs of the host and initiate the series of events leading to antibody production, provided that all components of ASU are available. A small fraction of the grafted cells, estimated between 2 and 15%, lodge in the recipient spleens. After a suitable time interval, grafted ASU will have generated antibody-forming progeny cells. We have enumerated in recipient spleens cells releasing hemolytic antibody (direct and indirect plaque-forming cells, abbreviated PFC) and the cells releasing hemagglutinating antibody (cluster-forming cells, abbreviated CFC). These three types of immunocytes belong to distinct cell populations, for CFC do not produce hemolytic plaques, PFC do not produce clusters, and individual PFC produce either direct or indirect plaques, but not both (14-16). We show in Figure 1, left side, the kinetics of appearance of direct and indirect PFC in the spleens of irradiated mice following transplantation of 2×10^7 spleen cells (from non-primed donor mice) mixed with 5×10^8 SRBC. Indirect PFC appear one or two days later than direct PFC and the times of maximum numbers are six days apart. The important observation is that all recipient spleens contained significant numbers of PFC. However, not all spleens, i.e., 28 of 40, contained CFC, the third type of immunocyte studied (data not shown in Figure 1). In order to find CFC in the spleens of all recipients, we had to transplant 10^8 spleen cells instead of 2×10^7. The appearance of CFC following the injection of 10^8 cells is illustrated in Figure 1, left side.

The described observations suggested that 2×10^7 grafted cells contained enough ASU lodging in the recipient spleens to provide them with direct and indirect PFC, but not enough ASU to also provide each recipient spleen with CFC. Experiments with grafts of 5×10^6 spleen cells confirmed the initial observations on the absence of CFC from recipient spleens containing PFC, and indicated, furthermore, that a fraction of spleens did not contain also indirect PFC, while almost all of them contained direct PFC (14). If there is more than one type of ASU, each of which is

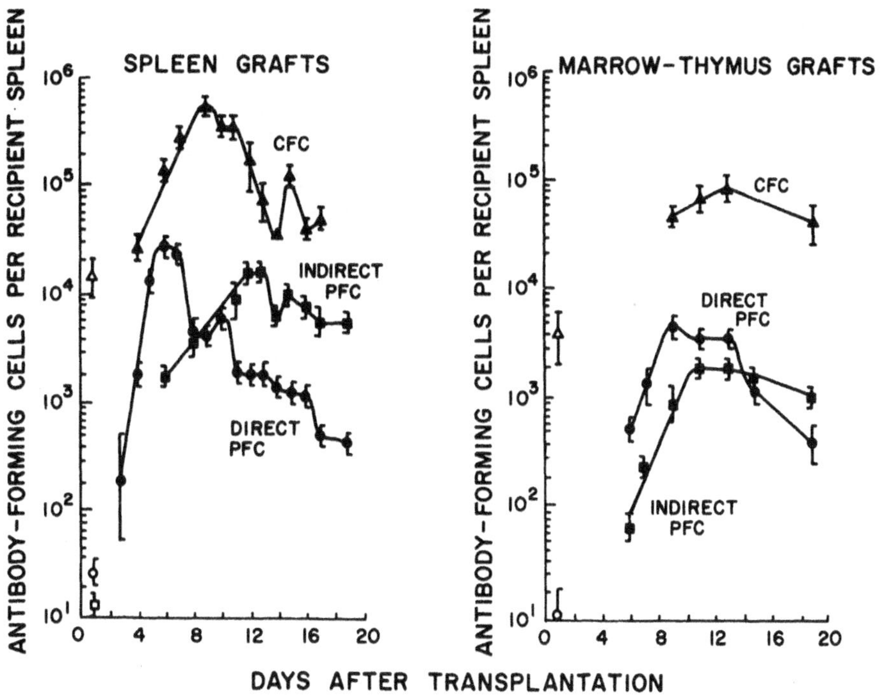

Figure 1. Mean number of immunocytes (with two standard errors) in spleens of (C3H x C57BL)F1 mice exposed to 900 R of X-rays at varying intervals after injection of either spleen cells (left side) or thymocytes and marrow cells (right side) mixed with SRBC. 3-9 spleens per point. 2×10^7 spleen cells were transplanted for plaque assays and 10^8 for cluster assays. 5×10^7 thymocytes and 5×10^7 marrow cells were transplanted for all assays.

endowed with the ability to generate a single type of anti-SRBC immunocyte in recipient spleens, then it should be possible to reduce further the size of spleen cell grafts so that a given inoculum contains only one or two ASU. This hypothesis was tested by limiting dilution assays, i.e., by establishing in groups of animals grafted with graded numbers of cells, the percentage of positive spleens for PFC and CFC. The data were then analyzed by Poisson statistics. The curves relating inoculum size to the observed and expected (Poisson) frequencies of positive spleens for direct and indirect PFC, and for CFC are shown in Figure 2. For each immunocyte population, a number of inoculum sizes yielded positive and negative spleens. The observed values were generally close to those calculated on the assumption of Poisson distribution of ASU reaching the recipient spleens on transplantation. However,

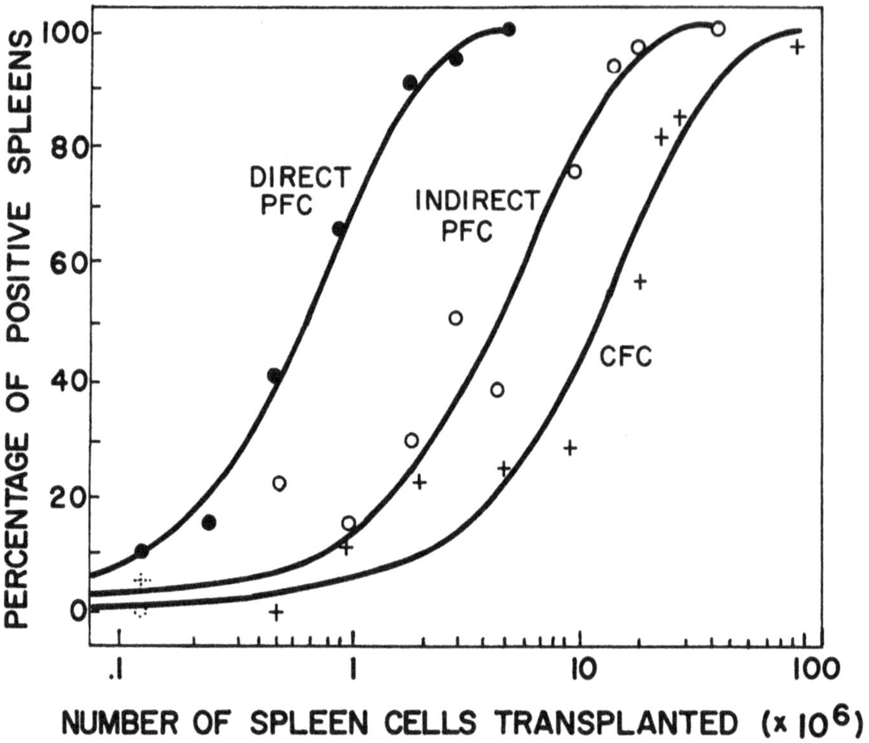

Figure 2. Percentage of recipient spleens positive for immunocytes following X-irradiation and injection of graded numbers of spleen cells (from non-primed donors) mixed with SRBC. Curves indicate expected percentages assuming Poisson distribution of antigen-sensitive units and symbols observed percentages. Dotted symbols, data obtained after statistical analysis was completed.

---

the relation between the percentage of positive spleens and the actual number of spleen cells grafted varied greatly for each type of immunocyte. This indicates that ASU in donor spleens differed with respect to the properties of their progeny cells and to their frequency. We calculated that in spleen cell suspensions of unprimed mice there was one ASU for direct PFC in $\sim 10^6$ nucleated cells, one ASU for indirect PFC in $\sim 7 \times 10^6$ cells, and one ASU for CFC in $\sim 19 \times 10^6$ cells.

Limiting dilution assays and frequency analysis for splenic immunocytes after transplantation of selected numbers of cells (14) have shown unequivocally that production of CFC occurred in the absence of PFC and vice-versa. Hence, ASU are restricted in their potential for differentaition to a single major type of functional progeny cell. Furthermore, unipotent ASU reach the recipient spleens independently. Direct PFC production by the smaller inocula was not followed by indirect PFC production for as long as 35 days (14), another indication that direct and indirect PFC descend from separate ASU. Results of current experiments with ASU from primed donors are consistent with this view, since indirect PFC were produced in the absence of direct PFC. In conclusion, splenic ASU are heterogeneous and unipotent, presumably as heterogeneous as antibody-forming cells.

ANTIGEN-SENSITIVE UNITS IN MARROW-THYMUS CELL MIXTURES

It has been demonstrated by Claman, et al. (11) that suspensions containing adult marrow and thymus cells and SRBC were active in producing anti-sheep PFC of the direct type when transferred to irradiated mice. The activity of the cell mixtures was far greater than that accounted for by the combined activities of each cell population alone containing few, if any, complete ASU. It seemed as if marrow and thymus contributed complementary cell types that were able to function as ASU in artificial mixtures, the thymus contributing the lymphoid antigen-sensitive cells, and the marrow the antibody-secreting cells (2-4). We verified that the thymus-marrow mixtures also contained antigen-receptor cells, presumably macrophages, by blocking the reticulo-endothelial system of recipient mice with a colloidal suspension of carbon, two days before the transplantation of thymus-marrow-SRBC mixtures. The PFC responses elicited by the grafts were equal in untreated and treated recipient mice; however, in intact animals, i.e., non-irradiated, non-grafted mice, carbon administration inhibited the PFC responses to SRBC (17 and unpublished observations). These experiments suggested that antigen-receptor cells were contained in marrow-thymus cell mixtures along with the other components of ASU; it remains to be established whether the antigen-receptor cell is thymus- or marrow-derived.

Following transplantation of 5×10^7 thymocytes, 5×10^7 marrow cells, and 5×10^8 SRBC into irradiated syngeneic mice, direct PFC, indirect PFC, and CFC appeared in the recipient spleens (Figure 1, right side). The spleens of all host mice were positive for PFC, but only 13 of 28 recipient spleens were positive for CFC. The lack of CFC in some of these mice resembled that encountered after transplantation of 2×10^7 spleen cells into irradiated mice (14). We interpret this to mean that the marrow-thymus mixtures used contained enough committed ASU to provide each recipient mouse with splenic PFC, but not with CFC. The number of splenic immunocytes produced by a unit number of grafted marrow-thymus cells was smaller than that produced by a unit number of grafted spleen cells (Figure 1). It is possible that one component of ASU is limiting the PFC and CFC responses. The most likely candidate for this role is the antigen-receptor cell or macrophage, since immunocytes were produced by grafts of as few as 3×10^5 marrow cells with 5×10^7 thymocytes (Table 1), or by 10^7 thymocytes grafted with 10^6 marrow cells (unpublished observations).

The kinetics of appearance of direct PFC were different after spleen and after marrow-thymus cell transplantation (Figure 1). In the latter case the time of peak response was delayed and the subsequent decline was slower. Consequently, direct and indirect PFC arose simultaneously and persisted at about equal values of cells per spleen until twenty days post-grafting. We do not yet fully understand the observed differences between cellular immune responses elicited by splenic and marrow-thymus ASU. However, the simultaneous production of direct and indirect PFC is hardly compatible with the view that a switch from secretion of one type of hemolysin to secretion of another occurs within a clonal population of immunocytes during the course of an immune reaction (18-20).

To investigate the potential for differentiation of marrow-thymus ASU, we have started limiting dilution assays of each separable component of ASU for anti-SRBC immunocyte production. The task is a difficult one because each ASU component has to be tested while other components are not limiting. In Table 1 we show immunocyte production in spleens of 13 animals grafted with

Table 1. Presence of immunocytes in individual spleens of irradiated (C3H x C57BL)F1 mice grafted with 5×10^7 thymocytes, 3×10^5 marrow cells, and 5×10^8 SRBC.

Days After Grafting	Immunocytes Per Spleen		
	Direct PFC	Indirect PFC	CFC
8	850	neg.	neg.
8	2,570	neg.	neg.
12	184	neg.	neg.
12	560	neg.	neg.
12	367	205	neg.
12	391	163	neg.
12	402	114	neg.
12	470	144	neg.
12	470	180	neg.
12	565	605	neg.
12	625	215	neg.
8	1,900	1,000	42,500
12	361	151	26,400

3×10^5 marrow cells and 5×10^7 thymocytes. All mice were positive for direct PFC, 9 for indirect PFC, and only 2 also for CFC. Preliminary results of a series of experiments such as the one presented, but with variable numbers of marrow and thymus cells, have led us to conclude that ASU resulting from marrow-thymus mixtures are similar, although not identical to the complete ASU of spleen cell suspensions. Both populations of ASU are heterogeneous, each ASU being committed to produce a given type of immunocyte progeny. Furthermore, the frequency of ASU types in spleen and marrow-thymus mixtures decreases in the same order from precursors of direct PFC, indirect PFC, and CFC (14).

CONCLUSIONS

The lymphopoietic pathway of differentiation produces an extremely specialized and heterogeneous population of functional immunocytes. The specialization conferred upon these end cells results i) in antibody production against a single antigenic determinant, or "specificity differentiation" (13,21); ii) in production of a single class of antibody molecule among several directed against the same antigen, or "class differentiation" (15, 16); iii) in production of one allotypic variant of a given molecular species of antibody, or "allotypic restriction" (22). Our experiments with anti-sheep immune reactions indicate that the immediate precursors of antibody-forming cells are also heterogeneous and that they display in the spleens of adult mice and in bone marrow-thymus cell mixtures the same type of differentiation detectable among their progeny cells. In other words, ASU reactive with the antigen complex of SRBC, are tooled up prior to antigen administration to generate either direct or indirect PFC (class differentiation), or CFC without PFC and vice-versa (specificity differentiation or class differentiation). According to Gell and Sell (23) also allotypic restriction occurs in precursors of antibody-forming cells. Due to the existence of natural anti-SRBC antibodies in mice, it was conceivable that differentiation of ASU was due to cross-reacting antigens of the intestinal flora. Experiments with germfree mice have not supported this idea (24).

If cells of the immune system of mice descend from undifferentiated (pluripotent) hemopoietic stem cells, we have to conclude that diversity-generating processes have taken place before the formation of ASU. Similar processes are responsible for differentiation along other pathways, e.g. erythropoietic and granulocytopoietic. Possibly the same general control mechanisms operate in differentiation of all hemopoietic cell lines. We would then expect to find potentially immunocompetent cells with intermediate properties of differentiation and maturation, particularly in adult or fetal marrow and thymus. Since marrow and thymus contain cells endowed with different functions, differentiation may occur independently in each cell line, i.e., in each component of ASU. One can visualize the interplay or random association of such heterogeneous and specialized cells (upon antigenic stimulation) as a source of sufficient variability to account for the great heterogeneity of antibody responses and for the spectrum of immune reactivities of mammals. Recently introduced techniques of cellular immunology, antibody responses in vitro, and dissociation of ASU in monofunctional cells render these concepts amenable to experimental test.

REFERENCES

1. Steiner, L.A. and H.N. Eisen. Bact. Rev., 30: 383, 1966.
2. Mitchell, G.F. and J.F.A.P. Miller. Proc. Natl. Acad. Sci. U.S., 59: 296, 1968.
3. Davies, A.J.S., E. Leuchars, V. Wallis, R. Marchant, and E.V. Elliott. Transplantation, 5: 222, 1967.
4. Davies, A.J.S., E. Leuchars, V. Wallis, and P.C. Koller. Transplantation, 4: 438, 1966.
5. Kennedy, J.C., L. Siminovitch, J.E. Till, and E.A. McCulloch. Proc. Soc. Exp. Biol. Med., 120: 868, 1965.
6. Nossal, G.J.V., K.D. Shortman, J.F.A.P. Miller, G.F. Mitchel, and J.S. Haskill. In "Antibodies," Cold Spring Harbor Symposia on Quantitative Biology, 32: 369, 1967.
7. Gallily, R., and M. Feldman. Immunology, 12: 197, 1967.
8. Mosier, D.E. Science, 158: 1573, 1967.
9. Fishman, M., and F.L. Adler. In "Immunity, Cancer, and Chemotherapy," E. Mihich, Ed., Academic Press, New York, N.Y., 1967, p. 177.
10. Pinchuck, P., M. Fishman, F.L. Adler, and P.H. Maurer. Science, 160: 194, 1968.
11. Claman, H.N., E.A. Chaperon, and R.F. Triplett. J. Immunol., 97: 828, 1966.
12. Radovich, J., H. Hemingsen, and D.W. Talmage. J. Immunol., 109: 756, 1968.
13. Mäkelä, O. In "Antibodies," Cold Spring Harbor Symposia on Quantitative Biology, 32: 423, 1967.
14. Shearer, G.M., G. Cudkowicz, M. St. James Connell, and R.L. Priore. J. Exptl. Med., 128: No. 3, 1968.
15. Shearer, G.M., and G. Cudkowicz. J. Immunol., in press.
16. Brahmi, Z., and B. Merchant. Federation Proc., 27: 735, 1968.
17. Sabet, T., C. Newlin, and H. Friedman. Proc. Soc. Exp. Biol. Med., 128: 274, 1968.
18. Nossal, G.J.V., A. Szenberg, G.L. Ada, and C.M. Austin. J. Exptl. Med., 119: 485, 1964.
19. Papermaster, B.W. In "Antibodies," Cold Spring Harbor Symposia on Quantitative Biology, 32: 447, 1967.

20. Sterzl, J. In "Antibodies," Cold Spring Harbor Symposia on Quantitative Biology, 32: 493, 1967.
21. Green, I., P. Vassalli, V. Nussenzweig, and B. Benacerraf. J. Exptl. Med., 125: 511, 1967.
22. Weiler, E. Proc. Natl. Acad. Sci. U.S., 54: 1765, 1965.
23. Gell, P.G.H., and S. Sell. J. Exptl. Med., 122: 813, 1965.
24. Shearer, G.M., G. Cudkowicz, and H.E. Walburg, Jr., This volume.

ACKNOWLEDGMENTS

This work was supported by the Damon Runyon Memorial Fund for Cancer Research through a post-doctoral fellowship to one of us (G.M.S.), and by the American Cancer Society, Inc., through Institutional Research Grant IN-54 and Research Grant T-476.

A COMPARATIVE STUDY OF THE IMMUNE REACTION IN GERMFREE AND CONVENTIONAL MICE

M. G. Hanna, Jr., P. Nettesheim and H. E. Walburg, Jr.

Biology Division, NCI-AEC Carcinogenesis Program

Oak Ridge National Laboratory, Oak Ridge, Tennessee

INTRODUCTION

In recent years considerable attention has been devoted to the germinal centers of lymphatic tissue, particularly as these structures relate to the immune reaction (see reviews by Thorbecke and Hurlimann, 1965, and Congdon and Hanna, 1967). Results of various studies have demonstrated 1) that an antigen-stimulated proliferation of large pyroninophilic cells occurs in germinal centers (Hanna, 1964; Cottier _et al._, 1967), and this proliferation is antigen dose dependent (Hanna _et al._, 1966); 2) that the sensitized immune cell compartment develops first in lymphatic nodules at the time of germinal center hyperplasia (Thorbecke _et al._, 1964); and 3) that during the course of an immune reaction, antigen is localized and retained extracellularly in plasma membrane infoldings of specialized reticular cells which constitute the fixed stroma of the germinal center (Szakal and Hanna, 1968, Nossal _et al._, 1968).

A complicating factor in these immunologic studies of conventional animals is a chronic stimulation by adventitious antigens, resulting in formation of germinal centers in the spleen, Peyers patches, and most lymph nodes. Ideally, detailed studies concerned with the growth, development, and functional significance of germinal centers require antigenically non-stimulated lymphatic tissue. The closest approximation to such ideal conditions is the germfree animal. It is imperative in such studies, however, to establish that the animals are not defective in their overall immune capacity. This is still an equivocal point regarding germfree mice. Many studies have suggested that the immune capacity of germfree animals, as defined either morphologically or serologically, is limited or under-developed in comparison with conventional animals (Bauer et al., 1964; Gustaffson and Laurell, 1959; Horowitz et al., 1964; Olson and Wustmann, 1965; Wustmann and Olson, 1964). This would make questionable the usefulness of germfree animals for immunologic studies. On the other hand, other studies suggest that there is no difference in immune potential between germfree and conventional animals (Wagner, 1955; Sell, 1965; Bosma et al., 1967).

The purpose of the present investigation was to compare the immune capacity of intact germfree and conventional mice, in particular with regard to the growth and development of lymphatic tissue germinal centers.

MATERIALS AND METHODS

Animals

Carworth Farm #1 (CF No. 1) mice were reared and maintained in the Biology Division of the Oak Ridge National Laboratory. Derivation and maintenance of the germfree colony have been described previously (Walburg et al., 1966). The mice were 10-12 weeks of age at the beginning of the experiment. During the course of the experiment, food, feces, and tissues of germfree mice were routinely tested for contamination with bacteria, fungi, mycoplasma, parasites, and detectable murine viruses. The results were consistently negative. Conventional mice used were CF #1, 10-12 weeks of age; they were originally germfree but were subsequently raised and maintained in conventional surroundings. The animals used were from the fourth, fifth and sixth generation of the first conventionalized breeder pairs.

Antigens

Sterile suspensions of sheep erythrocytes (SRBC) were purchased from Baltimore Biological Company (Baltimore, Maryland). Dilutions of the SRBC suspensions for injection into germfree mice were performed, using sterile procedures, in the isolators. Human gamma globulin (HGG) was purchased in crystalline form from Pentex Biochemical Company (Kankakee, Illinois) and labeled with ^{125}I at Abbott Laboratories (Chicago, Illinois). The soluble preparation was sterilized by millipore filtration. The specific activity of the labeled HGG used in these experiments was 40 microcuries per milligram. All injections of antigen were intraperitoneal (IP). Mice received either 10^8 SRBC or 1.5 mg ^{125}I-HGG.

Serology

The sera for each group of experimental animals were frozen until completion of the experiment, and all titrations for any one group were performed on the same date. The titration procedure involved a serial dilution using a 0.025 ml loop (Micro-titration kit, Alexandria Engineering Company, Alexandria, Virginia). Titers are expressed as log_2 of the reciprocal of the last dilution which shows a positive hemagglutinin pattern.

Procedures

At various intervals after antigen injection, animals were killed by cutting the throat, and blood was collected. One-tenth ml serum aliquots and whole spleen of animals receiving ^{125}I-HGG were counted for radioactivity in a Packard auto-gamma counter.

Sections of spleen, liver, kidney, and mesenteric lymph node were used for autoradiography with Kodak NTB-2 liquid emulsion. These sections were exposed for 40 days in the cold, then developed, fixed, and stained with Hematoxylin and Eosin (H & E). Other sections not prepared for autoradiography were also stained with H & E.

For determinations of surface areas of the white pulp germinal centers in each longitudinal spleen section, the image of the spleen section was micro-projected at a magnification of 150X. The outline of each germinal center in a longitudinal spleen section was traced on paper, and the respective areas were measured with a compensating polar Planimeter. At each interval the germinal center areas were plotted graphically, as the mean germinal center surface area per longitudinal spleen section. One longitudinal section was measured from each of the animals killed at each interval.

The *in vivo* culture system was used to follow the growth and development of the secondary antibody-forming potential. This assay enables one to estimate the size of the immune progenitor cell pool. Germfree and conventional mice were primed with 10^8 SRBC. At various intervals after priming, their spleens were assayed for secondary antibody-forming capacity by transferring 10^7 nucleated spleen cells, together with an optimum amount of antigen, into 800 r X-irradiated isologous recipients and by measuring the 6-day hemagglutinin response.

RESULTS

Primary Response to Sheep RBC

The latent and log phase of the hemagglutinin response was not significantly different in germfree and conventional mice (Fig. 1). From day 6 on, the antibody level in germfree mice was consistently 1 to 2 $\log_2$ units above that in the conventional animals. Once peak titer was achieved in both groups, the hemagglutinin levels were relatively stable for the following three months.

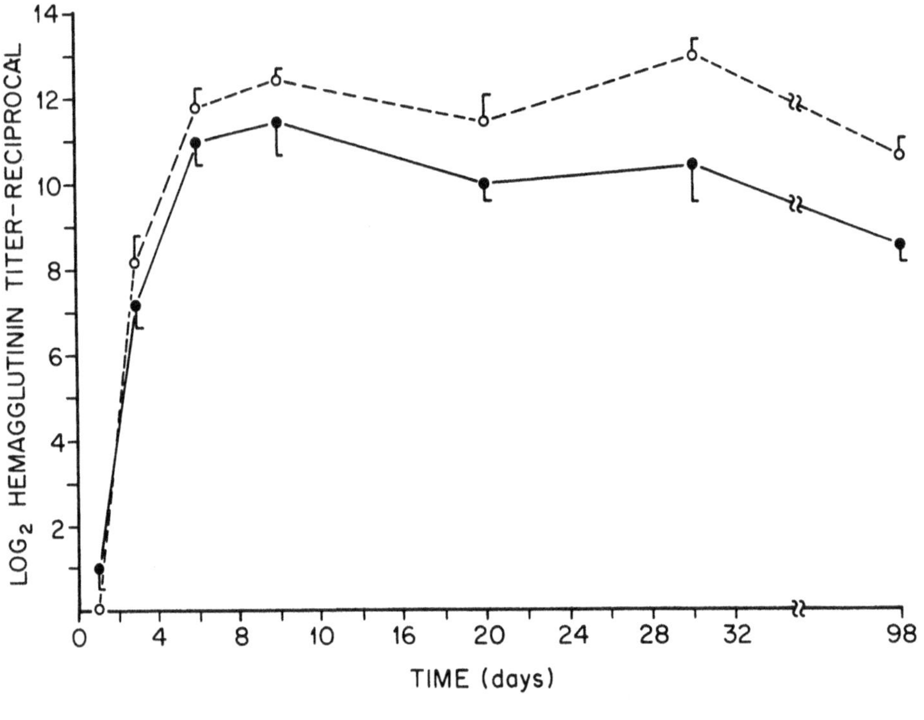

Fig. 1. Hemagglutin production to SRBC during the primary response. o---o germfree, ●——●, conventional. Four to five mice per point, perpendicular bars represent one standard error of the mean.

The mean germinal center surface area increased approximately 2 X 10^3 μ^2 in conventional animals during the first 10 days after injection of SRBC (Fig. 2). During the same interval, the change in the germfree animals was approximately 4 X 10^3 μ^2. Thus there was a difference of twofold in the overall growth of germinal centers in germfree, as compared with conventional, mice. The peak size of the germinal centers in germfree animals at day 10, however, was not significantly different from that of the conventional animals at the same interval. By day 20 the mean germinal center surface area in both groups decreased to 2 X 10^3 μ^2, which is approximately equal to the background level in conventional mice. However, the germinal centers of germfree animals continued to decrease during the following three months.

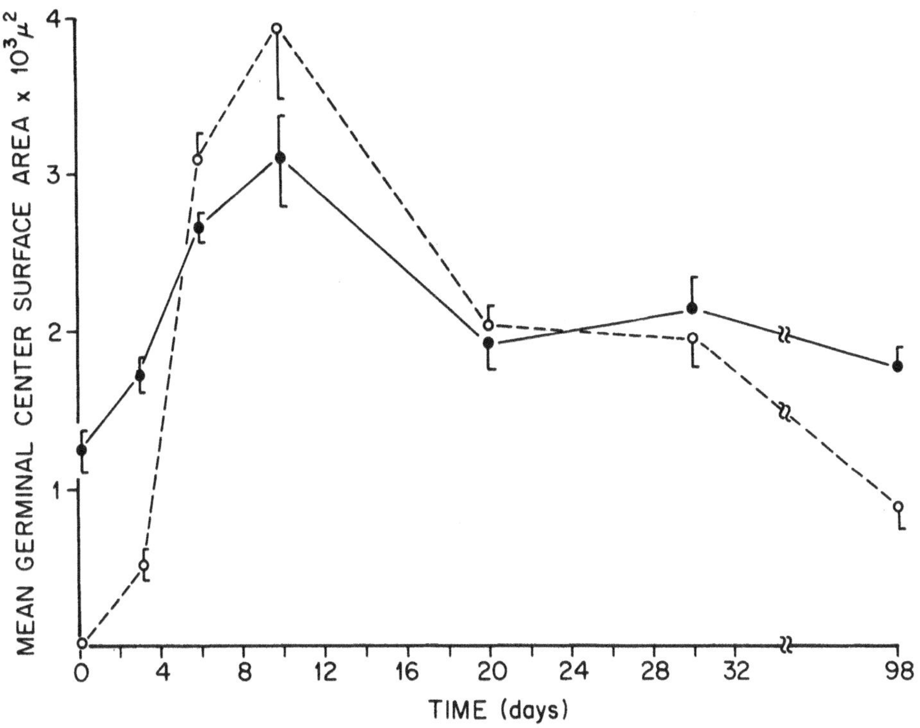

Fig. 2. Germinal center development in spleen to SRBC during the primary response. o---o germfree, ●—● conventional. Four to five mice per point, perpendicular bars represent one standard error of the mean.

Secondary Antibody-Forming Capacity in Germfree Mice

One week after injection of SRBC, the spleen cells of germfree donor mice gave a 6-day hemagglutinin response which was significantly above that produced by an equivalent number of spleen cells from unprimed donors, indicating that secondary antibody-forming potential was developing (Fig. 3). Maximum level of the secondary antibody-forming capacity, however, was not reached before 30-60 days after priming. This peak level was maintained until 5 months after antigenic stimulation, at which time the experiment was terminated.

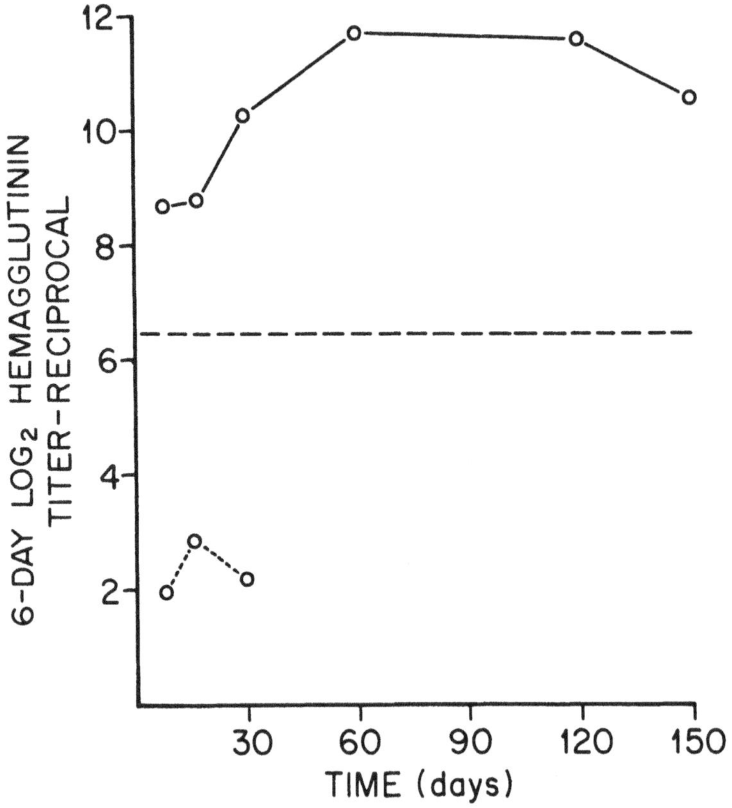

Fig. 3. Secondary antibody-forming potential in germfree mice. -----, primary antibody-forming potential of 10 X 10^6 cells.

Antigen Clearance from Spleen and Serum during the Primary Response to HGG

The clearance of ^{125}I-HGG during the primary response is shown in Fig. 4. There is no significant difference in the clearance pattern in spleen or serum during the first 8 days after injection in the germfree or conventional animals. Immune elimination became apparent in spleen and serum between days 8 and 10 in the germfree and between days 10 and 15 in the conventional mice, suggesting an earlier onset of antibody production in the former. Although not shown in Fig. 4, the same clearance profile for both groups was measured in the liver.

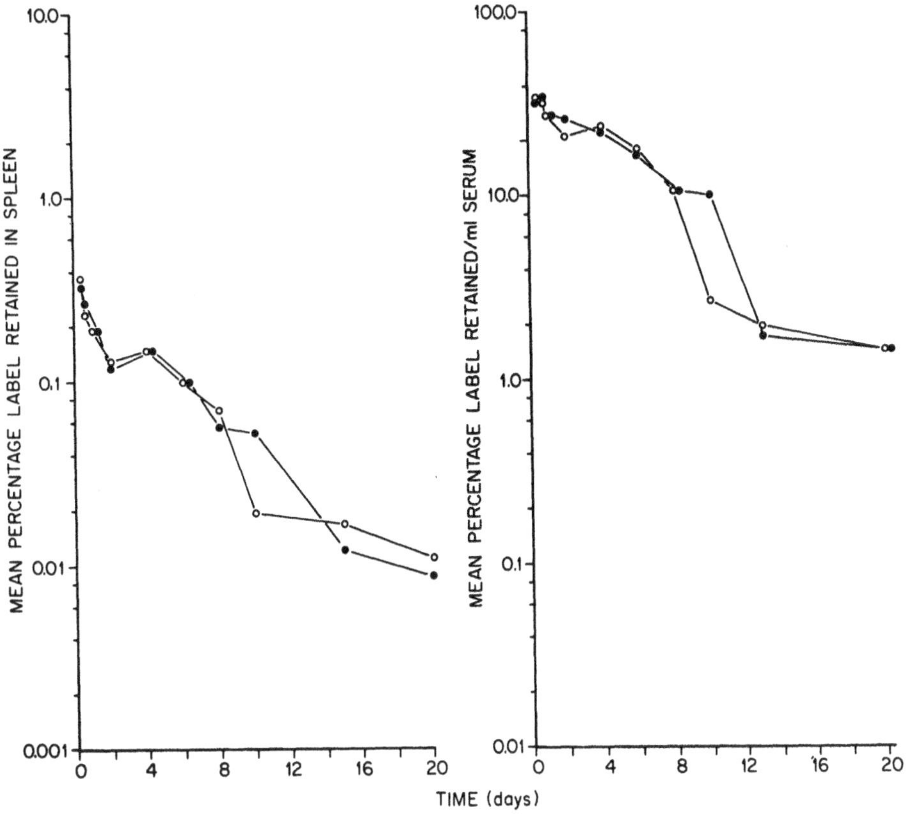

Fig. 4. ^{125}I-HGG clearance in spleen and serum during the primary response. o---o germfree, ●—● conventional.

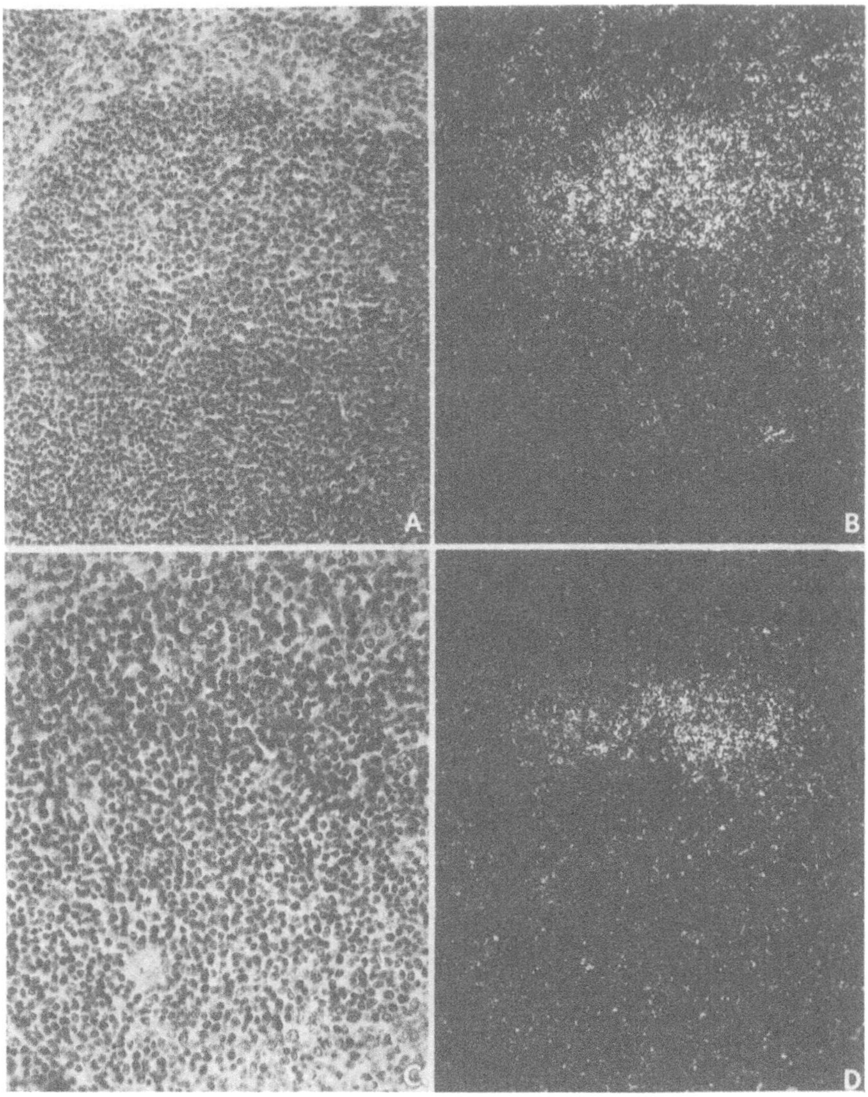

Fig. 5. ^{125}I-HGG localization in spleen lymphatic nodules during the primary response. A) Active germinal center in spleen 12 hours after injection (300X), B) dark field photomicrograph of the identical area in the adjacent section (300X), C) perifollicular region of lymphatic nodule in germfree spleen 12 hours after injection (500X), D) dark field photomicrograph of the identical area in the adjacent section (500X).

Histology and Autoradiography During the Primary Response to HGG

One hour after injection of 1.5 mg ^{125}I-HGG in conventional mice, label was found preferentially in the marginal zones of spleen lymphatic nodules. At 6, 12, and 24 hours, the **preexisting germinal** centers were heavily labeled while label decreased in other areas of the spleen (Fig. 5a & b). Between 2 and 20 days after antigen injection, the label was predominantly concentrated in germinal centers though the concentration decreased progressively with time. The overall pattern of antigen localization during the first 10 days was similar in the germfree mice as compared to the conventional mice. The major difference, however, was that although label concentrated in defined perifollicular areas of the lymphatic nodules, no active germinal centers existed (Fig. 5c & d). However, between 10 and 20 days germinal centers formed in these regions of antigen retention.

DISCUSSION

There appears to be a lack of conformity in comparative studies of germfree and conventional animals regarding their immune capacity (Gustafsson and Laurell, 1959; Horowitz et al., 1964; Bauer et al., 1964; Wustmann and Olson, 1964; Olson and Wustmann, 1965; Wagner, 1955; Sell, 1965; Bosma, 1967). These differences in findings could result from the use of different forms and types of antigens, routes of injection, and/or the techniques in environmental control. In the present study there was no deficiency in hemagglutinin-forming potential, development and growth of lymphatic tissue germinal centers, or antigen localization as well as antigen clearance in germfree mice compared to conventionally reared animals of the same strain. Rather, the data of these experiments suggest that the germfree mice may have an increased immune capacity. However, since the level of antigenic background stimulation in the conventional mouse is not controlled, the magnitude of this difference, if it is real, could be expected to vary among experiments. Since the immune competence of the germfree mouse was found to be at least equivalent to that of the conventional mouse, the former can be used to study the initial morphologic changes in lymphatic tissue after antigen stimulation.

A significant feature of the antigen localization in the lymphatic nodules of germfree mice is the deposition of antigen in areas in which no active germinal centers were detected. It was apparent, however, that germinal centers formed later in these specific regions in the nodules. Anatomically, the antigen-trapping device in germfree mice corresponds to the germinal center region in conventional mice. It is reasonable to assume, based on the ultrastructure studies of antigen localization in conventional mice (Szakal and Hanna, 1965; Nossal et al., 1968), that there is a pre-existing stroma capable of localizing antigen and that the characteristic germinal center cells are grouped in relation to this structure. Thus these studies support the contention of Menzies (1965) that the stromal compartment is constant even though its contained lymphoid population is variable. Further, these studies indicate that the reticulum cells of the "centron" of the germinal center are capable of initially trapping and localizing antigen in the absence of the characteristic pyroninophilic, parenchymal cells of the active germinal centers. These findings further support the contention of Nossal et al. (1964) that there is a causal relationship between antigen localization and the development of active germinal centers.

An interesting point but one unresolved in this study is the origin of the parenchymal lymphoid cells of the germinal centers. It may be assumed that the lymphoid cell population of the germinal center is an immunologically competent cell compartment since the development of these centers corresponds closely with the rise in hemagglutinin titer and since the proliferation of cells in the centers seems to be antigen-dependent. The proliferation which occurs in germinal centers during the interval of the immune reaction in the presence of specific antigen suggests an expansion of immunologically competent cells in this location. This contention is further supported by studies which demonstrate that suppression of SRBC hemolysin-producing cells with specific passive antibody infusion also results in marked depression in the hyperplasia of the germinal centers during the primary response (Hanna et al., unpublished data, 1968).

These present data in germfree mice show, however, that the hemagglutinin response can be initiated in the absence of active germinal centers. Nevertheless, to our knowledge, no evidence has been presented which demonstrates, in mammals, a maintained primary immune reaction with development of the sensitized cell compartment without the development or presence of active germinal centers.

Thorbecke et al. (1964) have proposed that the role of germinal centers is primarily related to the development of immunologic memory and that the sensitized cells are first found in lymphatic nodules concurrently with the hyperplasia of the germinal center. The present study shows that the memory cell compartment in the SRBC primed germfree mice is maximum between 30 and 60 days after priming and remains at this level for the subsequent three months. During this period there is a marked decrease in the size of germinal centers. Although germinal centers may contribute to the development of the pool of sensitized cells, they do not remain the seat of this compartment of cells.

Based on the sequence of histologic and serologic events observed in this serial study of the primary response in germfree mice, we would suggest the following: Lymphatic tissue has two unique components, dispersed immunologically competent progenitor cells and defined areas of reticular cells, capable of localizing antigen extracellularly. After the initial antigenic stimulation, one phase of the proliferative expansion of the immunologically competent progenitor cells occurs in the region of the antigen-retaining reticular cells ("centron" as described by Menzies, 1964). This results in the accumulation of large pyroninophilic cells in this area and ultimately in the formation of a histologically well-defined germinal center. At the same time antibody becomes detectable in the serum. The serum antibody response and the growth of the germinal center simultaneously develop to peak levels. During this time the sensitized cell compartment is being established in the lymphatic nodules. Therefore the proliferation of large pyroninophilic cells in germinal centers is probably related to both the development of antibody-producing cells as well as memory cells. Since the largest proportion of antibody-producing cells and memory cells are eventually found outside the germinal centers (Thorbecke et al., 1964), the lymphoid cell population of the centers must be considered transitory. This is also supported by kinetic studies of active germinal centers (Hanna, 1964, Cottier et al., 1967). The later decrease in the size of the centers probably is a manifestation of the diminishing growth rate of the antibody response as well as the memory formation. This may be a consequence of the elimination of antigen from these areas.

SUMMARY

Systematic studies using sheep erythrocytes and ^{125}I-human gamma globulin as antigens were performed to compare the immune capacity of germfree and conventional mice, particularly in regard to the growth and development of lymphatic tissue germinal centers. The results showed no deficiency in overall immune capacity in germfree mice in terms of hemagglutinin forming potential, development and growth of germinal centers, or antigen localization as well as antigen clearance. Further, these studies support the idea that the germinal center consists of a constant antigen-retaining reticulum cell stroma and a variable and/or transient population of immunologically competent lymphoid cells. Although germinal centers may contribute to the development of the sensitized immune cell compartment, they do not remain the seat of this compartment of cells.

ACKNOWLEDGEMENT

The authors acknowledge the capable assistance of Mrs. Leona C. Peters, Miss Edna I. Mynatt, and Miss Mary Lou Williams.

REFERENCES

1. Bauer, H., Horowitz, R. E., Watkins, K. C., and Popper, H. J., J. Amer. Med. Assoc., 87:715, 1964.
2. Bosma, M. J., Makinodan, T., and Walburg, H. E., Jr., J. Immunol., 99:420, 1967.
3. Congdon, C. C and Hanna, M. G., Jr., In: H. Cottier, N. Odartchenko, R. Schindler and C. C Congdon, Eds., Germinal Centers in Immune Responses, New York, Springer-Verlag, 1967, p. 1.
4. Cottier, H., Keiser, G., Odartchenko, N., Hess, M., and Stoner, R. D., In: H. Cottier, N. Odartchenko, R. Schindler and C. C Congdon, Eds., Germinal Centers in Immune Responses, New York, Springer-Verlag, 1967, p. 270.
5. Gustafsson, B. E., and Laurell, G. B., J. Exptl. Med. 110:675, 1959.
6. Hanna, M. G., Jr., Lab. Invest., 13:95, 1964.
7. Hanna, M. G., Jr., Congdon, C. C, and Wust, C. J., Proc. Soc. Exptl. Biol. Med., 121:285, 1966.
8. Hanna, M. G., Jr., Francis, Mary W., and Peters, Leona C., Immunology (in press), 1968.
9. Hanna, M. G., Jr., Nettesheim, P., and Francis, Mary W., unpublished data, 1968.
10. Horowitz, R. E., Bauer, H., Paronetto, F., Abrams, G. D., Walkins, K. C., and Popper, H., Amer. J. Pathol., 44:747, 1964.
11. Menzies, D. W., Nature, 208:163, 1965.
12. Nossal, G. J. V., Ada, G. L., and Austin, Caroline M., Aust. J. Exptl. Biol. Med. Sci., 42:311, 1964.
13. Nossal, G. J. V., Abbot, Judith Mitchell, and Lummus, Z., J. Exptl. Med., 127:277, 1968.
14. Olson, G. B., and Wustmann, B. S., Fed. Proc., 14:1400, 1965.
15. Sell, S., J. Immunol., 95:300, 1965.
16. Szakal, A. K., and Hanna, M. G., Jr., Exp. Mol. Pathol., 8:75, 1968.
17. Thorbecke, G. J., Jacobson, E. B., and Hochwald, G. M., In: J. Sterzl, Ed., Molecular and Cellular Basis for Antibody Formation, New York, Academic Press, 1964, p. 587.
18. Thorbecke, G. J., and Hurlimann, J., Nouvelle Rev. Francaise d'Hematologie, 5:383, 1965.
19. Wagner, M., Ann. N. Y. Acad. Sci., 31:236, 1955.
20. Walburg, H. E., Jr., Mynatt, Edna, and Robie, D. M., Rad. Res., 27:616, 1966.
21. Wustmann, B. S., and Olson, G. B., J. Immunol., 92:41, 1964.

DEVELOPMENT OF IMMUNOLOGIC COMPETENCE IN GERMFREE AND CONVENTIONAL MICE*

M. J. Bosma

The Institute for Cancer Research, 7701 Burholme Avenue

Fox Chase, Philadelphia, Pennsylvania

The ability of an animal to resist the vast variety and multitude of disease-causing pathogens is primarily a function of its immune potential. How this immune potential is programmed with specificity against so many prospective immunogens is one of the great mysteries of immunology.

Perhaps relevant to this problem is the logarithmic amplification of the immune potential during the first few months after birth. If the postnatal increase of specific immune competence were dependent on antigenic programming, then in germfree (GF) animals which do not experience the bulk of naturally occurring antigen, one would expect a suboptimum rate of increase in specific immune competence compared to conventionally-reared (CVN) animals. This possibility was tested in mice in terms of immunocompetent precursors of hemagglutinin-synthesizing cells.

EXPERIMENTAL TEST SYSTEM

A valid comparative study of GF and CVN mice required that experimental conditions not prejudice the results. This requirement was satisfied by a lymphoid cell transfer system, a system in which passive recipients served as standard environments for the antibody synthesis of transferred spleen cells. This involved the following: (a) known numbers of spleen cells and sheep RBC antigen

* Research sponsored by the United States Atomic Energy Commission under contract with the Union Carbide Corporation while the author was at the Biology Division of the Oak Ridge National Laboratory, Oak Ridge, Tennessee (1964-1966).

(ratio of 1/10) were mixed in vitro, (b) 1 ml aliquots of the suspension were injected intravenously into lethally-irradiated and therefore immunologically incapacitated barrier recipients,* and (c) antibody synthesis was assessed by serum hemagglutinin analysis (1).

The biological integrity of this test system was assured by the following:

1) The use of an autologous cell transfer system showed that the presence of histoincompatibility factors in randombred Carworth Farm #1 mice did not significantly impair hemagglutinin formation of transferred spleen cells (1).

2) The choice of sheep RBC as the test antigen seemed to be a good one in that the frequency of mice having natural hemagglutinin was virtually nil among GF animals and very low among barrier and CVN animals (see Figure 1). The frequency of natural sheep RBC hemagglutinins in very young CVN mice decreased with age which would suggest that they may be of maternal origin. Thus, immunologic experience with the test antigen was minimal as reflected by the level of natural hemagglutinin.

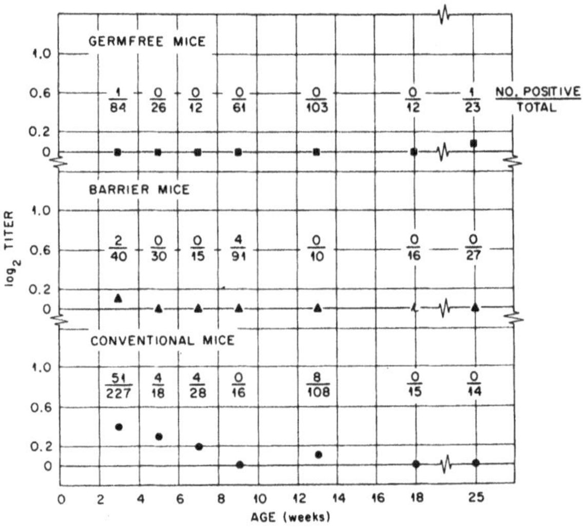

Fig. 1. Natural Sheep Agglutinin as a Function of Age. The mean $\log_2$ titer of the total sample size is plotted on the ordinate.

---

* Barrier recipients refer to a germfree-derived colony of mice raised within an isolation facility according to standard husbandry techniques. The mice used throughout this study were of the Carworth Farm #1 strain.

3) Naturally occurring organisms present in the barrier recipient did not cause any apparent adverse or adjuvant effect toward immunocompetent cells of GF mice. This was demonstrated by transferring, aseptically, spleen cells of GF mice into GF or CVN irradiated recipients. As can be seen in Figure 2, the kinetics and magnitude of hemagglutinin response were identical in either case. These results were also representative of the _in situ_ performance of spleen cells because studies with intact mice revealed an equal hemagglutinin or hemolysin responsiveness of GF and CVN mice (1).

4) Hemagglutinins synthesized by spleen cells of GF or CVN 14-week-old donors were catabolized without preference in barrier recipients. This is shown in Figure 3 for passively-infused 19S or 7S hemagglutinating antibody. This, coupled with the finding that spleen cells of GF or CVN mice expressed a qualitatively similar hemagglutinin response at all ages (i.e. the proportion of 19S to 7S hemagglutinins), meant that hemagglutinin titers were meaningful indices for a comparative study.

5) Optimum "culture" conditions were established for spleen cell populations of GF and CVN donors (1).

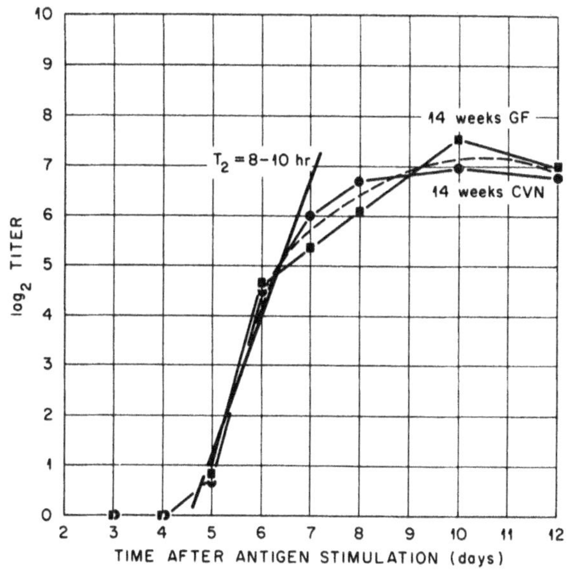

Fig. 2. Kinetics of Hemagglutinin Formation for Spleen Cells Transferred into Germfree (GF) or Conventionally-reared (CVN) recipients. Hemagglutinin response expressed by 1 ml inocula of 80 x 10^6 spleen cells plus 80 x 10^7 sheep RBC injected intraperitoneally into X-irradiated (800 R) recipients. Sample size per point, 10 mice.

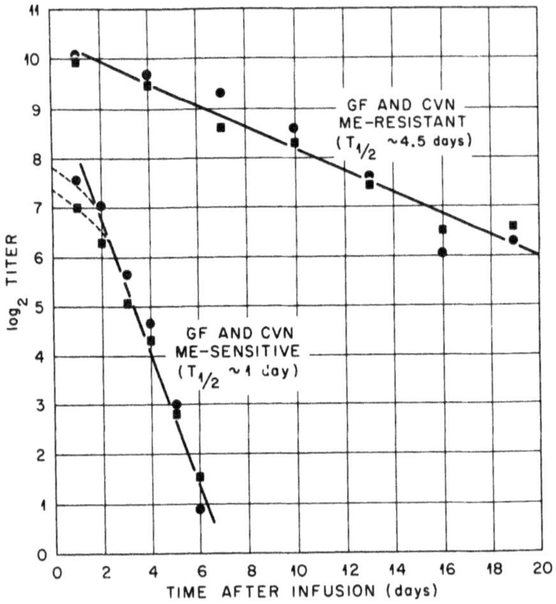

Fig. 3. Catabolism of Passively-Infused Hemagglutinins. Undiluted anti-sheep RBC serum (0.35 ml) was administered intravenously at time T_0 into unirradiated recipients (appropriate control experiments revealed that unirradiated recipients catabolize immunoglobulins at the same rate as 800 R-treated mice, passively receiving either hemagglutinating antibody only or both spleen cells and hemagglutinating antibody). Sera were produced by spleen cells of germfree (GF - ■) or conventionally-reared (CVN - ●) mice. Prior to infusion, hemagglutinins were segregated into 19S and 7S classes according to their sensitivity or resistance to 2-mercaptoethanol, respectively. The basis of this classification was checked by sucrose density centrifugation (1). Sample size per point, 7-9 mice.

IMMUNOLOGIC COMPETENCE

In order to define immunologic competence on a cell basis, the relationship of the relative number of antigen-stimulated cells to serum antibody output had to be established. It was necessary to determine, then, the time after cell transfer when the hemagglutinin response was maximal. As seen in Figure 4, following the

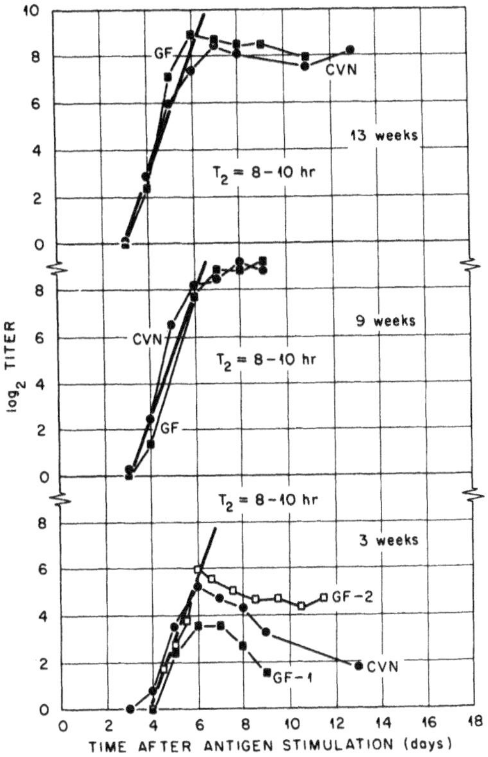

Fig. 4. Hemagglutinin Response of Spleen Cell Populations from Donors of Various Ages -- expressed by 1 ml inocula of 80 x 10^6 spleen cells and 80 x 10^7 sheep red blood cells injected intravenously into 800 R recipients. T_2 is the time required for the serum antibody concentration to double. GF, germfree mice; CVN, conventional mice. GF-1 is a response expressed by spleen cells of 2-week-old germfree donors. Sample size per point, 8-10 mice.

transfer of spleen cell populations from donors of various ages, peak hemagglutinin response occurred at 6-7 days in all instances. Also, during the log phase of response, serum hemagglutinin titer doubled every 8-10 hours at all three age levels. Allowing for the known differences in the number of immunocompetent precursor cells at various ages (1), these data suggested that the serologic performance of all spleen cell populations was comparable.

Having established the optimum time for a point assessment of immunologic competence, varying numbers of spleen cells from donors of various ages were intravenously injected into recipient mice.

As shown in Figure 5, the 7-day hemagglutinin output was a linear function of the number of spleen cells transferred and all data were described by the same general equation. This was interpreted to mean that independent units, presumably immunocompetent precursor cells, control hemagglutinin formation. Accordingly, the immune response would be expected to assume an all-or-none character at limiting spleen cell doses.

Limiting cell dilution studies were therefore carried out for spleen cell populations of 3- and 13-week-old GF or CVN mice. A bimodal distribution of two classes resulted and recipient mice were classified as responders (1 or more precursor cells) or non-responders (no precursor cells), depending upon whether or not they exhibited a significant hemagglutinin titer. By the use of Poisson

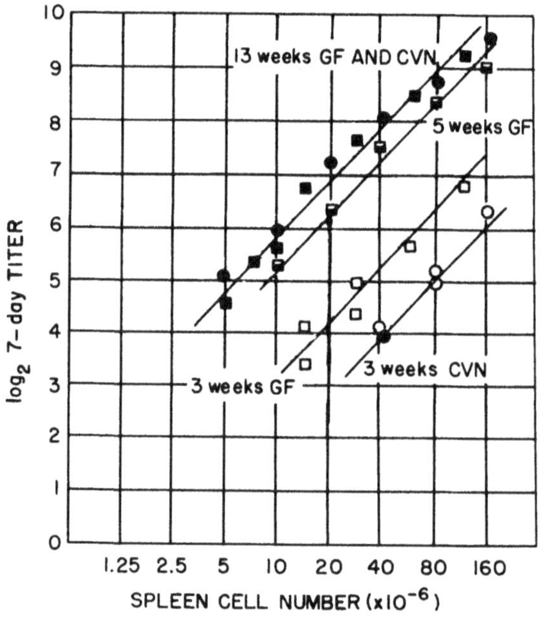

Fig. 5. Hemagglutinin Response as a Function of the Number of Spleen Cells Transferred (sheep RBC/spleen cell ratio held constant at 10/1). The relationship between hemagglutinin titer and number of spleen cells transferred is described by:

$$AbR = k\log_2(\#N_o) + \log_2(\text{constant})$$

where AbR is $\log_2$ 7-day titer, $\#N_o$ is the number of spleen cells injected, and k is 1. Sample size per point, 10-15 mice.

statistics, the number of precursor cells in the different spleen cell populations was calculated and then the frequency of non-responder recipient mice was plotted as a function of the number of precursor cells (Figure 6). In this manner a straight line was obtained.* It was reasonable to conclude, then, that for all spleen cell populations tested a minority population of immuno-competent precursor cells acted as the controlling units of hemagglutinin formation.

Hence, the linear dependence of hemagglutinin formation upon spleen cell number (Figure 5) shows that the 7-day hemagglutinin

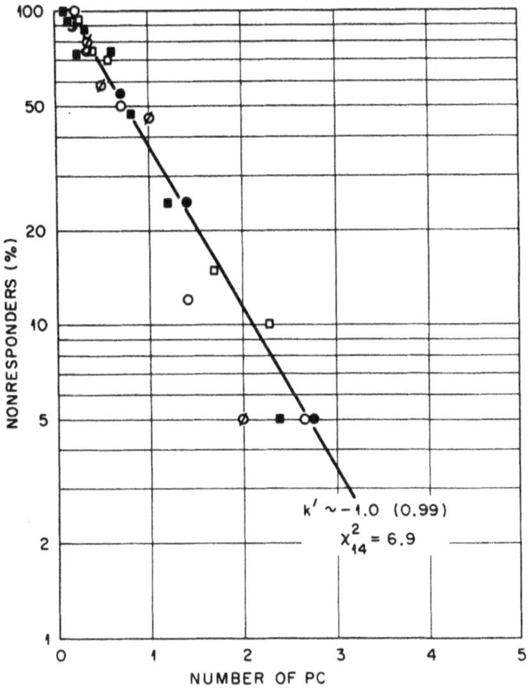

Fig. 6. Immunologic "Equivalence" of Immunocompetent Precursor Cells Independent of Age or Source. Data were normalized by use of the general equation $\log_{10}P_o = (-k'/2.3)$ (PC), where P_o is the probability of nonresponders, k' is about 1, and PC is the number of precursor cells.

* The small "shoulder" is probably due to homograft factors since it was not observed in an isologous cell transfer system (2). Single-hit behavior implies a step-limiting cell, but does not preclude cell-cell interaction in the immune process.

response is mainly a function of the number of antigen-stimulated precursor cells and therefore a measure of immunologic competence.

INCREASE OF IMMUNOLOGIC COMPETENCE

Determination of the increase of immunologic competence in the spleen of GF and CVN mice was based on the 7-day hemagglutinin response generated by a fixed number of spleen cells (80×10^6) from mice of various ages (1-25 weeks). This was expressed on a spleen basis by using the following relationship: immunologic competence of the spleen = (7-day hemagglutinin titer/80×10^6 spleen cells) x (number of cells/spleen) x (1/maximum 7-day hemagglutinin titer of the entire study). The number of cells per spleen was determined by multiplying the observed wet weight of the spleen by the number of cells/unit weight of spleen. This latter conversion factor was constant and independent of the age of mice.

The results of such an analysis, shown in Figure 7, revealed that the changes in immunologic competence described typical growth curves which were identical for GF and CVN mice. This development was exponential during the first five weeks after birth, doubling about every 4 days, followed by a slower growth phase which tended to plateau after about a 500-fold increase (at 9 to 12 weeks).

The growth of the spleen mass was substantially less. During the first 5 weeks after birth, it doubled every 2 weeks, and after 9 weeks showed only about a 4-fold increase. The lower hemagglutinin titers per unit number of transferred spleens for CVN donors vs. GF donors at 3 and 5 weeks of age (see Figure 5) were compensated by the greater spleen mass (2-fold) of CVN mice over that of GF mice. Therefore per spleen equivalent the growth of immunologic competence of GF and CVN mice was identical.

CONCLUSIONS

These studies demonstrated that the postnatal increase of immunologic competence to sheep RBC antigen was identical in GF and CVN mice. This finding was based on an unbiased experimental test system which allowed us to relate hemagglutinin synthesis to a cellular definition of immunologic competence.

Thus, the very low levels of serum γ-globulin and the relative lack of any morphological signs of immune responsiveness in GF animals (3, 4, 5) do not imply an underdeveloped immune capacity, but rather the absence of immunologic expression normally due to antigenic stimuli or subliminal infection. Immune competence, on the other hand, is not dependent on extra-uterine exposure to the normal bacterial load. Also the very low level of γ-globulin

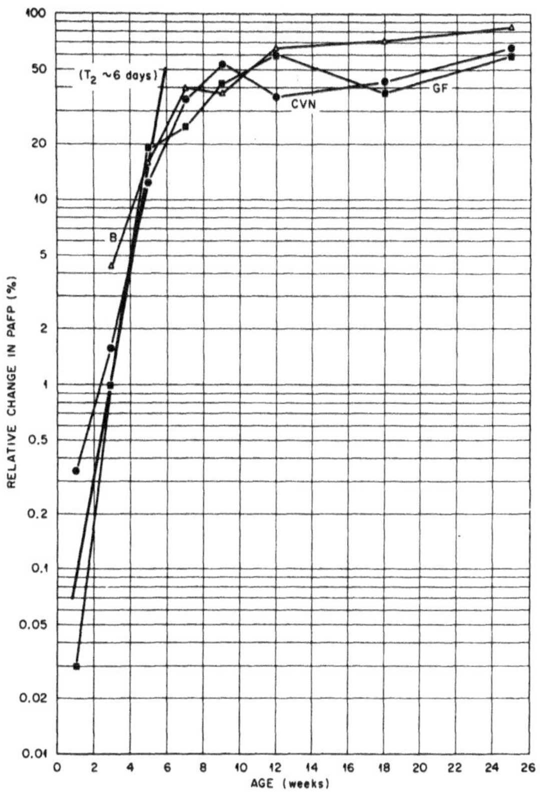

Fig. 7. The Increase of the Primary Antibody-Forming Potential of the Spleen (PAFP). This was expressed as follows: PAFP = (7-day hemagglutinin titer/80 x 10^6 spleen cells) x (number of cells/spleen) x (1/maximum 7-day hemagglutinin titer of the entire study). GF, germfree mice; CVN, conventionally-reared mice; and B, barrier mice.

---

in GF mice and consequently the relative lack of maternal antibodies are apparently of no consequence to the development of immunologic competence.

To reconcile the apparent inconsistency of these results to those (6, 7, 8, 9) who reported a suboptimum immune response for GF animals, we need to emphasize the distinction between immunologic competence and immunologic expression. In the cell transfer system, one can relate hemagglutinin synthesis (expression) to immunologic competence with reasonable accuracy. However, the expression of immune competence of intact individuals is subject to homeostatic regulatory factors to a greater degree than in a cell transfer system (10) and it is possible that these factors are influenced by changes in the environment (11, 12). The effects of this regulation may be apparent in degree, depending on differences in type and dose of antigen and of its route of injection, time and manner of antibody assay.

Finally, the logarithmic increase of lymphoid precursor cells sensitive to sheep RBC antigen was independent of the bulk of naturally occurring antigen, and also was apparently not affected by the normal level of γ-globulins or by maternal antibodies. It would seem, therefore, that the ontogeny of immunologic precursor cells is an intrinsic property of the organism whereby: (1) precursor cells are programmed to elaborate an immunologic specificity by some mechanism independent of antigenic input, or, alternatively, (2) antigen-processing cells (13) that are not step-limiting cellular events, program precursor cells with immunologic specificity.

REFERENCES

1. Bosma, M. J., Makinodan, T. and Walburg, H. E., Jr., J. Immunol., 99:420, 1967.
2. Brown, R. A., Albright, J. F. and Makinodan, T. Nature (Lond.), 210:1383, 1966.
3. Gustafsson, B. E. and Laurell, G. B. J. Exp. Med., 110: 675, 1959.
4. Sell, S. J. Immunol., 93:81 and 122, 1964.
5. Miyakawa, M. Ann. N. Y. Acad. Sci., 78:221, 1959.
6. Bauer, H. et al. J.A.M.A., 187:715, 1964.
7. Wostmann, B. S. and Olson, G. B. J. Immunol., 92:41, 1964.
8. Horowitz, R. E. et al. Amer. J. Pathol., 44:747, 1964.
9. Olson, G. B. and Wostmann, B. S. J. Immunol., 97:267 and 275, 1966.
10. Makinodan, T. et al. Med. Clin. N. Amer., 49:1569, 1965.
11. Mark, R. and Dixon, F. J. J. Immunol., 91:614, 1963.
12. Albright, J. F. and Makinodan, T. J. Cell Physiol., 67 (suppl):185, 1966.
13. Fishman, M. and Adler, F. L. Cold Spring Harbor Symposium on Quantitative Biology, XXXII:343, 1967.

THE TRUE PRIMARY IMMUNE RESPONSE IN GERMFREE COLOSTRUM-DEPRIVED PIGLETS

Yoon Berm Kim and Dennis W. Watson

Department of Microbiology, University of Minnesota

Minneapolis, Minnesota

One of the most challenging and difficult problems in the field of immunology today concerns the nature of the mechanism by which the antigen induces the initiation of antibody synthesis. Many theories have been proposed but this apparently unique mechanism remains unresolved. In our earlier studies on the mouse-actinophage system (1) it became apparent that what was a primary response to us as investigators was, indeed, secondary to the mouse; the presence of "normal" antibody in the serum of these animals prior to the injection of actinophage suggested that the animal had already been exposed to this ubiquitous antigen sometime during its lifetime. These antigens are known to be present in food; therefore, maintaining an animal in a germfree environment does not preclude the presence of all of the antigens present in the normal environment (2). This is not only true of actinophage but also applies to most natural antigens such as erythrocytes and bacterial antigens (3). A review of the recent literature relative to the so-called "normal background" which concerns those cells present in "normal" spleens capable of making antibody against sheep erythrocytes detected by hemolytic plaque techniques (4, 5) indicates the wide distribution of these antigens in nature. It became obvious, therefore, that one must distinguish between the primary response commonly observed in conventional animals and a true primary immune response. The prerequisite for an ideal model in which to study the true primary immune response would involve: (a) a host which has never been in contact with an antigen; (b) its tissues should be free of "background" antibody-forming cells and the serum should be completely devoid of immunoglobulins; (c) it should be immunologically competent.

In searching for such a model, the colostrum-deprived germfree piglet attracted our attention. In a series of papers Segre et al. (6, 7) used the model to prove Jerne's theory (8) of antibody formation. According to this concept, there is present in all animals a low level of "natural" antibody which is specific for a wide variety of antigenic determinants. These low levels of "natural" antibody combine with the antigen and essentially are selected out; the clonal selection theory of Burnet (9) is an extension of this theory at the cellular level. Segre et al. (6, 7) had demonstrated greater immunocompetence with observed low levels of antibodies in this model and emphasized their importance in the mechanism of antibody synthesis.

In contrast to these findings, Sterzl et al. in 1962 (10) reported that germfree, colostrum-deprived piglets were free of detectable antibody as determined by a sensitive bactericidal test using strains of Escherichia coli and Salmonella in the S phase. The sensitivity of his assay was 0.00001 to 0.000001 μg antibody N/ml. When these piglets were artificially colonized with a given type of E. coli, bactericidal antibody was produced within 4-7 days after the administration of bacteria. The antibody was specific for the strain with which the piglets had been colonized. These conflicting results made it necessary to re-examine this model and determine whether or not our specific model, the Minnesota miniature piglet (11), fulfilled the prerequisites outlined above.

1. "Immunologically Virgin" Piglets

Germfree, colostrum-deprived piglets obtained by aseptic hysterectomy 3 to 5 days prior to term were devoid of detectable immunoglobulins (12-14). By the combined use of hyperimmunization and immunoelectrophoretic analysis it was possible to conclude that these piglets could be obtained which had at least less than 1/20,000,000 the concentration of immunoglobulins of the mother. In addition, the actinophage neutralizing assay provides one of the most sensitive methods for the detection of antibody. For instance, it is possible to detect antibody activity in hyperimmune sow serum, with a K value of 100, at a 1:1,000,000 dilution. In piglet sera or its fractions concentrated 20 times, it was not possible to detect specific activity. Again, it is possible to conclude that piglet serum has antibody concentrations of less than 1/20,000,000 of the maternal sow.

There are, however, conditions which permit the transfer of maternal antibody into these piglets. Placental damage can be brought about by infection or by the injection of antigens containing Gram-negative bacterial endotoxins. In addition, these piglets have an efficient mechanism for intestinal absorption within the first few hours after birth. Maternal fluids entering the mouth are rapidly transmitted to the blood. For this reason we have never obtained immunoglobulin-free piglets by natural farrow (12). It is obvious, therefore, that the technique of hysterectomy is critical in obtaining immunoglobulin-free piglets. In addition, these animals were devoid of any "background" hemolytic plaque-forming cells (15, 16) for sheep erythrocytes in the spleen, thymus, lymph nodes when >90% of the total organs were plated out by hemolytic plaque assay technique (4, 5).

2. Immunologic Competence

Piglets taken by hysterectomy 3-5 days prior to term, maintained germfree, were highly immunologically competent when tested by a single injection of 10^{12} actinophage (MSP-2). Antibody was detected within 48 hrs after the injection of antigen (14). In addition other antigens such as keyhole limpet hemocyanin (KLH) and sheep erythrocytes (SRC) (16-19) have been used. Hemolytic plaque-forming cells in the spleen were detected within 60 hrs after the injection of 5×10^9 SRC, i.p., and reached a maximum on the 6th day. In contrast, uninjected littermate controls maintained on Mullsoy diet (Borden Company, Pharmaceutical Division, New York) showed no detectable anti-MSP-2 for 6 days and then gradually developed a detectable amount of anti-MSP-2 indicating that Mullsoy diet contains phage or their cross-reacting antigens; however, there were no hemolytic plaque-forming cells at least for 2 months, suggesting that Mullsoy diet does not contain SRC or their cross-reacting antigens. To determine the specificity of this immune response, 20 mg KLH and 10^{12} MSP-2 injected into these animals did not stimulate development of hemolytic plaque-forming cells at least for 34 days and those receiving 20 mg KLH or 5×10^9 SRC did not produce anti-MSP-2 above that observed in the uninjected controls. These results indicated that SRC, KLH and MSP-2 antigens were devoid of cross-reacting antigens.

These data support our working hypothesis that so-called "natural antibodies" or "background" plaque-forming cells do not develop spontaneously without specific antigenic stimulation. It is also evident that these results do not support Jerne's natural selection theory of antibody formation (8) because in the piglet model, antibody is not required for the initiation of antibody response and antigen is the sole inducer of the antibody.

3. Sequential Synthesis of 19SγG to 7SγG

The absence of immunoglobulins in the sera of this model (12-14) makes it ideal to study the ontogenic development of the immunoglobulins after antigenic stimulation.

The first antibody formed in the true primary response to specific antigenic stimulation was a 19S immunoglobulin which appeared to be distinct from the so-called γM-macroglobulin as defined immunoelectrophoretically. Later the 7SγG antibody was formed which appeared to be antigenically identical to the 19S. For this reason the 19S antibody was designated 19SγG (20). Since γM and γA do not appear early after antigenic stimulation, it was possible to isolate the 19SγG free from 19SγM macroglobulin and 7SγG free from γA. Purified 19SγG had a sedimentation constant of 18.5 ($S^0_{20 \cdot w}$) and a molecular weight of 900,000 (19). Purified 7SγG-immunoglobulin of both the piglet and sow had a sedimentation constant of 6.6 S ($S^0_{20 \cdot w}$) and a molecular weight of 170,000 (19). The use of a radioimmunoelectrophoretic technique with I^{131} labeled MSP-2 and KLH revealed specific antibody activity only in 19SγG and 7SγG-immunoglobulins and not in 19SγM (19). The existence of 19S antibodies has been reported by many investigators (21-28). In addition, the macroglobulin (β_2M) of the Walderström type (29) has been well characterized (30-32) and designated γM (33). This latter macroglobulin is characterized by a distinct heavy chain μ. Both macroglobulins are modified by mercaptoethanol (1, 20, 30, 34, 35). The early observations of Deutsch and Morton on the effect of mercaptoethanol on the Walderström type macroglobulin (30) has resulted in the use of this technique by many investigators to identify the early 19S antibody as a γM (36, 37). Since the antibody activity of 19SγG is also destroyed by mercaptoethanol (1, 20), this method is not specific for γM. It is possible that many of the early antibodies observed in other species and designated γM are 19SγG.

It may be concluded, therefore, that the first antibody formed in the true primary response in the piglet model is a 19SγG immunoblobulin induced only by specific antigenic stimulation with the sequential synthesis of 7SγG.

4. Antigenic Competition

The rapid rate of antibody production in the true primary response and the greater response to a single antigen than to a multiple antigen (14) suggested the presence of a large number of uncommitted multipotent immunocompetent cells (X-cells, Fig. 1) in these immunologically virgin piglets; in contrast, in conventional animals comparable cells may have been committed and transformed to monopotent committed cells (Y-cells, Fig. 1) by the ubiquitous antigens in the environment. If this is true, these multipotent uncommitted cells will be largely committed to the first antigen to which they are exposed resulting in antigenic competition (38, 39). Evidence for this hypothesis was obtained by sequential injection of KLH and MSP-2 or MSP-2 and SRC in germfree, colostrum-deprived piglets maintained on Mullsoy diet (17). When piglets were injected with 20 mg KLH, i.p., 6 hours prior to the injection of 10^{12} MSP-2, i.p., the anti-MSP-2 response was reduced 75 to 85% compared with MSP-2 alone. When piglets were injected with 10^{12} MSP-2 i.p. 6 hrs before the injection of 5×10^9 SRC, i.p., hemolytic plaque-forming cells were not detected until 96 hrs; in the controls receiving SRC alone, plaque-forming cells were detected within 60 hrs and at 48 hrs in piglets injected with SRC 6 hrs before MSP-2. In addition, the peak response in plaque-forming cells was reduced greater than 50% when MSP-2 was injected before SRC. There is significant antigenic competition which favors the hypothesis that immunologically virgin piglets have uncommitted multipotent competent cells and that the antigenic competition is at this cellular level. These results are less compatible with the hypothesis that the X-cells (Fig. 1) are monopotential as suggested in the clonal selection hypothesis.

More recent unpublished experiments using a longer time interval between the injection of the competing antigens gave nearly complete inhibition of anti-SRC response; therefore, additional experiments will be required to determine the optimal interval for complete antigenic competition. Other factors could contribute to the apparent incomplete inhibition of antibody response to a second antigen such as physiological and anatomical barriers which prevent antigens from coming in contact with all of the uncommitted immunocompetent cells. This may be a basic biological control mechanism to prevent complete commitment and preserve a portion of their immune mechanism for other antigens. It is also possible that a number of uncommitted X-cells arise by differentiation from the omnipotent stem cells (S-cells, Fig. 1) during the neonatal period.

5. X→Y→Z Concept

In recent reviews and publications the concept of a sequential series of cell differentiations has been used to designate immunologically functioning cells (40-43). A conceptual model for an "immunocompetent system" is given in Fig. 1. Stem cells (S) are omnipotent and may differentiate to immunocompetent X-cells, which are uncommitted multipotent cells (UC), as a physiological differentiation independent of specific antigenic stimulation. This step may be influenced by thymus, Bursa equivalents (44, 45). Antigen priming induces a differentiation of the X-cells (UC) to committed monopotent Y-cells (C, I). The priming role of antigen (properly processed antigen by "macrophages") is instructive (as

Figure 1.

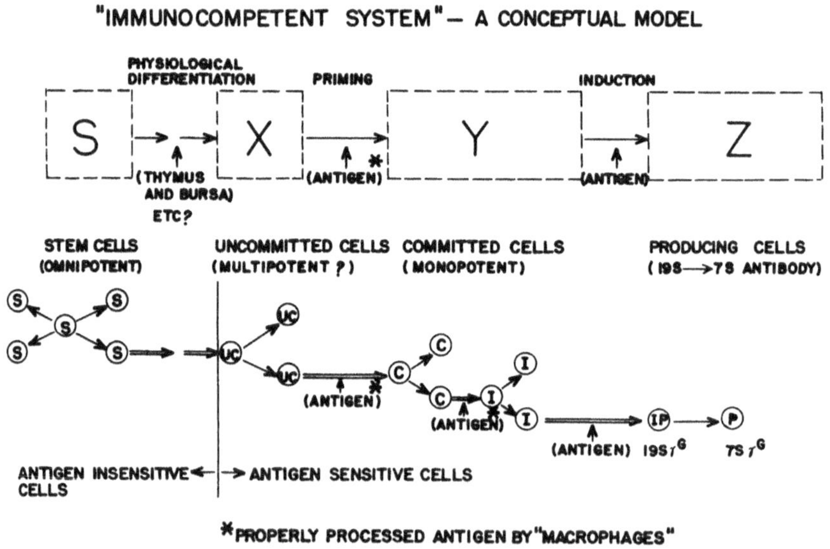

template or derepressor), not selective. The committed monopotent Y-cells can be induced by antigen to the producing Z-cells (IP, P). It is also possible that the immature plasma cells (IP) produce the 19SγG and the mature plasma cells (P) synthesize the 7SγG. The Y-cells (particularly C-cells) serve as memory cells and the Y→Z step could involve clonal selection in the sense of Burnet (9). In conventional animals continually exposed to ubiquitous antigens, many of their immunocompetent cells are already committed. When these antigens are used by the investigator to study mechanisms of antibody formation, the mechanisms studied may indeed support the clonal selection hypothesis because they are studying mostly the Y to Z step.

We suspected, therefore, that the immunologically virgin piglets would have only X-cells, and therefore the probability of eliciting an immune response in vitro would be greater than in tissues of conventional animals where the X-cells have been committed to Y-cells by ubiquitous antigens. To test this, the organ culture method of Auerbach et al. (46) was adopted. In contrast to our expectations, organs from immunologically virgin piglets gave a poor in vitro response when compared with organs from conventional animals. If, however, organs were removed from the piglets, 6 hrs after exposure to the antigen in vivo, excellent in vitro responses were obtained. These observations suggested that in the true primary response proper processing of antigen (47, 48) is required for the conversion of X→Y cells. If this can be accomplished in vitro, this model should be ideal to study the mechanism of the true primary event in antibody production at the cellular level.

ACKNOWLEDGMENT

We are grateful to Dr. Almut Dettmers, Department of Animal Science, and Dr. Eldon Hill and his staff of the Hormel Institute, University of Minnesota, who have made this work possible by their generous and expert assistance in the procurement of the animals used in this investigation.

REFERENCES

1. Kim, Y.B., Bradley, S.G. & Watson, D.W., J. Immunol., 93:798, 1964.

2. Reyniers, J.A., Ann. N.Y. Acad. Sci., 78:1, 1959.
3. Springer, G.F., Angew. Chem. (Eng.), 5:909, 1966.
4. Jerne, N.K., Nordin, A.A. & Henry, C., In Cell bound antibodies, p. 109, Wistar Institute Press, Philadelphia, 1963.
5. Ingraham, J.S. & Bussard, A., J. Exp. Med., 119:667, 1964.
6. Segre, D. & Kaeberle, M.L., J. Immunol., 89:782, 1962.
7. Myers, W.L. & Segre, D., J. Immunol., 91:697, 1963.
8. Jerne, N.K., Proc. Nat. Acad. Sci., 41:849, 1955.
9. Burnet, M., The clonal selection theory of acquired immunity. Vanderbilt University Press, Nashville, 1959.
10. Sterzl, J., Kostka, J. & Lanc, A. Folia Microbiol. 7:162, 1962.
11. Dettmers, A.E., Rampel, W.E. & Comstock, R.E., J. Animal Sci., 24:216, 1965.
12. Kim, Y.B., Bradley, S.G. & Watson, D.W., In Swine in Biomedical Research, p. 273, Frayn Printing Co., Seattle, Wash., 1966.
13. Kim, Y.B. & Watson, D.W., Ann. N.Y. Acad. Sci., 133: 727, 1966.
14. Kim, Y.B., Bradley, S.G. & Watson, D.W., J. Immunol., 97:52, 1966.
15. Sterzl, J., Vesely, J., Jilek, M. & Mandel, L., In Molecular and cellular basis of antibody formation, p. 463, Publishing House of Czech. Acad. Sci., Prague, 1965.
16. Kim, Y.B. & Watson, D.W., Fed. Proc., 27:493, 1968.
17. Kim, Y.B. & Watson, D.W., Bact. Proc., p. 75, 1968.
18. Kim, Y.B., Bradley, S.G. & Watson, D.W., J. Immunol., 99:320, 1967.
19. Kim, Y.B., Bradley, S.G. & Watson, D.W., J. Immunol., (in press, 1968).
20. Kim, Y.B., Bradley, S.G. & Watson, D.W., J. Immunol., 97:189, 1966.
21. Heidelberger, M. & Pedersen, K.O., J. Exp. Med., 65: 393, 1937.
22. Kabat, E.A., J. Exp. Med., 69:103, 1939.
23. Bauer, D.C. & Stavitsky, A.B., Proc. Natl. Acad. Sci., 47:1667, 1961.
24. Benedict, A.A., Brown, R.J. & Ayengar, R., J. Exp. Med., 115:195, 1962.
25. Fahey, J.L. & Humphrey, J.H., Immunology, 5:104, 1962.
26. Uhr, J.W. & Finkelstein, M.S., J. Exp. Med., 117:457, 1963.

27. Berlin, B.S., Proc. Soc. Exp. Biol. Med., 113:1013, 1963.
28. Nossal, G.J.V., Szenberg, A., Ada, G.L. & Austin, C.M., J. Exp. Med., 119:485, 1964.
29. Waldenström, J., Advan. Intern. Med., 5:398, 1952.
30. Deutsch, H.F. & Morton, J.I., J. Biol. Chem., 231:1107 and 1119, 1958.
31. Mannik, M & Kunkel, H.G., J. Exp. Med., 116:859, 1962.
32. Cohen, S., Biochem. J., 89:334, 1963.
33. Nomenclature for Human Immunoglobulins. Bull. W.H.O. 30:447, 1964.
34. Kim, Y.B., Bradley, S.G. & Watson, D.W., J. Immunol., 98:868, 1967.
35. O'Connor, J.E. & Snyder, I.S., J. Bact. 94:815, 1967.
36. Sahiar, K. & Schwartz, R.S., J. Immunol., 95:345, 1965.
37. Williams, G.M., J. Exp. Med., 124:57, 1966.
38. Cremer, N.E., J. Immunol., 90:685, 1963.
39. Adler, F.L., Progr. Allergy, 8:41, 1964.
40. Sercarz, E. & Coons, A.H., In Mechanisms of immunological tolerance, p. 73, Publishing House of Czech. Acad. Sci., Prague, 1962.
41. Sterzl, J., Advan. Biol. Sci., p. 149, 1962.
42. Perkins, E.H. & Makinodan, T., J. Immunol., 92:192, 1964.
43. Sterzl, J. & Silverstein, A.M., Advan. Immunol., 6:337, 1967.
44. Good, R.A. & Papermaster, B.W., Advan. Immunol., 4:1, 1964.
45. Glick, B., In The thymus in immunobiology, p. 343, Hoeber-Harper, New York, 1964.
46. Globerson, A. & Auerbach, R., J. Exp. Med., 124:1001, 1966.
47. Fishman, M. & Adler, F.L., J. Exp. Med., 117:595, 1963.
48. Adler, F.L., Fishman, M. & Dray, S., J. Immunol., 97:554, 1966.

HEMOLYTIC PLAQUE-FORMING CELLS AND ANTIGEN-SENSITIVE UNITS IN SPLEENS OF GERMFREE AND CONVENTIONAL MICE*

G. M. Shearer,** G. Cudkowicz and H. E. Walburg, Jr.

Department of Experimental Biology, Roswell Park Memorial Institute, Buffalo, N.Y., and Biology Division, Oak Ridge National Laboratory, Oak Ridge, Tenn.

Immunocytes producing so-called natural antibodies are not detected in the spleens of germfree, colostrum-deprived animals reared on diets lacking bacterial and other antigens (1). Upon exposure of such animals to sheep erythrocytes, potentially immunocompetent cells generate, by proliferation and differentiation, antibody-forming cells. The number of antibody-forming cells increases with the animal's age, but is not influenced by pretreatments with nonspecific stimulants of the lymphoid or reticuloendothelial system (1). It seems that antigen is not necessary for the differentiation of the primitive lymphoid cells not producing antibody, but capable of interacting with antigens and of initiating an immune response. The role of antigen is to trigger these primitive cells, called antigen-sensitive units (2) into the path of differentiation leading to mature lymphoid cells synthesizing a specific immunoglobulin.

Antibody-forming cells are extremely heterogeneous and differentiated, since each cell produces, as a rule, only one of several possible molecular species of antibody formed during an im-

---

* Research jointly sponsored by Institutional Grant IN-54 of the American Cancer Society, Inc. to Roswell Park Memorial Institute, and by the U.S. Atomic Energy Commission under contract with the Union Carbide Corporation.

** Postdoctoral Fellow of the Damon Runyon Memorial Fund for Cancer Research.

mune response (3). We have found, for example, that in mice given sheep erythrocytes (SRBC), spleen cells producing hemagglutinins do not produce hemolysins and vice versa (4, 5). Furthermore, cells producing hemolysins of the direct type (presumably IgM) do not produce hemolysins of the indirect or facilitated type (presumably IgG) (5-7). The precursors of antibody-forming cells, i.e., antigen-sensitive units (ASU), are also heterogeneous and differentiated, since they generate, upon interaction with SRBC or other antigens, only one population class of immunocyte (2, 5,8).

Sheep erythrocytes possess antigenic determinants that cross-react with bacterial antigens commonly found in the intestinal flora of mice (9). It was conceivable, therefore, that the differentiation of splenic and thymus-marrow ASU reactive with SRBC was enhanced in unprimed conventional mice (2, 5) by cross-reacting bacterial antigens. In other words, the intestinal flora could have played a role in "programming" the immune system of mice by way of antigens cross-reacting with SRBC and other cellular antigens (9, 10). It was of interest, therefore, to quantitate antibody-forming cells and antigen-sensitive units in the spleens of germfree and conventional mice given SRBC to test this possibility.

DIRECT PLAQUE-FORMING CELLS IN NON-IRRADIATED MICE

SRBC were injected into the peritoneal cavity of germfree and conventional Carworth Farm No. 1 (CF1) mice of both sexes at the age of 11-14 weeks. Animals were sacrificed at daily intervals and spleen cell suspensions were prepared for direct hemolytic plaque assay. The methods used for rearing the mice, for immunization, and for preparing and assaying the cell suspensions have been described elsewhere (4, 5, 11). The results are presented in Figure 1. The number of plaque-forming cells (PFC) increased rapidly in the spleens of conventional and germfree mice from day 2 to day 4, reached a maximum on days 4 and 5, and then declined. There were no meaningful differences between germfree and conventional mice in the time course of the direct PFC response even when the concentrations of splenic PFC (not shown in Figure 1) were compared rather than the total numbers of PFC per spleen. Therefore, germfree CF1 mice were as competent as normal mice to produce, upon antigenic stimulation, functional antibody-forming cells detected by the direct plaque assay. These observations are consistent with previous findings that the serum titers of anti-sheep hemolysins and hemagglutinins, and the numbers of direct PFC per spleen are equal in germfree and conventional mice (12, 13).

In spite of the similar PFC responses in intact animals,

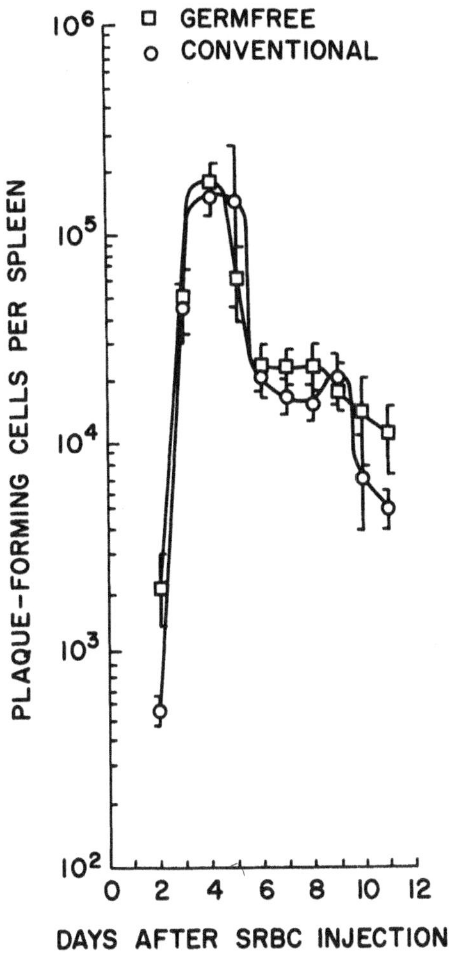

Figure 1. Mean number of plaque-forming cells in the spleen of germfree and conventional mice after intraperitoneal injection of $\sim 5 \times 10^8$ SRBC. Two standard errors of the mean are indicated by vertical bars.

---

germfree mice may not possess as many ASU as conventional mice. This may be so because the PFC found in the spleens of intact mice represent only a fraction of the potential of the ASU present. In fact, the expression of immunological competence in intact mice is under depressing homeostatic influences, so that only the ASU contained in $\sim 12 \times 10^6$ spleen cells of sensitized mice are expressed at any given time (14).

ANTIGEN-SENSITIVE UNITS IN IRRADIATED MICE

The number of antigen-sensitive units in spleens or other lymphoid organs is usually estimated by transplantation assays (5). We were unable to resort to a transfer system because the only germfree animals available for this study were noninbred CF1 mice. We attempted therefore, to approximate the conditions of transplantation by exposing mice to sublethal whole-body irradiation. This treatment reduces the population of immunocompetent cells to a minimum, and induces endogenous regeneration of new ASU (endorepopulation).

In a pilot experiment, conventional CF1 mice were exposed to sublethal doses of X-radiation. Antigen was administered to groups of irradiated mice 2 hours, or 7 days, or 14 days after irradiation, and plaque assays were then made at daily intervals. We intended to measure the surviving fraction of ASU 2 hours after X rays, and the regeneration of ASU that would eventually occur during the following two weeks. In particular, we aimed at a treatment schedule (X-ray dose and timing of antigen) that would yield a number of regenerated splenic ASU close to that contained in grafts of 2×10^7 spleen cells of unprimed donors. Under the latter conditions (exorepopulation) the magnitude of the PFC response is a linear function of the number of grafted ASU, i.e., the PFC response is not depressed by homeostatic controls. We found that the suitable treatment schedule was 550 R of whole-body X rays followed by SRBC injection 14 days later (Table I).

Table I. Plaque-forming cells in spleens of conventional irradiated mice undergoing endorepopulation or exorepopulation by spleen cell grafts. Assays were made at several intervals after SRBC injection, but only results on peak day of response are given.

Treatment and strain	SRBC Postirradiation Interval	Day of peak response (after SRBC)	PFC per spleen ± standard error (No. of mice)
CF1 mice 550 R	2 hrs.	4	138 (2)
CF1 mice 550 R	14 days	7	20,500 ± 4,380 (6)
(C3H x C57BL)F_1 mice 900 R and 2×10^7 spleen cells	2 hrs.	7	26,400 ± 3,750 (6)

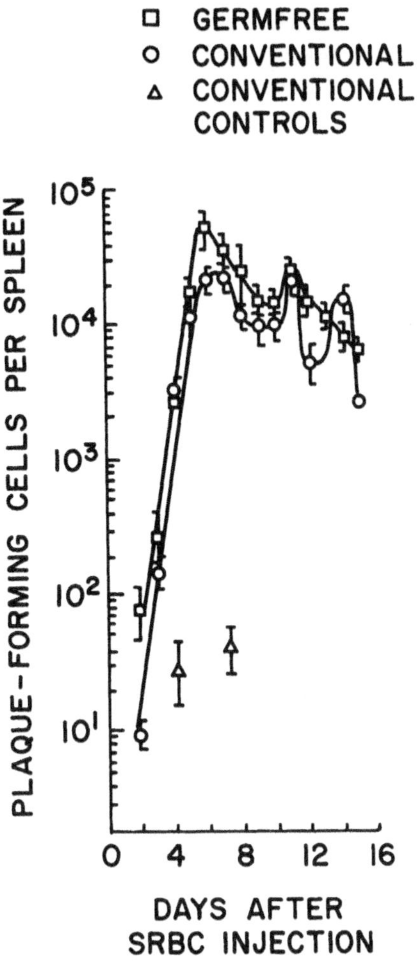

Figure 2. Mean number of plaque-forming cells in the spleens of germfree and conventional mice exposed to 550 R of X rays. SRBC were given 14 days postirradiation. Two standard errors of the mean are indicated by vertical bars.

Germfree and conventional mice of similar age were then exposed to 550 R of X rays and injected with SRBC 14 days later to compare their direct PFC responses (Figure 2). The number of PFC increased exponentially in the spleens of both types of mice from day 2 to day 6 after antigen administration, reached a maximum on days 6 and 7, and then fluctuated irregularly. The number of PFC per spleen in germfree mice was as great as that in conventional mice at every time interval studied. It follows that regeneration of the lymphoid system and differentiation of ASU reactive with SRBC are quantitatively undistinguishable in germfree and conventional mice.

CONCLUSION

The direct plaque response of intact germfree mice to SRBC antigens is equivalent to the response of conventional mice in terms of numbers of PFC and kinetics of PFC formation. After radiation injury (550 R), the number of surviving ASU, the regenerative potential of the immune system, and the kinetics of PFC formation do not differ in germfree and conventional mice. This implies that the unirradiated mice were alike also with respect to the total number of ASU and of more primitive lymphopoietic progenitor cells. Hence, the differentiation of ASU from more primitive elements, presumably contained in marrow and thymus (2), and the differentiation of PFC from antigen-sensitive units, were not detectably affected by bacterial antigen load.

The germfree mice could have been exposed to relatively small amounts of bacterial antigens (cross-reacting with SRBC) contained in the sterilized food. It is unlikely, however, that these amounts of antigens were sufficient to induce immunocompetent cell differentiation quantitatively undistinguishable from that which occurred in conventional mice, since the magnitude of an immune response is a function of small amounts of antigen administered rather than a threshold antigen-dose effect. Furthermore, some enteric bacteria such as Escherichia coli and Aerobacter aerogenes are capable of inducing the appearance of anti-sheep PFC in the spleens of specific pathogen-free mice upon intraperitoneal injection (9). For these considerations, it is likely that cellular differentiation of the immune system of mice is not only independent of antigen load, but of cross-reacting antigens altogether.

REFERENCES

1. J. Sterzl, "Factors determining differentiation pathways of immunocompetent cells," in Antibodies, Cold Spring Harbor Symposium on Quantitative Biology, Cold Spring Harbor, New York, 32: 493, 1967.
2. G. Cudkowicz and G. M. Shearer, "Differentiation of potentially immunocompetent cells in the mouse," this volume.
3. O. Mäkelä, "The specificity of antibodies produced by single cells," in Antibodies, Cold Spring Harbor Symposium on Quantitative Biology, Cold Spring Harbor, New York, 32: 423, 1967.
4. G. M. Shearer and G. Cudkowicz, "Cluster formation in vitro by mouse spleen cells and sheep erythrocytes," J. Immunol., in press.

5. G. M. Shearer, G. Cudkowicz, M. St. James Connell, and R. L. Priore, "Cellular differentiation of the immune system of mice. I. Separate splenic antigen-sensitive units for different types of anti-sheep antibody-forming cells," J. Exptl. Med., in press.
6. F. Celada and H. Wigzell, "Immune responses in spleen colonies. II. Clonal assortment of 19S- and 7S-producing cells in mice reacting against two antigens," Immunology, 11: 453, 1966.
7. Z. Brahmi and B. Merchant, "Exclusive secretion by rabbit immunocytes of either γM or γG hemolysins,"Fed. Proc., 27: 735, 1968.
8. G. J. V. Nossal, K. D. Shortman, J. F. A. P. Miller, G. F. Mitchell, and J. S. Haskill, "The target cell in the induction of immunity and tolerance," in Antibodies, Cold Spring Harbor Symposium on Quantitative Biology, Cold Spring Harbor, New York, 32: 369, 1967.
9. V. Cheng and J. J. Trentin, "Enteric bacteria as a possible cause of hemolytic antibody-forming cells in normal mouse spleens," Proc. Soc. Exptl. Biol. Med., 126: 467, 1967.
10. J. J. Trentin, Ed., "Cross-Reacting Antigens and Neoantigens," Williams and Wilkins Co., Baltimore, Maryland, 1967, entire volume.
11. H. E. Walburg, Jr., E. I. Mynatt, and D. M. Robie, "The effect of strain and diet on the thirty-day mortality of X-irradiated germfree mice," Rad. Res., 27: 616, 1966.
12. J. F. A. P. Miller, P. Dukor, G. Grant, N. R. St. C. Sinclair, and E. Saquet, "The immunological responsiveness of germ-free mice thymectomized at birth. I. Antibody production and skin homograft rejection," Clin. Exptl. Immunol., 2: 531, 1967.
13. M. Bosma, T. Makinodan, and H. E. Walburg, Jr., "Development of immunologic competence in germfree and conventional mice," J. Immunol., 99: 420, 1967.

SEVERITY OF THE PARENT TO F_1 GRAFT-VERSUS-HOST REACTION IN IRRADIATED GERMFREE MICE

H. E. Walburg, Jr. and G. E. Cosgrove

Biology Division, Oak Ridge National Laboratory

Oak Ridge, Tennessee

Lethally irradiated mice protected with allogeneic hemopoietic cells do not survive as long as those protected with syngeneic hemopoietic cells due to a "secondary disease." This observation was first made by Barnes and Loutit in 1954 (1,2), and "secondary disease" was more fully characterized by Congdon and Urso in 1957 (5). This disease is primarily due to an immune reaction of grafted cells against the host tissues, the graft-versus-host (GVH) reaction (10). Uphoff (15,16) demonstrated a similar lesion in lethally irradiated F_1 hybrid mice which had been treated with parental strain bone marrow. Schwartz, Upton, and Congdon (13) first described the "foreign spleen reaction," a GVH disease resulting from injection of parental spleen cells into sublethally irradiated F_1 hybrid mice.

Although the pathology of the GVH reaction and its associated secondary bacterial invasion have been well characterized (5,7) it is not known what role this microbial invasion (i.e., secondary infection) plays in the pathogenesis of the lesions. With the development of germfree techniques and the availability of germfree animals, the role of secondary infection can be more clearly defined. Earlier reports on GVH reaction in germfree animals are contradictory; some suggest a decreased GVH reaction in germfree animals (6,14) and others describe similar pathology and mortality from the GVH reaction in germfree and conventional animals (11,17). This study was undertaken to help clarify the role of microorganisms in the GVH reaction and to compare quantitatively the immunological reaction in germfree and conventional mice.

Materials and Methods. All germfree and conventional mice were reared and maintained at the Biology Division of the Oak Ridge

National Laboratory. Inbred mice of the C3H strain, whose ancestors were germfree breeders supplied by the Lobund Laboratory, University of Notre Dame, were cross-mated with inbred mice of the RFM strain (19) to produce C3RF (C3H ♀ X RFM ♂) and RFC3 (RFM ♀ X C3H ♂) F_1 hybrid mice. Germfree mice were consistently negative on periodic examination for bacteria, fungi, mycoplasma, internal and external parasites, and for all viruses for which they were tested, i.e., polyoma, Reo. 3, Theiler's, "K", Sendai, PMV, mouse hepatitis, and mouse adenovirus. Conventional mice were derived from germfree ancestors which had been removed from the germfree environment, placed in a breeder room with conventional mice, and reared there for at least two generations. Mice were exposed to various amounts of whole-body X-radiation with a G.E. Maxitron X-ray Machine. Radiation factors were: 300 kVp, 20 mA, hvl 0.98 mm Cu, and exposure rate of approximately 100 R/min. Irradiation procedures for germfree mice have been described previously (18). Within 2 hr. after irradiation, hybrid mice of both sexes were injected intraperitoneally with 0.5 ml of a suspension of spleen cells from 9-week-old females of the maternal parent strain (i.e., RFC3 received cells from female RFM). Spleen cell suspensions were prepared by gentle extrusion of the cellular pulp into a .85% saline solution in a Potter-Elvehjem glass homogenizer in the cold. The dose of cells was estimated on the basis of spleen equivalents, one spleen equivalent being equal to the average number of cells contained in one donor spleen.

At various times after treatment mice were killed and the thymus, lymph nodes (axillary, inguinal, and mesenteric), duodenal Peyer's patch, sternum (for bone marrow), spleen, and liver were removed, fixed in Bouin's fixative, embedded in paraffin, sectioned, and stained with hematoxylin and eosin. Slides were examined for histologic changes characteristic of the foreign spleen reaction (7), and graded for lesion severity, with 1+ representing a mild granuloma, 2+ a moderate granuloma, and 3+ a severe granuloma often accompanied by necrosis. Cages were observed daily for dead mice. Survival data were corrected for withdrawal of mice from the experiment by a life table analysis (12). Cumulative severity, obtained by adding the mean graded severity of the lesion at each observation time after treatment, was plotted as a function of the days after treatment. The lesions in lymph nodes were selected as a representative of those in other hemopoietic organs.

RESULTS

Several preliminary experiments were conducted with conventional mice to establish optimal conditions for production of the GVH reaction in this strain combination. Studies on the effect of recipient age where irradiated C3RF F_1 mice were injected with 1/10

TABLE 1. EFFECT OF RECIPIENT AGE ON GVH REACTION

X-ray Dose(R)	Recipient age(wks)	No. Mice	Survival (%)	Cumul. granuloma severity at 21 days after treatment
	1	25	<10	18
	4	15	<10	21
300	5-6	65	<10	19
	7	15	55	19
	24	10	75	15
	1	25	<10	17
	4	15	60	20
200	5-6	30	65	19
	7	15	90	7
	24	10	100	5
	1	25	<10	16
	4	15	60	10
100	5-6	30	70	4
	7	15	100	3
	24	10	100	0

spleen equivalent from 9-week-old female C3H mice showed that the older the animal, the less sensitive it was to the production of a GVH reaction, both as measured by survival and by the development of a lymph node granuloma (Table 1). This effect of recipient age was less evident at high doses of X-rays than at low doses, especially in the production of lymph node granuloma, i.e., granuloma development was similar for all age groups exposed to 300 R but an age effect was noted when these groups were exposed to 200 or 100 R.

Studies on the effect of X-ray exposure where conventional irradiated 5- to 6-week-old C3RF F_1 recipients were injected with 1/10 spleen equivalent from 9-week-old female C3H mice revealed that the greater the exposure the more profound the reaction (Table 2). C3RF F_1 hybrid mice injected with 1/10 spleen equivalent showed a high mortality after exposure to 300 R or more of X-rays and severe lymph node granuloma after exposure to 200 R or more. No mortality or lymph node granuloma was observed in animals exposed to 50 R of X-rays and injected with 1/10 spleen equivalent.

To study the effect of spleen cell dose on the development of GVH reaction, 5- to 6-week-old C3RF F_1 recipients were exposed to 300 R of X-rays and injected with from 1/10 to 1/500 spleen-equivalents from 9-week-old female C3H mice. Both mortality and severity of lymph node granuloma were proportional to cell dose, with 1/10 spleen-equivalent giving a maximal response and 1/500 spleen equivalent showing only slight lymph node granuloma and no mortality (Table 3).

TABLE 2. EFFECT OF X-RAY EXPOSURE ON GVH REACTION

X-ray Dose(R)	No. Mice	Survival (%)	Cumulative granuloma severity at 21 days after treatment
500	85	<10	Not Done
400	15	<10	15
300	29	<10	15
200	67	65	14
100	25	70	4
50	15	100	0

TABLE 3. EFFECT OF SPLEEN CELL DOSE ON GVH REACTION

Cell dose in Spleen equiv.	No. Mice	Survival (%)	Cumulative granuloma severity at 21 days after treatment
1/10	29	<10	15
1/50	14	30	9
1/100	14	50	6
1/500	15	100	2

The development of GVH reaction in germfree mice was first studied using lethal combinations of X-ray exposure and spleen cell dose. Five- to six-week-old germfree and conventional recipients were exposed to 500 or 200 R of X-rays and injected with 1/10 spleen-equivalent from 9-week-old female donors. Several donor-recipient combinations were studied: 1) all germfree (GF)(GF donor cells into GF recipients), 2) all conventional (CONV) (CONV donor cells into CONV recipients and 3) GF-CONV (GF donor cells into CONV recipients). When recipients were of the RFC3 strain and donors were of the RFM strain, the all-GF combination had as high or higher mortality than the all-CONV combination (Fig. 1). When the recipients were of the C3RF strain and the donors were of the C3H strain the all-GF combination had a somewhat lower mortality than the all-CONV combination (Fig. 2). When spleen cells from germfree donors were injected into irradiated conventional recipients (GF-CONV combination) the mortality was similar to that observed in the all-CONV combination (Fig. 2). The development of lymph node granulomas was examined in these groups and no differences were observed, in that all groups developed severe lesions (Fig. 3). Radiation controls and spleen-injected controls in all groups showed 100% survival and were negative for the development of GVH reaction in all hemopoietic organs.

Further studies on the GVH reaction in GF mice using sublethal combinations of X-ray exposure and spleen cell dose were conducted in an attempt to quantitate more accurately the histological response. Five- to six-week-old germfree and conventional C3RF

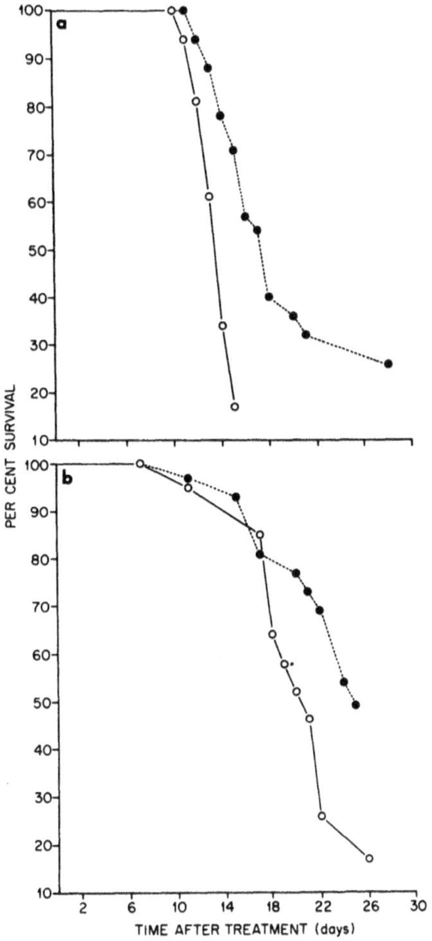

Fig. 1. Influence of germfree environment on foreign spleen reaction as measured by survival. (a) 500 R or (b) 200 R ●.....●, CONV donors and CONV recipients (30); O——O GF donors and GF recipients (20). Number of mice entering experiment in parentheses.

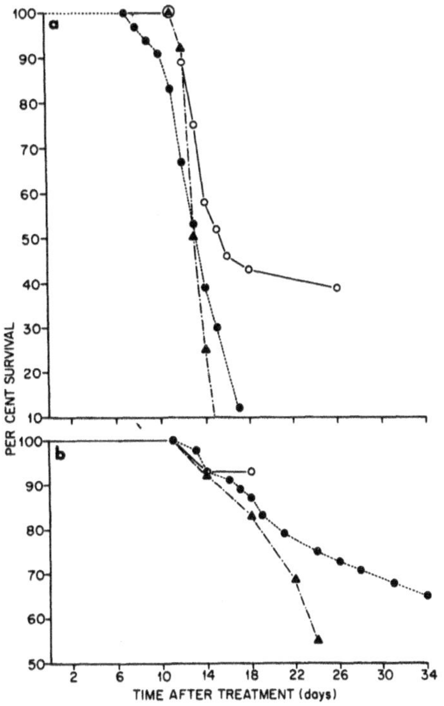

Fig. 2. Influence of germfree environment on foreign spleen reaction as measured by survival. (a) 500 R (b) 200 R. ●......● CONV donors and CONV recipients (65-85); O——O, GF donors and GF recipients (35-40); ▲——·——▲, GF donors and CONV recipients (15). Number of mice entering experiment in parentheses.

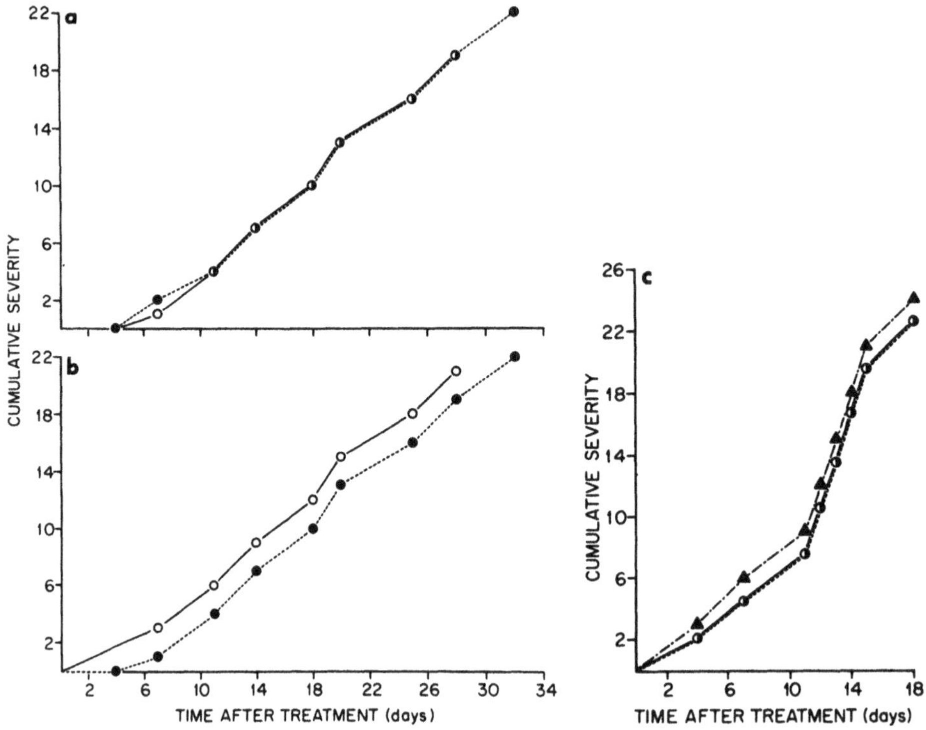

Fig. 3. Influence of germfree environment on foreign spleen reaction as measured by severity of lymph-node granuloma. (a) RFC3 recipient, RFM donor, 500 R; (b) RFC3 recipient, RFM donor, 200 R; (c) C3RF recipient, C3H donor, 500 R. ●.....●, CONV donors and CONV recipients (3); O———O, GF donors and GF recipients (2); ▲———·———▲, GF donors and CONV recipients (1). Number of mice examined per point in parentheses.

recipients were exposed to 300 R X-rays and injected with 1/40 or 1/80 spleen equivalents from C3H donor females raised in the same environment (i.e., GF mice received cells from GF donors and CONV mice from CONV donors) (Fig. 4). Similarly, RFC3 recipients were injected with 1/100 or 1/500 spleen equivalents from RFM donors (Fig. 5). In all cases, the development of lymph node granulomas was as severe, or more severe, in the GF system as in the CONV system.

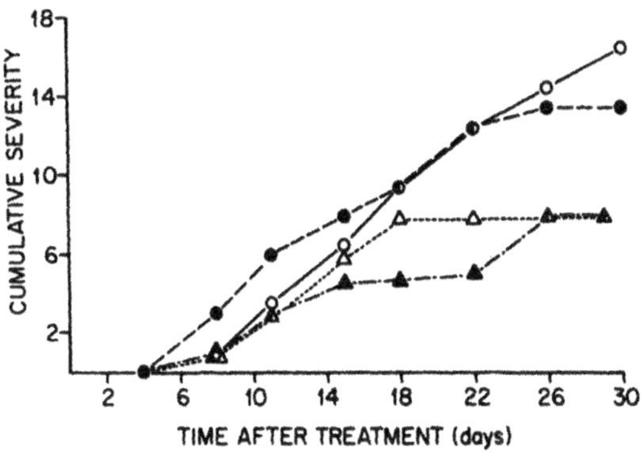

Fig. 4. Influence of germfree environment on foreign spleen reaction as measured by severity of lymph-node granuloma. O——O, GF recipients injected with 1/40 spleen equivalent from GF donors (4); ●— —●, CONV recipients injected with 1/40 spleen equivalent from CONV donors (2); △.....△, GF recipients injected with 1/80 spleen equivalent from GF donors (2); ▲——·——▲, CONV recipients injected with 1/80 spleen equivalent from CONV donors (2). Number of mice examined per point in parentheses.

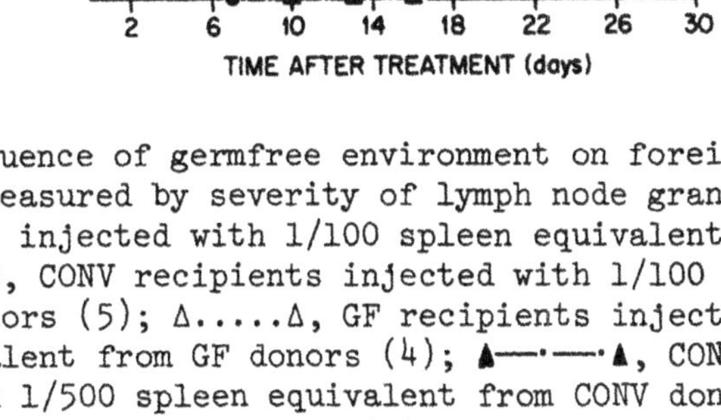

Fig. 5. Influence of germfree environment on foreign spleen reaction as measured by severity of lymph node granuloma. O——O, GF recipients injected with 1/100 spleen equivalent from GF donors (4); ●— —●, CONV recipients injected with 1/100 spleen equivalent from CONV donors (5); △.....△, GF recipients injected with 1/500 spleen equivalent from GF donors (4); ▲—·—·▲, CONV recipients injected with 1/500 spleen equivalent from CONV donors (5). Number of mice examined per point in parentheses.

DISCUSSION

Our data support conclusions of other workers that different combinations of X-ray exposure (8) and spleen cell dose (4,9) affect the severity of the foreign spleen reaction, and further, demonstrate that young recipients develop a more severe foreign spleen reaction than do adults which are more commonly used as recipients in experiments of this type (i.e., 12-24 weeks). Iossifides et al. (9) have also reported an age related decrease in severity of the foreign spleen reaction when the dose of spleen cells was small. The resistance of adults to GVH reaction, especially after exposure to small amounts of radiation may reflect an interference with establishment of the graft in older recipients. Our choice of 5- to 6-week-old recipients reflects both the sensitivity of this age and the shorter holding time in germfree isolators. Thus a susceptible strain combination, a sensitive recipient age, and a system of varying the severity of the response through manipulation of X-ray exposure and spleen cell dose provide a method for quantitative analysis of the severity of the GVH reaction.

Previous studies have demonstrated that a GVH reaction can occur in germfree animals. Germfree mice made chimeric with allogeneic or xenogeneic bone marrow cells do not always develop the wasting, early mortality, and severe dermatitis seen in their conventional counterparts (6,17). However, when such chimeric animals are examined histologically the GVH reaction is quite evident (17). At the same time, some strains of germfree mice do develop all the signs and the gross pathology of the more acute GVH reaction which occurs when newborn mice are injected with allogeneic spleen cells (11).

Our results demonstrate again that a severe GVH reaction can occur in mice reared in a germfree environment. Studies with lethal combinations of X-radiation and parental spleen cells did not clearly establish an influence of microbial environment on the mortality associated with the foreign spleen reaction, since experiments using the RFC3 cross demonstrated decreased mortality in the conventional system whereas those using the reciprocal cross showed an increased mortality when conventional and germfree systems were compared. On the other hand, there was no demonstrable influence of the microbial flora on the development of lymph node granulomas in these mice, suggesting that mortality and such parameters as mean survival time (14) may not always reflect the true severity of the GVH reaction.

Further study of the foreign spleen reaction in germfree mice employed sublethal combinations of X-ray exposure and spleen cell dose, which produce less severe and less rapidly developing lymph node granulomas, a system which might be more sensitive to effects of the microbial flora than those used previously. With this technique, lymph node granulomas caused by the foreign spleen reaction proved as severe in germfree mice as in their conventional counterparts. These data suggest that the microbial flora which is absent in germfree mice plays no role in the development of GVH reaction in sublethally irradiated F_1 hybrid mice injected with parent spleen cells. The ability of spleen cells from conventional and germfree donors to produce similar GVH pathology suggests that they are equally immunocompetent when stimulated by foreign tissue antigens. This is consistent with earlier studies in our laboratory which demonstrated no difference in immunocompetence of spleen cells from germfree and conventional mice challenged with sheep erythrocytes (3).

ACKNOWLEDGEMENTS

We wish to thank W. D. Gude and his staff, for the histological preparation of the experimental material, and W. E. Jennings for technical assistance.

(Research sponsored by the U. S. Atomic Energy Commission under contract with the Union Carbide Corporation.)

REFERENCES

1. Barnes, D. W. H.; Loutit, J. F. 1954. Nucleonics 12: (5)68.
2. Barnes, D. W. H.; Loutit, J. F. 1955. p. 134 In Bacq, Z. M. and Alexander, P. (eds.). Radiobiology Symposium. Butterworths Scientific Publications, London.
3. Bosma, M. J.; Makinodan, T.; Walburg, H. E., Jr. 1967. J. Immunol. 99: 420.
4. Cole, L. J.; Ellis, M. E. 1958. Science 128: 32.
5. Congdon, C. C.; Urso, I. S. 1957. Am. J. Pathol. 33: 749.
6. Connell, S. J.; Wilson, R. 1964. Radiat. Res. 25: 181.
7. Cosgrove, G. E.; Upton, A. C.; Congdon, C. C. 1962. Am. J. Pathol. 40: 455.
8. Cosgrove, G. E.; Upton, A. C.; Congdon, C. C. 1962. Transplant. Bull. 30: 141/29.
9. Iossifides, I. A.; Rabotti, G. F.; Brand, M. 1964. Transplant. 2: 33.
10. Koller, P. C.; Davies, A. J. S.; Doak, S. M. A. 1961. pp. 220-223 In Haddow, A. and Weinhouse, S. (eds.). Advances in Cancer Research, Volume 6. Academic Press, New York.
11. McIntire, K. R.; Sell, S.; Miller, J.F.A.P. 1964. Nature 204: 151.
12. Sachs, R. 1959. Toxicol. Appl. Pharmacol. 1: 203.
13. Schwartz, E. E.; Upton, A. C.; Congdon, C. C. 1957. Proc. Soc. Exptl. Biol. Med. 96: 797.

14. Solomon, J. C.; Lecourt, J. C. 1966. Proc. Soc. Exptl. Biol. Med. 122: 640.
15. Uphoff, D. E. 1957. J. Natl. Cancer Inst. 19: 123.
16. Uphoff, D. E.; Law, L. W. 1957. Radiation Res. 7: 456.
17. Van Bekkum, D. W.; De Vries, M. J.; Van der Waay, D. 1967. J. Natl. Cancer Inst. 38: 223.
18. Walburg, H. E., Jr.; Mynatt, E. I.; Robie, D. M. 1966. Radiation Res. 27: 616.
19. Walburg, H. E., Jr.; Upton, A. C.; Tyndall, R. L.; Harris, W. W.; Cosgrove, G. E. 1965. Proc. Soc. Exptl. Biol. Med. 118: 11.

DIET AND IMMUNE MECHANISMS

B.S. Wostmann, J.R. Pleasants and Patricia Bealmear

The Lobund Laboratory

University of Notre Dame, Notre Dame, Indiana

I. Introduction

Germfree animals generally have a less well developed RE system, characterized by lower weight of especially those tissues which normally would be exposed to microbial stimulation (1). In germfree mice and chickens the concentration of immuno competent cells in these tissues is lower than in conventional controls (2,3). Gamma globulin levels of germfree animals are lower than in comparable conventional controls, but appear to be quite diet dependent in certain species (4). Immuno electrophoresis of the serum of germfree CFW mice reared on the "antigen free", low molecular weight, water soluble diet formula developed by Pleasants et al (5) indicates total absence of the IgG line in young adults, while the pattern of comparable animals reared on solid diet always demonstrates at least some IgG.

Although antibody response can be considered adequate in all germfree species investigated (mouse, rat, guinea pig, chicken), the primary response often reflects the lower RE content (6,7). However, especially in the germfree mouse, different authors report different primary response patterns, which may or may not differ from the response of the comparable conventional animal (7, 8,9). In all of these experiments, as in the studies of the specific bactericidal action of the serum of a number of germfree animal species (10), dietary stimulation appears to be a major variable.

II. Diet and pathogenicity of Salmonella typhimurium in germfree rats.

In our studies on the biochemistry of the anti-microbial defense mechanisms, adult germfree rats were monoassociated with *S. typhimurium* (N.D. #750A). In earlier series the animals were maintained on diet L 462 (11). Upon monoassociation a weight loss of 25-30% occurred, and approximately ¼ of the animals died within the first 2 weeks. Later experiments in which diets 5010 C* and L 485 (12) were fed showed much less loss in weight, while no death occurred. To verify these results, within one isolator half of the rats to be associated were fed diet L 462, and the other half diet 5010 C. In another series half of the animals were fed L 462, and the other half diet L 485. Fig. 1 depicts body weights and number of deaths before and after monoassociation.

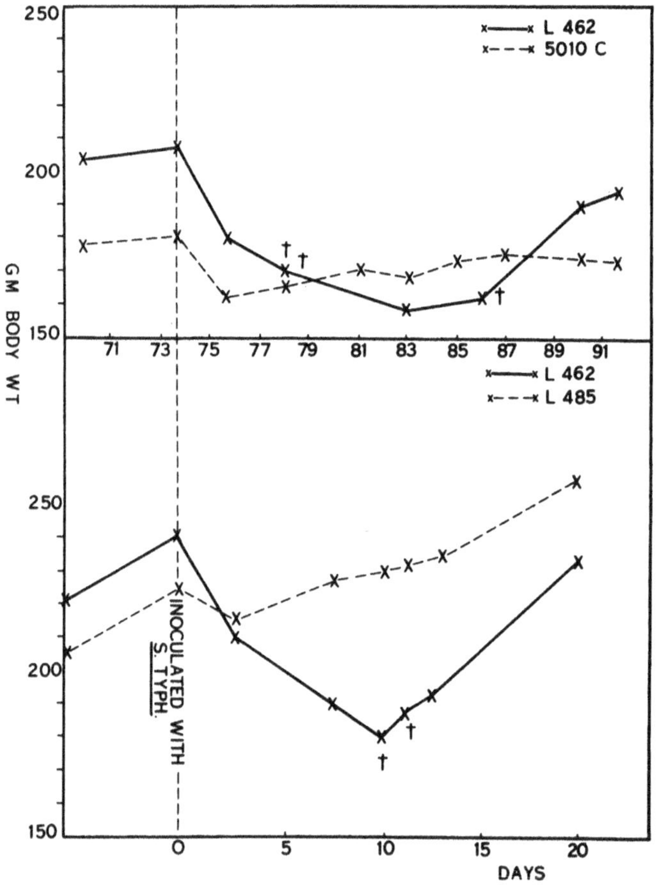

Fig. 1. Body weight and mortality of germfree rats maintained on different diets and monoassociated with *Salmonella typhimurium*.

* Ralston Purina Co. St. Louis, Missouri.

The data show that, while in both experiments the animals maintained on diet L 462 demonstrated severe weight loss and death, monoassociated rats fed either 5010 C or L 485 showed little weight loss and no mortality. Before monoassociation, however, diet L 462 had appeared superior to diets 5010 C and L 485 judged on the basis of body weight. These experiments suggest a pronounced influence of diet on resistance to pathogenic microorganisms.

III. Effect of diet on elements of the reticulo-endothelial system of germfree mice.

To obtain further information on the effect of diet and dietary stimulation on the immune mechanisms of the germfree animal, germfree CFW mice were fed 4 different solid diets: L 462, L 474 E_{12} (13) 5010 C and L 485. The animals were sacrificed at 65 days of age and total serum protein, leukocyte counts, weight and histology of mesenteric lymph nodes and liver, bone marrow cellularity and anti-sheep RBC plaque forming cells (PFC background value (14) in the spleen determined. These values were compared to those obtained from comparable germfree mice reared on the "antigen-free", low molecular weight, water-soluble diet formula described by Pleasants (5), and to conventional mice maintained on diet L 485.

Table 1 shows part of these data. They indicate no difference in total serum protein between any of the groups; weight of the mesenteric nodes of all germfree groups are comparable, and are lower than those found in the conventional mice. Background anti-sheep RBC PFC values were also similar in all germfree groups, and lower than in the conventional mice. The major differences among the germfree groups, were found in the leukocyte pattern. Total leukocyte counts in germfree mice fed diets L 462 and L 474 E_{12} were comparable and approximately twice as high as in animals maintained on diets 5010 C and L 485. The latter values were again approximately two to three times higher than those found in the mice fed the water-soluble formula. Lymphocyte values followed this general pattern. Total granulocyte counts in animals reared on diet L 462 were statistically higher than in the animals reared on 5010 C or L 485 diet; counts in mice fed diet L 474 E_{12} were higher, but not significantly higher. Total granulocyte counts in mice maintained on the water soluble formula tended to be even lower than those in animals reared on solid diets 5010 C and L 474 E_{12}. Comparison with conventional mice reared on diet L 485 showed that the leukocyte pattern is affected far more by diet than by status.

Surprisingly little of the above difference was reflected by other elements of the reticulo-endothelial system (RES). In the bone marrow the only significant difference noted between the four

Table 1. Results obtained from germfree and conventional mice maintained on solid and water soluble, low molecular weight diets (see text). Number of animals in parenthesis. Mean and standard error given.

	GERMFREE										CONV.
	L 474 E_{12}		L 462		5010 C		L 485		H_2O-I	H_2O-II	L 485
	♂(9)	♀(10)	♂(10)	♀(10)	♂(10)	♀(10)	♂(10)	♀(10)	♂(4)	♂(4)	♂(20)
Body weight gm	26.8 ±0.5	21.2 ±0.7	28.8 ±0.8	21.7 ±0.3	28.3 ±0.9	22.9 ±0.5	26.8 ±0.4	22.4 ±0.7	-- --	21.3 ±1.0	24.7 ±0.8
Serum protein %	6.1 ±0.1	5.8 ±0.1	6.1 ±0.1	6.0 ±0.2	5.9 ±0.1	5.8 ±0.1	6.0 ±0.1	5.8 ±0.1	5.9 ±0.3	6.2 ±0.1	6.0 ±0.147
WBC x 10^3 Total	18.4 ±1.6	19.0 ±1.3	20.0 ±1.7	18.5 ±1.7	8.6 ±0.4	11.0 ±0.6	8.4 ±0.7	8.2 ±0.8	4.6 ±0.7	2.2 ±0.2	9.9 ±0.11
Agran.	15.3 ±1.6	16.8 ±1.3	17.0 ±1.5	16.7 ±1.6	7.0 ±0.6	9.7 ±0.6	6.0 ±0.5	7.3 ±0.7	3.1 ±0.4	1.8 ±0.2	7.5 ±0.2
Gran.	3.1 ±0.3	2.2 ±0.4	3.0 ±0.4	1.8 ±0.2	1.6 ±0.3	1.3 ±0.3	2.4 ±0.5	0.9 ±0.1	1.5 ±0.4	0.4 ±0.1	2.4 ±0.2
Weight mg % Mesenteric	47 ±3	69 ±7	38 ±4	82 ±5	45 ±5	73 ±5	45 ±6	67 ±6	--	77 ±3	--
Node			42±3[1]								134±11[1]
PFC	29 ±6	-- --	33 ±5	-- --	24 ±4	-- --	32 ±7	-- --	-- --	32 ±7	91 ±22[2]

1) Data from earlier experimental series (2)
2) Data from earlier experimental series (15)

solid diet groups was a significantly higher percentage of granulocyte precursors in germfree mice reared on diet L 474 E_{12}. Granulocyte levels on this diet approached the values found in the conventional animals. Histologically the lymph nodes of all four groups were quite comparable, with dense nodules predominating and an occasional inactive center visible in the nodule. Some plasma cells were present in the inactive centers, but few, if any, mitotic figures were in evidence. There were large areas of white pulp in all the spleens examined, but no active centers, few plasma cells, and few mitotic figures.

There was, however, a pronounced increase in Kupffer cells in the livers of the mice fed diet L 474 E_{12}, L 462 and the water soluble diet.

IV. Conclusions

The above data indicate that diet may be a major factor in the functional development of the RES. Comparison of the components of the various solid diets fed to germfree mice shows that both diets L 462 and L 474 E_{12} contain substantial amounts of casein and other bovine milk protein, while the other two (5010 C and L 485) contained little or none. Analysis of three different brands of casein routinely used for the production of rat and mouse diets demonstrated the presence of 7×10^8 to 5×10^9 nonviable microorganisms in these commercial products. In our opinion the incorporation of commercially produced casein into the diet of germfree animals may lead to extensive stimulation of the RES by nonviable microbial forms and possibly by the protein itself (16).

However, the data mentioned in section II indicate that dietary factors other than casein may have a strong influence on anti-microbial defense mechanisms. In the case of germfree rats monoassociated with *S. typhimurium* the rearing on diets 5010 C and L 485 appeared advantageous in comparison with the results obtained with diet L 462.

The water-soluble, low molecular, chemically defined and "antigen free" diets, which are sterilized by Millipore filtration without loss or chemical change would appear most ideally suited for immunological studies which, besides control of the microbial factor, require total absence of specific and non-specific dietary stimulation. However, these diets are expensive and few laboratories are, at present, equipped to use them. Whenever solid diets are fed it appears mandatory that these diets be well defined and reported in detail, in order to make an evaluation of dietary influences possible and to ascertain reproductibility of results. (This research was supported by a grant from the National Science Foundation #GB 6561.)

References

1. Gordon, H.A., E. Bruchner-Kardoss, T.C. Staley, M. Wagner and B.S. Wostmann. 1966. Acta Anatomica 64:301.

2. Olson, G.B. and B.S. Wostmann. 1966. J. Immunol. 97:267.

3. Thorbecke, G.J., H.A. Gordon, B.S. Wostmann and J.A. Reyniers. 1957. J. Inf. Dis. 101:237.

4. Wostmann, B.S. 1961. Ann. N.Y. Acad. Sci. 94:272.

5. Reddy, B.S., J.R. Pleasants and B.S. Wostmann. The Germfree Animal in Research (M.E. Coates, ed.) Academic Press, London, 1968. In Press.

6. Wostmann, B.S. and G.B. Olson. 1964. J. Immunol. 92:41.

7. Olson, G.B. and B.S. Wostmann. 1966. J. Immunol. 97:275.

8. Bosma, M.J., T. Makinodan and H.E. Walburg. 1967. J. Immunol. 99:420.

9. Baker, P.J. and M. Landy. 1967. J. Immunol. 99:687.

10. Michael, J.G., J.L. Whitley and M. Landy. 1961. J. Expt. Med. 115:131.

11. Wostmann, B.S. 1959. Ann. N.Y. Acad. Sci. 78:175.

12. Kellogg, T.F. and B.S. Wostmann. 1967. Gnotobiotics Newsletter 3:1.

13. Wostmann, B.S. and T.F. Kellogg. 1967. Lab. Animal Care 17:542.

14. Hege, J.S. and L.J. Cole. 1967. J. Immunol. 99:61.

15. Nordin, A.A. 1968. Proc. Soc. Exptl. Biol. Med. In Press.

16. Sell, S. 1964. J. Immunol. 93:122.

THE RESPONSE OF GERMFREE RATS TO DIETARY CHOLESTEROL

Thomas Kellogg and Bernard Wostmann

Department of Microbiology

University of Notre Dame, Notre Dame, Indiana

Male germfree (G.F.) and conventional (Conv.) rats 90 to 120 days of age were fed purified diets (1) (casein-rice starch with 5% corn oil) containing 0%, 0.05%, 0.1% and 0.5% cholesterol. After a minimum of two weeks, the animals were sacrificed. Serum, liver, and intestinal tract total cholesterol were determined by a modification of the $FeCl_3$ procedure of Caraway (2).

The results are shown in table 1. The serum cholesterol levels were similar in all groups regardless of dietary cholesterol or microbiological status. Kellogg and Wostmann (3) have recently reviewed the available data on serum and liver cholesterol levels of comparable germfree and conventional mammals and birds. Although there is some disagreement, the best controlled experiments indicate little if any consistent difference between serum cholesterol levels of germfree and conventional groups.

Table 1 shows that liver cholesterol levels were subject to an interaction between microbiological status and dietary cholesterol levels. Conventional rats receiving 0.1% cholesterol with the diet showed no increase in liver cholesterol concentration over their sterol free diet fed counterparts, whereas in comparable G.F. rats, liver cholesterol levels increased by 360%. Conventional rats fed the 0.5% cholesterol diet had liver cholesterol concentrations similar to those in the germfree group fed 0.1% cholesterol in the diet (Approx. 1100 mg/100g liver) while the germfree rats fed at the 0.5% cholesterol level reached much higher values. These data are portrayed graphically in fig. 1. The G.F. liver cholesterol values are approximately 3 times as high as the Conv. values for each level of dietary cholesterol above 0%.

Table 1

SERUM AND LIVER CHOLESTEROL CONTENT OF GERMFREE AND CONVENTIONAL RATS FED DIET 474-E12 WITH VARYING AMOUNTS OF CHOLESTEROL

PERCENT CHOLESTEROL ADDED TO DIET	LIVER CHOLESTEROL mg/100 g		SERUM CHOLESTEROL mg/100ml	
	GERMFREE	CONVENTIONAL	GERMFREE	CONVENTIONAL
none	304	287	117	118
0.05	579	---	105	---
0.10	1119	296 425	118	89
0.50	2957	1070	120	109

Figure 1

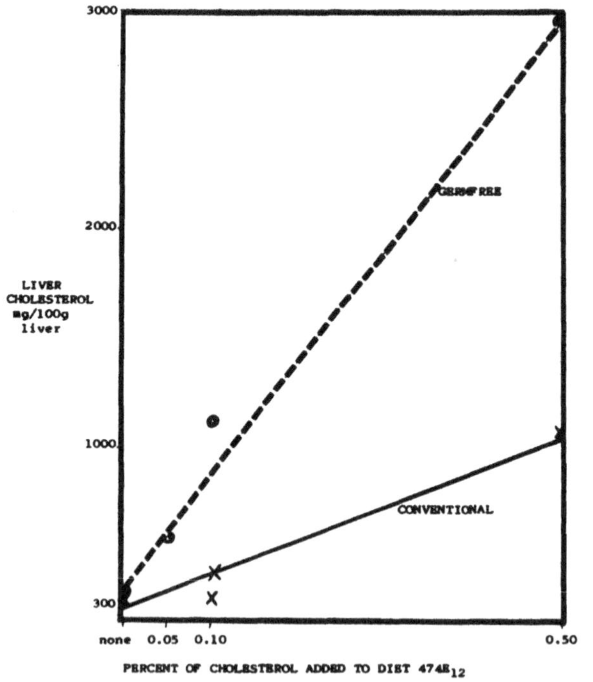

REGRESSION OF CHOLESTEROL CONCENTRATION IN LIVER VS. AMOUNT OF DIETARY CHOLESTEROL FED MALE GERMFREE AND CONVENTIONAL RATS

These data suggest a marked protective effect of the intestinal microflora on liver cholesterol accumulations resulting from dietary cholesterol. They also point out that, in the rat, liver and blood can not be considered as one pool with regard to response to exogenous cholesterol since liver cholesterol in the various groups varied over a range of a factor of ten with no appreciable variation in serum cholesterol. (This research was supported by grants from the National Institutes of Health #HE 08351 and from the American Heart Association.)

References

1. Wostmann, B.S., and T. F. Kellogg. Purified starch-casein diet for nutritional research with germfree rats. J. Lab. Animal Care. 17:589. (1967)

2. Caraway, W.T. Microchemical Methods of Blood Analysis pp. 82-85 Charles C. Thomas, Springfield, Illinois, 1960.

3. Kellogg, Thomas F. and Bernard S. Wostmann. The Metabolism of Lipids in Germfree Animals. in The Germfree Animal in Research. M.E. Coates Ed. Academic Press, London, 1968.

THE ABSORPTION OF FAT BY GERMFREE AND CONVENTIONALIZED RATS[1]

Mario Reina-Guerra, B. Tennant, D. Harrold, and M. Goldman

Department of Clinical Sciences and Radiobiology Laboratory, School of Veterinary Medicine, University of California, Davis, California

Numerous differences in the morphology of the gastrointestinal tract have been demonstrated between germfree and conventional animals (1-7). Differences have also been observed in the intestinal absorption of xylose (8), thiamine (9), and methionine (10), all of which are more rapidly absorbed in germfree animals. The balance studies of Evrard, et al. (11), suggest that utilization of dietary fat is accomplished more efficiently in germfree rats although Wiech, et al. (12) were unable to demonstrate differences between germfree and conventional rats in the absorption of linoleic acid.

We became interested in fat absorption following observations by Levenson and Tennant (13) which indicated germfree rats had significantly less body fat than conventionalized rats. To evaluate the importance of possible differences in fat absorption, we have studied the intestinal absorption of oleic acid ^{131}I and triolein ^{131}I in germfree and conventionalized rats. A summary of these investigations is presented below.

[1]This investigation was conducted as part of Western Regional Research Project W-88 and was supported in part by RRF-2279.

MATERIALS AND METHODS

Germfree Fischer strain rats were obtained[2] at weaning and were divided according to weight and sex into 2 equal groups. One group remained germfree. The other was transferred to a separate sterile isolator and was "conventionalized" using a procedure similar to that previously described (13). This procedure consisted of contaminating the drinking water with fresh cecal contents obtained from a rat out of an ordinary open animal room. Subsequently open animal room rats were introduced into the cages of the conventionalized rats and were allowed to live in association with these rats for approximately 1 month.

Both germfree and conventionalized groups were raised in Trexler type isolators and were fed diet L-356 (14)[3] and distilled water ad libitum. Diet and water were sterilized for 1 hour at 122° C and were entered into the isolators through peracetic acid sterilocks. A multivitamin preparation[4] containing 5 mg. thiamine HCl per ml was also entered through locks sterilized with peracetic acid and was added to the drinking water at the rate of 0.1 ml per 100 ml.

At 4 1/2 to 5 months of age, germfree and conventionalized rats were transferred into individual stainless steel cages and food and water were withheld over night. The following morning each rat received eith 1 μC of oleic acid -^{131}I in 0.2 ml of oleic acid carrier or 1 μC of triolein ^{131}I in peanut oil carrier by intragastric tube. Six hours following administration, the rats were killed and the amounts of radioactivity in the stomachs, small intestines, ceca, and colons were determined.

RESULTS

A greater percentage of the administered triolein ^{131}I was found in the stomachs of germfree rats (45.0%) than in conventionalized rats (23.4%), a difference which was significant at the 0.01 level. Germfree rats also retained a greater amount of oleic acid ^{131}I (36.0%) than did conventionalized rats (26.2%) although this difference was not signficnant at the 0.05 level.

---

2 Charles River Breeding Laboratories, Wilmington, Massachusetts.
3 General Biochemicals, Chagrin Falls, Ohio.
4 Vi-syneral, U.S. Vitamin and Pharmaceutical Corp., New York. N.Y.

No significant differences were observed between germfree and conventionalized rats in the amounts of either triolein ^{131}I or oleic acid ^{131}I found in the small intestine. In the oleic acid trial, 8.7% was found in the small intestine of germfree rats and 7.9% in the conventionalized. In the triolein trial, 9.9% of the administered dose was present in the small intestine of the germfree rats and 12.1% in the conventionalized.

The ceca of germfree rats contained 0.8% of the oleic acid and 0.3% of the triolein. In conventionalized rats, 3.5% of the oleic acid and 4.4% of the triolein were found in the ceca. These differences in the amounts of oleic acid and triolein found in the ceca were significant at the 0.05 and 0.01 levels respectively.

In order to compare the amounts of oleic acid ^{131}I and triolein ^{131}I absorbed from the intestine of germfree and conventionalized rats, it was necessary to correct for differences in gastric emptying. When these corrections were made using the procedure of Makela, et al. (15), no differences were found in the net absorption of either oleic acid or triolein. Of the administered oleic acid which left the stomach during the experimental period, 84% was absorbed by germfree rats and 85% by conventionalized rats. Germfree rats absorbed 78% of the triolein which cleared the stomach, conventionalized rats absorbed 77%.

DISCUSSION

Investigations with germfree rats have demonstrated several important influences which the intestinal microflora has on lipid metabolism. Germfree rats have less body fat than conventionalized rats (13) but usually have greater serum and liver cholesterol concentrations (16). Cholesterol turnover rate and bile acid turnover rate (17) are significantly increased in germfree rats.

Wiech, et al., observed that the rate of intestinal absorption of linoleic acid was the same in germfree and conventional rats (12). The results of our studies are similar and show that oleic acid is absorbed by germfree and conventionalized rats at the same rate. This information along with the data demonstrating similar absorption of triolein suggests that triglyceride lipolysis also occurs at the same rate in germfree and conventionalized rats.

Abrams has recently demonstrated that gastrointestinal tract motility is significantly reduced in germfree mice (18). We also observed a decreased rate of gastric emptying and an apparent delay in intestinal transit in germfree rats. These data suggest that microorganisms have a significant role in determining normal gastrointestinal motility.

SUMMARY

The intestinal absorption of oleic acid ^{131}I and triolein ^{131}I was studied in germfree and conventionalized rats. Gastric emptying of both compounds was delayed in germfree rats and correspondingly, the amount of radioactivity which reached the cecum was reduced significantly. When corrections were made for differences in gastric emptying, there were no significant differences between germfree and conventionalized rats in net intestinal absorption of either oleic acid ^{131}I or triolein ^{131}I. In these experiments, intestinal microorganisms did not appear to influence the rate of fat absorption directly but significantly influenced the transit of fat in the gastrointestinal tract.

REFERENCES

1. Abrams, G. D., Bauer, H., and Sprinz, H., Lab Invest., 12: 355, 1963.
2. Dupont, J., Jervis, H. R., and Sprinz, H., J. Comp. Neurol., 125: 11, 1965.
3. Geever, E. F. and Levenson, S. M., Experientia 20: 391, 1964.
4. Gordon, H. A. and Bruckner-Kardoss, E., Am. J. Physiol., 201: 175, 1961.
5. Nakao, K. and Levenson, S. M., Experientia, 23: 494, 1967.
6. Sprinz, H., Kundel, D. W., Dammin, G. J., Horowitz, R. E., Schneider, H., and Formal, S. B., Am. J. Path., 39; 681, 1961.
7. Wostmann, B. and Bruckner-Kardoss, E., Am. J. Physiol. 197: 1934, 1959.
8. Heneghan, J. B., Am J. Physiol., 205: 417, 1963.
9. Gordon, H. A., Bruckner-Kardoss, E., and Kan, D., p. 21, In abstracts, 5th International Congress of Nutrition, September 1-7, 1960, Washington, D.C.
10. Herskovic, T., Katz, J., Floch, M. H., Spencer, R. P., and Spiro, H. M., Gastroenterology, 52: 1136 (Abstract).
11. Evrard, E., Hoet, P. P., Eyssen, H., Charlier, H., and Sacquet, E., Brit. J. Exp. Path., 43: 409, 1964.
12. Wiech, N. L., Hamilton, J. G., and Miller, O. N., J. Nutr., 93: 324, 1967.
13. Levenson, S. M. and Tennant, B., Fed. Proc. 22: 109, 1963
14. Larner, J. and Gillespie, R. E., J. Biol. Chem., 225: 279, 1957
15. Makela, T. E., Hakkila, J., and Siurala, M., Ann. Med. Exper. Biol. Fenn 40: 231, 1962.
16. Wostmann, B.S. and Wiech, N. L., Am. J. Physiol., 201, 1027, 1961.
17. Wostmann, B. S., Wiech, N. L., and King, E., J. Lipid Res., 7: 77, 1966.
18. Abrams, G. D. and Bishop, J. E., Proc. Soc. Exp. Biol. Med., 126: 301. 1967.

PROTEIN METABOLISM IN GERMFREE RATS FED CHEMICALLY DEFINED, WATER-SOLUBLE DIET AND SEMISYNTHETIC DIET

Bandaru S. Reddy, B.S. Wostmann and J.R. Pleasants

Lobund Laboratory, Department of Microbiology,

University of Notre Dame, Notre Dame, Indiana

The importance of germfree animals and chemically defined, water-soluble diets developed at Lobund Laboratory in nutritional and immunological studies was discussed in detail by Pleasants (1). Nutritional adequacy of this diet has been demonstrated by the fact that it supported reproduction of germfree mice into fifth generation and the growth of germfree rats from birth through maturity (2). Although these results indicate that the diet is at least qualitatively adequate for germfree rats and mice, it is of considerable importance to determine if the animal is physiologically "normal" in a chemical environment so limited and so defined. So, as a further test of nutritional adequacy and physiological normality, germfree and conventional rats fed this diet were compared to similar rats fed a semisynthetic diet based on casein-starch using morphology, serum amino nitrogen, pancreatic enzyme levels, liver protein, nitrogen balance and protein efficiency ratio (PER) as criteria. In addition, these studies would indicate to what extent the intestinal microflora could influence the protein metabolism of the host.

METHODS AND MATERIALS

Germfree and conventional male albino rats of Lobund stock (Wistar origin) used in this investigation were genetically closely related. All rats were housed in individual screen bottom cages and maintained as described in our earlier report (2). Following weaning, germfree and conventional rats were fed ad libitum one of the following experimental diets: a Millipore-filtered, chemically defined, water-soluble diet (L-474E9) based on amino acids and glucose (2) and a steam-sterilized, semisynthetic diet (L-474E15) based on purified casein and starch (2). At 80 days of age, the animals were transferred to individual metabolism cages, and feces and urine collected for 10 days after an adjustment period of 7 days.

After the experimental period, the rats were starved for 4 hours with free access to water and bled from the heart under ether anesthesia. Then the organs were removed and weighed. Pancreatic glands were used for histopathology.

Amino acid nitrogen and glucose in plasma were determined by the methods of Frame et al. (4), and Washko and Rice (5) respectively. Plasma protein was measured using the Atago refractometer. Liver protein was determined by biuret method (6), and total nitrogen in the diet, feces and urine and true protein in urine (after TCA precipitation) by micro-Kjeldhal method. Lipase (7) amylase (8) trypsinogen and chymotrypsionogen (9) activities were measured in the supernatant fraction from the pancreas homogenized with ice-cold 0.15 M NaCl and centrifuged at 20,000 X g. Trypsinogen and chymotrypsinogen in pancreas were activated with enterokinase.

RESULTS AND DISCUSSION

The body weights of both germfree and conventional rats fed liquid diet were slightly lower than in those fed semisynthetic diet. Organ weights (liver, spleen, kidney, pancreas, lungs and heart) expressed per unit of body weight, were not affected by diet except for a slightly lower liver weight in germfree rats fed liquid diet. The free amino acid pool, as represented by free amino acid N (ranged from 6.06 to 6.36 mg N/100 ml), and glucose (ranged from 113 to 124 mg/100 ml) in the plasma were similar in all groups. Plasma protein levels were slightly lower on liquid diet (Gf.5.89; Conv.6.11 g/100 ml) than on semisynthetic diet (GF.6.00; Conv.6.49 g/100 ml) under both germfree and conventional conditions. Germfree status and diet had no appreciable effect on liver protein levels (ranged from 553 to 623 mg/100g body weight). However, these levels were slightly increased in germfree rats fed semisynthetic diet as compared to germfree rats fed liquid diet (623 vs. 553 mg/100g body weight).

The results of nitrogen balance studies are summarized in table 1. Germfree animals showed higher fecal N excretion and lower urinary N excretion than their conventional controls on either diet. Since some part of the fecal N is of microbial origin, this data indicate that the germfree rats excreted more non-microbial N than the controls. Increased fecal N excretion was unlikely to be the result of decreased absorption of amino acids, because Herskovic et al. (10) showed an increase in methionine absorption in germfree animals compared to conventional animals. But it appears probable that increased fecal N excretion might be due to excretion of partially degraded endogenous protein, since in germfree rats, there was no microbial degradation of protein and subsequent microbial deamination of amino acids.

Table 1

Nitrogen retention and protein efficiency ratio in germfree and conventional rats

	Germfree		Conventional	
	L-479E9	L-474E15	L-479E9	L-474E15
PER[a]	0.79	0.72	0.62	0.60
	mg N/ day/ 100 g body weight			
Intake	126	192	163	179
Feces	10	34	7	14
Urine	77	116	126	133
Retention	34	42	30	32
Urinary true protein N	1.00	0.85	0.76	0.57

[a]Protein efficiency ratio = $\frac{\text{gram weight gain}}{\text{gram protein consumed}}$

Germfree status and also feeding of semisynthetic diet slightly increased nitrogen retention compared to their respective counterparts. On the other hand, germfree status and also feeding of liquid diet increased protein efficiency ratio (PER) compared to their respective controls. An interesting finding of this investigation is that there was an increase in urinary excretion of protein (TCA precipitable) in germfree rats fed either diet as compared to their controls. This finding is suggestive of changes in filtration and resorption of proteins (probably low molecular weight) through the kidney of germfree animal.

The results presented in table 2 show that lipase and amylase levels in the pancreas were not affected either by germfree status or by diets. Germfree status had no effect on trypsinogen and chymotrypsinogen levels. However, there was an increase in trypsinogen and chymotrypsinogen activities in rats fed semisynthetic diet over those animals fed liquid diet.

Table 2

Pancreatic enzymes in germfree and conventional rats

Enzyme	Germfree		Conventional	
	L-479E9	L-474E15	L-479E9	L-474E15
	Units/ pancreas/ 100 g BW			
Lipase[a]	493	560	599	545
Amylase[b]	1374	1545	1556	1407
Trypsinogen[c]	720	943	749	955
Chymotrypsinogen[d]	896	1194	913	1072

[a] Unit of activity is expressed as micromoles of fatty acid liberated per minute
[b] Unit of activity is expressed as micromoles x 10^3 of maltose liberated per minute
[c] Unit of activity is expressed as micromoles of TAME hydrolyzed per minute
[d] Unit of activity is expressed as micromoles of BTEE hydrolyzed per minute

In contrast to the observations of Geever et al. (11) histopathological examination of pancreas from the germfree and conventional rats fed liquid diet showed no evidence of microscopic lesions indicative of pancreatic dysfunction. However, these investigations (11) fed a liquid diet lacking in chromium and containing Tween 80. That neither visible and microscopic pathological changes in the pancreas nor any major changes in the levels of enzymes in the pancreas were detected in both germfree and conventional rats fed liquid diet indicates no impairment of the functional capacity of the pancreas as a result of feeding liquid diet.

These results clearly provide further evidence that the liquid diet developed in our laboratory is nutritionally adequate for germfree and conventional rats. Germfree rats utilized nitrogen more efficiently than the conventional animals whether it is presented in the form of amino acids or intact protein.

(This study was supported by NIH grants HD 00855 and AM 11300).

REFERENCES

1. Pleasants, J.R. 1966 Development of chemically defined, water-soluble diets nutritionally adequate for germfree rats and mice. Ph.D Thesis, University of Notre Dame, Notre Dame, Indiana.

2. Reddy, B.S., B.S. Wostmann and J.R. Pleasants 1968 Nutritionally adequate diets for germfree animals. In: The Germfree Animal in Research, ed., M.E. Coates. Academic Press, London (in press).

3. Reddy, B.S., and B.S. Wostmann 1966 Intestinal disaccharidase activities in the growing germfree and conventional rats. Arch. Biochem. Biophys., 113: 609.

4. Oser, B.L. 1965 Hawk's Physiological Chemistry. McGraw-Hill Book Company, New York, p. 1048.

5. Washko, M.E., and E.W. Rice 1961 Determination of glucose by an improved enzymatic procedure. Clin. Chem., 7: 542.

6. Inchiosa, Jr., M.A. 1964 Direct biuret determination of total protein in tissue homogenates. J. Lab. Clin. Med., 63: 319.

7. Roe, J.H., and R.E. Byler 1963 Serum lipase determination using one-hour period of hydrolysis. Anal. Biochem., 6: 451.

8. Dahlqvist, A. 1962 A method for the determination of amylase in intestinal content. Scand. J. Clin. Lab. Invest., 14: 145.

9. Hummel, B.C.W. 1959 A modified spectrophotometric determination of chymotrypsin, trypsin and thrombin. Canadian J. Biochem. Physiol., 37: 1393.

10. Herskovic, T., J. Katz, M.H. Floch, R.P. Spencer and H.M. Spiro 1967 Small intestinal absorption and morphology in germfree, monocontaminated and conventionalized mice. Gastroenterology, 52: 1136 (abstract).

11. Geever, E.F., F.S. Daft and S.M. Levenson 1965 Pancreatic atrophy and fibrosis in the germfree rat on a chemically defined diet. Federation Proc., 24: 246 (abstract).

SUDDEN DEATH IN GERMFREE MICE REARED THROUGH SUCCESSIVE GENERATIONS ON CHEMICALLY DEFINED LIQUID DIET

J.R. Pleasants, B.S. Reddy and B.S. Wostmann

Lobund Laboratory, University of Notre Dame

Notre Dame, Indiana

A diet for germfree animals which consists of chemically defined, low-molecular-weight nutrients, and which can be prepared in liquid form for sterilization by filtration, brings the experimental animal one step farther toward complete control of nutritional and antigenic variables. Such a diet, consisting of purified amino acids, sugars, ethyl linoleate, vitamins and minerals, all dissolved in a single water solution, has been fed to germfree rats (1) and to germfree mice (2,3). The fat-soluble portion of the diet had been solubilized in the water solution by use of Polysorbate 80. These preliminary experiments had implicated the nonionic detergent as a cause of persistent diarrhea in some mice. Changes in the lymph nodes of germfree mice fed the detergent diet also resembled those observed by Mori and Kato (4) in conventional rats fed polysorbate. For these reasons the detergent was eliminated and the fat-soluble portion of the diet was prepared and fed separately from the water-soluble portion, according to a procedure reported at the 1966 meeting of this Association. Diet prepared in this way has now been subjected to long-term testing of its nutritional adequacy.

METHODS

Animal Maintenance. Germfree CFW mice from the Lobund closely bred colony were offered the two components of the diet from weaning throughout life. They were housed singly except during mating. Housing consisted of stainless wire cages, with filter paper nesting material provided only for nursing mothers. The cages were maintained in Trexler-type isolators sterilized with peracetic acid. The young born to mice on this diet were allowed to suckle their mothers until weaning to defined diet. Thus, the defined diet and mother's milk were the only two foods ever consumed by mice of the second and later generations. The water-soluble component was offered in overhead brown bottles having holes 1/16" in diameter drilled in the plastic lid. The fat component was offered in overhead tubes or in small dishes, but consumption of the fat component was very erratic. Various flavors were added to increase consumption, but without success. On the basis of recent reports (5,6) that rodents develop a taste for foods because of the feeling of well-being that occurs within an hour or less after consumption of the food, it was decided to mix the fat component with some dry dextrose, which could provide such an immediate sense of well-being. Three times weekly, 0.15 gm of the fat component (Table 1) was absorbed into just enough dry dextrose to form a paste. The dry dextrose was prepared by evaporating to dryness some of the filtered 30% dextrose solution already in the isolator.

Diet composition. The composition of the fat component is shown in Table 1. The water-soluble component consisted of 3 solutions, one of amino acids and B vitamins (Solution 1, Table 2) one of sugars and minerals (Solution 2, Table 3) and one of pure 30% dextrose. Solutions 1 and 2 were filtered and stored separately because this prevented most of the browning which had occurred when the entire diet was stored as a single solution. The dextrose solution was filtered separately so that the sugar content of the diet (as percent of solids) could be varied for weanling mice. Experience had shown that weanlings adjusted better to the diet when it contained less sugar and less total solids than the final diet. The changing proportions of the 3 water solutions are shown in Table 4. These were mixed as needed for feeding, so that the complete mixture was not more than 2 days old when fed. The individual solutions, however, were stored in the isolator as long as 2 weeks. Water was always available to the animals from separate overhead bottles.

Except where noted, the organic ingredients of the diet were of National Research Council grade and were obtained from General Biochemicals, Inc., Chagrin Falls, Ohio. The inorganic ingredients were Fisher Certified Reagents or equivalent quality. The procedure for preparing the various water solutions may be unnecessarily complex at this time, but there is still an occasional problem of precipitation in the separate solutions. All diet solutions were filtered into the isolator through a Millipore pressure filter holder with 0.22 micron pore size filter. After autoclaving of the filter holder, it was inserted into a Tygon flare (sprayed with peracetic acid) on top of the isolator.

Table 1. Composition of fat supplement LADEK-64 fed germfree CFW mice (0.15 gm three times weekly = 0.065 gm/day)

	mg/gm	IU/gm	mg/day
Ethyl linoleate >99% pure*	970		63
DL-α-tocopheryl acetate	20	20	1.3
DL-α-tocopherol	10	11	0.65
Vitamin K_1 (phytyl)	1.1		0.07
	μgm/gm	IU/gm	IU/day
Vitamin A palmitate, pure**	650	1200	78
Vitamin D_3, pure	0.7	28	1.8

*Hormel Institute, Austin, Minn.
**Hoffman-La Roche, Nutley, N.J.

Table 2. Solution 1 of water-soluble component (diet L-479E10)

To 70 ml glass-distilled water just below 100° C add:

	gm		gm
L-Phenylalanine	0.90	L-Asparagine	1.20
L-Leucine	0.80	L-Lysine HCl	1.25
L-Methionine	0.85	L-Arginine HCl	0.75
L-Tryptophan	0.40	L-Threonine	0.50
L-Isoleucine	0.50	L-Histidine HCl · H_2O	0.55
L-Valine	0.70		

Cool solution to room temperature and add:

Glycine	0.50	L-Serine	1.55
L-Proline	3.00	L-Alanine	0.75
L-Tyrosine ethyl ester HCl		in 1 ml water	1.00
Monosodium L-glutamate			6.00
NaCl-KI mixture (0.377 gm NaCl + 0.003 gm KI)			0.38
Choline Cl			0.25
B vitamin mix 111E2*			0.065
Potassium acetate dissolved in 2 ml water			1.06

Add water to solution 1 to total volume of 83 ml.

*B vitamin mixture (0.065 gm) contains:

	mg		mg
Thiamin HCl	0.50	Calcium pantothenate	5.00
Riboflavin	0.75	p-Aminobenzoic acid	30.00
Pyridoxine HCl	0.63	Biotin	0.10
Niacin	3.75	Folic acid	0.15
i-Inositol	25.00	Cyanocobalamin (B_{12})	0.03

Table 3. Solution 2 of water-soluble component, for diet L-479 variants E10, E12, and E16

Make following solutions separately in glass-distilled water:

	ml water	gm E10	E12	E16
a) D-Dextrose	44	21.50	21.70	22.40
b) Ferrous gluconate	2.5	0.035	0.035	0.035
c) Salts 31*	2.5	0.036	0.036	0.036
d) Mono Ca fructose$(PO_4)_2$	12	5.0	4.5	3.2
e) $CaCl_2 \cdot 2\ H_2O$	1.5			0.6
f) Mono Mg fructose$(PO_4)_2$	2.0	0.5	0.75	
$MgCl_2 \cdot 6\ H_2O$				0.75

Add b to c, mix and add at once to a, then add d, e, and f to abc. Finally add water to toal volume of 83 ml.

Salts 31 (0.036 gm) contains:

	mg		mg
$Mn(acetate)_2 \cdot 4\ H_2O$	26.0	$(NH_4)_6Mo_7O_{24} \cdot 4H_2O$	0.6
$ZnSO_4 \cdot H_2O$	5.5	Na_2SeO_3	0.011
$Cu(acetate)_2 \cdot H_2O$	2.5	$Cr(acetate)_3 \cdot H_2O$	0.48
$Co(acetate)_2 \cdot H_2O$	0.9		

Table 4. Proportions of different water solutions used during weaning of germfree mice to water-soluble diet.

Days of Age	ml Sol. 1	ml Sol. 2	ml 30% Dext.*	ml H_2O	% Solids	% Sugar in Solids
21 - 25	1	1	-	1	20	45
25 - 30	1	1	1	1	22.5	63
30 - 35	1	1	1.5	0.5	26	69
After 35	1	1	2	-	30	72

* The 30% dextrose solution is prepared by dissolving 50 gm D-Dextrose in 100 ml glass-distilled water and adding water to a total volume of 167 ml.

RESULTS

From 7 female germfree mice started on the diet at weaning, 12 litters of 2nd generation mice were weaned. Five litters of 3rd generation, 2 litters of 4th generation, and 1 litter of 5th generation mice were eventually weaned. By the stringent test of reproduction into successive generations, it thus appears that the diet must contain at least minimal levels of every essential nutrient. These findings are extremely encouraging for their demonstration that there are no unknown nutrients required for the maintenance of germfree mice. The immunological results were also encouraging in that 2-months-old germfree mice of the second and later generations showed no detectable gamma globulin by immunoelectrophoresis. This suggests that dietary antigens have been the main stimulus for the residual gamma globulins seen in germfree mice on solid diets.

Despite the evidence of qualitative nutritional adequacy, however, an imbalance of some nutrients, or a marginal level of some nutrient(s) was indicated by the occurrence of sudden deaths among many members of the later generations. These deaths tended to occur around 4-5 months of age in the first 2 litters of the 2nd generation. Among later litters and in the later generations, the age at death reached a plateau at a mode of 65 days. The first deaths occurred while the fat intake was very erratic, and a deficiency of fat or fat-soluble vitamins was suspected. The sudden death syndrome continued, however, after fat intake had been stabilized by the incorporation of the fat component of the diet into dry dextrose.

Suddenness of death was the most consistent observation in 73 out of 85 spontaneous deaths observed in the later generation animals. The other 12 spontaneous deaths were associated with cecal volvulus. The animals dying suddenly had shown no weight decline or loss of appetite in the week preceding death. Only one animal was ever observed in a moribund condition (hind leg paralysis). The dead mice were found lying face down with hind legs fully extended behind them. At necropsy, nearly all animals showed hyperemia of the lungs. Many also showed congestion of the abdominal vasculature. A histological study of hearts and kidneys in these animals was made by Dr. Daniel Farnell, of the Department of Veterinary Pathology, Michigan State University. Most of the animals showed mild degrees of calcification in the kidneys and in heart muscle. A few showed severe calcification of the kidneys. These changes suggested a magnesium deficiency. In four mice sacrificed at 2 months of age while still in good health, serum Mg levels did not differ significantly from the level found in conventional mice fed commercial diets (about 3 mg Mg per 100 ml serum). However, the one mouse sacrificed while moribund had a serum Mg level of 1.95 mg/100 ml.

Among the 3rd and 4th generation females, all 9 spontaneous deaths occurred when the female was either in late pregnancy or was nursing a litter. In the 2nd generation animals, only 5/30 had died near the time of parturition. Although necropsy showed that the mice dying near parturition had stomachs moderately well filled with the filter paper bedding supplied to them at that time, it was not obvious how this might have affected the animals.

Although male mice died at about the same age as the female mice, some of the males showed gross changes which were never observed in the females. Seven males had severe hydronephrosis, and 21 showed at least moderate distention of the bladder with urine. When the urinary tract was further dissected, by splitting the symphysis pubis, it was found that the urethra was filled and distended with a translucent, rubbery cast of proteinaceous material. Out of 38 males which died suddenly and which had the entire tract dissected out, 32 (84%) showed the urethral casts. Two-thirds of these 32 showed some distention of the bladder with urine, and 22% had hydronephrosis.

In order to determine if this condition was peculiar to male mice fed the defined diet, a spot check was made of both conventional and germfree males fed either natural or semi-synthetic solid diets. Such casts were observed in 50/51 mature conventional males, and retention of moderate amounts of urine in the bladder was observed in about one third of them. Yet the health and fertility appeared unimpaired. One old male, however, had 6 ml of urine in his bladder and was suffering from hydronephrosis. Among germfree males, animals with casts made up at least 40% of each group examined, with one exception. Among 4 young adults fed diet L-462, consisting of grains and milk proteins (7), none showed urethral casts.

DISCUSSION

As mentioned above, the diet must be qualitatively adequate, i.e., it cannot be totally lacking in any essential nutrient. Any unknown nutrients stored in the bodies of the first generation animals (whose mothers were fed natural-type diets) could hardly suffice to maintain reproduction into the 5th generation. However, such continued reproduction is not incompatible with a marginal level of some nutrient. The fact that the sudden death syndrome occurred earlier in later generations, and the fact that it occurred at the end of the period of rapid growth, or during the nutritionally demanding periods of pregnancy and lactation, suggest such a marginal deficiency. Sudden death and the occurrence of soft tissue calcification have been reported in animals fed magnesium deficient diets (8,9).

The level of Mg in the diet L-479E10 on which the colony was first maintained is the same as that used by Greenstein group who devised the first water-soluble diet and fed it to conventional rats through 3 generations (10). The Mg level corresponds to that considered adequate for rats (11). The Mg requirement for mice has not been worked out. Part way through the experiment, the Mg level was increased by 50% over its previous level, (diet L-479E12, Table 3), but the sudden death syndrome continued to occur. A recent publication by Dubos reported sudden deaths, among pregnant and nursing mice only, in a colony of conventional mice fed a diet containing 50 mg Mg/100gm solids (12). These deaths were prevented by doubling the Mg level. A further increase in the Mg level of the defined diet was therefore made (L-479E16, Table 3) and the levels of Ca and P were reduced to prevent possible interference with Mg utilization by the previous high levels of Ca and P. In E16, the magnesium fructose diphosphate, whose Mg content varied from batch to batch, was replaced by magnesium chloride. The actual levels of the elements are shown below:

Table 5. Levels of Mg, Ca, and P in 3 variants of L-479 diet.

	mg/100 gm solids			Ratios	
	Mg	Ca	P	Ca/Mg	Ca/P
E10	35	590	850	17	0.69
E12	53	530	810	10	0.65
E16	90	540	500	6	1.08

The E16 diet has been only recently adopted. The colony was depleted by deaths and by accidental contamination, so that no animals have reached the susceptible age range since adoption of the new formula. An inbred C3H colony is also being started on the diet in order to reduce further the antigenicity of the environment by reducing the antigenicity of inter-animal contacts. Results with the old and new colonies should help determine the adequacy of diet L-479E16.

The urethral casts observed in male mice cannot be the cause of the sudden death syndrome since this occurred in both sexes. Similar casts have been reported in conventional rats (13) as resulting from the premature coagulation within the male urinary system of the material which forms a vaginal plug in the female after mating. This material results from the secretions of the seminal vesicles and the coagulating glands. It has been reported (14) that the coagulation process is sensitive to ionic concentrations. Thus the mineral balance of the diet may be important in the coagulation process. Although the induction of these urethral casts is not peculiar to the defined diet, the associated degree of urine retention and hydronephrosis was more severe than in germfree and conventional mice fed solid diets. The fact that germfree, but not conventional, male mice fed diet L-462 showed no urethral casts, would suggest that the germfree mouse may provide a good experimental system for determining the factors which control cast formation. Further studies are planned to discover these factors, so that urethral cast formation can be prevented in germfree mice fed defined diets.

It may be concluded that diet L-479E10 or E12 is qualitatively but not quantitatively adequate for maintenance of the germfree mouse. Sudden death, and a mild degree of soft tissue calcification in young adult animals, suggest a magnesium deficiency or an imbalance of other minerals which might affect magnesium utilization. Occurrence of urethral casts in adult male mice may also reflect a mineral imbalance. Further studies with a new formulation, diet L-479E16 are continuing, but definitive results are not yet available.
(This research was supported by NIH grants HD00855 and AM11300)

REFERENCES

1. Pleasants, J.R., B.S. Wostmann and D.R. Zimmerman 1964 Improved hand rearing methods for small rodents. Lab. Animal Care 14:37-47.

2. Pleasants, J.R., G.B. Olson, B.S. Reddy and B.S. Wostmann 1964 Growth, reproduction and serum proteins in germfree rats and mice fed chemically defined, water-soluble diet. Fed. Proc. 23:408 (abstract).

3. Wostmann, B.S., J.R. Pleasants and B.S. Reddy 1967 Water-soluble, non-antigenic diets. In: Husbandry of Laboratory Animals, ed. M.L. Connally. Academic Press, London.

4. Mori, T., and S. Kato 1959 Effects of orally administered "Tween 80" on the mesenteric lymph nodes of intact and adrenalectomized mice. Tohoku J. Exptl. Med. 69:197-209.

5. Garcia, J., F.R. Ervin, C.H. Yorke and R.A. Koelling 1967 Conditioning with delayed vitamin injections. Science 155: 716-718.

6. Lepkovsky, S. 1966 Discrimination and selection of food. Food Technol. 20:1430-1437.

7. Zimmerman, D.R., and B.S. Wostmann 1963 Vitamin stability in diets sterilized for germfree animals. J. Nutrition 79:318-322.

8. Kruse, H.D., E. Orent and E.V. McCollum 1932 Studies on magnesium deficiency in animals. J. Biol. Chem. 96:519-539.

9. Goldsmith, L.A. 1967 Relative magnesium deficiency in the rat. J. Nutrition 93:87-102.

10. Greenstein, J.P., S. M. Birnbaum, M. Winitz and M.C. Otey 1957 Quantitative nutritional studies with water-soluble, chemically defined diets. Arch. Biochem. Biophys. 72:396-416.

11. National Research Council 1962 Nutrient Requirements of Laboratory Animals. National Academy of Sciences, Washington, D.C.

12. Dubos, R., R.S. Schaedler and R. Costello 1968 Lasting biological effects of early environmental influences. J. Exptl. Med. 127:783-799.

13. Vulpe, M., D. Usher and C.P. LeBlond 1956 Influence of the extirpation of male accessories on the occurrence of a "soft calculus" in the urinary bladder of the rat. Endocrinology 59:257-259.

14. Gotterer, G.S., and H.G. Williams-Ashman 1957 Some factors which influence vesiculase action. Proc. Soc. Exptl. Biol. Med. 94:60-64.

THE NEPHROPATHY OF ACUTE CHOLINE DEFICIENCY IN GERMFREE, CONVENTIONALIZED AND OPEN ANIMAL ROOM RATS

Arnold L. Nagler[1,2], Eli Seifter[1,3], Erving F. Geever[1,2]
Wolf-D. Dettbarn[4] and Stanley M. Levenson[1]

Departments of Surgery[1], Pathology[2] and Biochemistry[3], Albert Einstein College of Medicine, Yeshiva University, and Department of Neurology[4], College of Physicians and Surgeons, Columbia University, New York, N.Y.

Weanling rats fed diets deficient in choline develop fatty livers promptly and bilateral renal cortical hemorrhagic necrosis within a period of 6-8 days. Impairment of the mobilization of fat from the liver because of the low availability of methyl donor compounds appears to be the cause of the liver fat accumulation. The mechanism of the development of the nephropathy is poorly understood.

Our goals have been (a) to determine the role of microorganisms in the development of the kidney pathology and (b) to determine the mechanism by which dietary choline deficiency produces the nephropathy.

To clarify the first goal, we subjected germfree, conventionalized and open animal room weanling rats to dietary choline deficiency; to explore the second aspect of this work, we measured the levels of acetylcholine and acetylcholinesterase activity of brain, intestine and kidney of normal and choline deficient open animal room, conventionalized and germfree rats.

MATERIALS AND METHODS

The diet used was basically the Salmon and Newberne (1) diet modified by the addition of certain vitamins, 0.4% cystine, 5% cholesterol and autoclaving, as described in a previous publication (2).

Male Fischer rats, 20-23 days old were used for all experiments; housing, microbiologic testing, feeding regimens, purposeful microbial contamination (conventionalization), handling, killing routines, tissue morphologic studies and acetylcholine (bioassay) and acetylcholinesterase (chemical) analyses have been described elsewhere (2,3,4). The choline deficient diet was fed ad libitum; when choline supplements were given, choline chloride

TABLE 1

ACUTE CHOLINE DEFICIENCY IN GERMFREE, CONVENTIONALIZED AND OPEN ANIMAL ROOM RATS[1]

STATUS	No. rats	Fatal Nephropathy +++	Non-fatal Nephropathy +++	++	+	0	Liver fat at death %	At Killing liver fat %	BUN mg/100 ml blood
OAR[2] Water	11	10	1	0	0	0	15± 1	17 (1)	426 (1)
Conv[3] Water	33	19	6	5	3	0	17± 1	26± 2	92± 20
GF[4] Water	33	1	1	4	2	25	31 (1)	36± 1	25± 1
OAR Choline	4	0	0	0	0	4	-	13± 1	22± 1
Conv. Choline	12	0	0	0	0	12	-	13± 1	23± 2
GF Choline	12	0	0	0	0	12	-	16± 1	23± 2

1. Male Fischer rats, 21 days old (weanling) at start of experiment
2. OAR - open animal room rats
3. Conv. - conventionalized rats - germfree rats contaminated with cecal contents of open animal room rats on the day of weaning, the day before the experimental diet was started.
4. Gf - germfree rats - rats free from viable bacteria and fungi according to methods described elsewhere. (9)

was dissolved in the drinking water (1.5 mg/ml). For survival experiments, the rats were followed for 12 days; for acetylcholine assay experiments, the rats were killed on the fifth day.

Statistical comparisons of the data were made using the Student "t" test.

RESULTS

Our first experiments (published elsewhere (2)) were conducted with open animal room, conventionalized and germfree rats. The results are summarized briefly in Table 1.

By 12 days, 10 of 11 open animal room rats, and 19 of 33 conventionalized rats died with severe nephropathy but only 1 of the 33 germfree rats developed fatal nephropathy. Growth rates and concentrations of blood urea nitrogen paralleled these findings, the germfree choline deficient rats showing good growth and no increase in blood urea nitrogen concentration during the entire experimental period as contrasted with the poor growth and marked rise in blood urea nitrogen concentration of the conventionalized or open animal room rats. While almost all the conventionalized and open animal room rats showed severe nephropathy at death or killing, only 6 of the 33 germfree rats showed any signs of nephropathy, and only 1 of these was severe.

The concentrations of acetylcholine in the proximal half of the small intestine, in the brain and kidneys of choline deficient male weanling open animal room rats fed the choline deficient diet for 5 days was respectively about 32%, 34% and 60% lower than in rats fed the same diet but supplemented with choline chloride in their drinking water; all these differences were statistically significant. The concentration of kidney acetylcholinesterase was significantly lower in the choline deficient rats (about 18%). These results have been presented elsewhere (3) and are summarized in Table 2.

The levels of small intestine acetylcholine of open animal room and germfree choline supplemented rats were similar. In both germfree and open animals room rats fed choline deficient diet, the small intestine acetylcholine levels fell to similar degrees, 24% and 32% respectively; these changes were statistically significant. The activity of intestinal acetylcholinesterase of choline deficient or choline supplemented rats was the same whether the animals were germfree or not. The kidney acetylcholine levels of choline supplemented germfree rats were similar to those of choline supplemented open animal room rats. When the choline deficient diet was fed for 5 days without choline supplementation, the levels of acetylcholine in the kidneys of the open animal room rats fell significantly as found previously, but the level in the germfree did not change (3). The body weights, gross and microscopic appearance of the kidneys, plasma total solid concentration, BUN's and hematocrits were similar in all these rats. The results are currently in press (4) and are summarized in Table 3 of this publication.

TALE 2

TISSUE LEVELS OF ACETYLCHOLINE AND ACETYCHOLINESTERASE IN CHOLINE DEFICIENT RATS[1]

STATUS	TISSUE	ACETYLCHOLINE LIKE MATERIAL µg/g	BUN (blood) mg/100 ml	ACETYLCHOLINESTERASE µg/g/hr
OAR[2] Water	Intestine	0.99± 0.10		
OAR Choline	Intestine	1.46± 0.14		
OAR Water	Kidney	0.07± 0.02	15 (3)	15.5± 1.1
OAR Choline	Kidney	0.27± 0.01	20 (3)	19.5± 1.4
OAR Water	Brain	0.71± 0.05		302 ± 8
OAR Choline	Brain	1.12± 0.07		318 ± 16

1. Male Fischer rats, 21 days old (weanling) at start of experiment
2. OAR - open animal room rats

TABLE 3

TISSUE LEVELS OF ACETYLCHOLINE AND ACETYLCHOLINESTERASE IN GERMFREE RATS[1]

STATUS		TISSUE	ACETYLCHOLINE LIKE MATERIAL µg/g	BUN (blood) mg/100 ml	ACETYLCHOLINESTERASE µg/g/hr
OAR[2]	Water	Intestine	0.54± 0.03		219± 12
OAR	Choline	Intestine	0.79± 0.04		232± 23
OAR	Water	Kidney	0.16± 0.03	23	
OAR	Choline	Kidney	0.26± 0.02	21	
GF[3]	Water	Intestine	0.63± 0.04		190±9
GF	Choline	Intestine	0.82± 0.03		195±13
GF	Water	Kidney	0.27± 0.02	27	
GF	Choline	Kidney	0.24 0.02	15	

1. Male Fischer rats, 21 days old (weanling) at start of experiment
2. OAR - Open animal room rats
3. GF - Germfree rats - rats free of viable bacteria and fungi according to methods described elsewhere (9)

DISCUSSION

Our investigations of acute choline deficiency in germfree, conventionalized and open animal room weanling male Fischer rats have shown that the germfree rat is markedly resistant to the development of the renal pathology typically seen in open animal room or conventionalized rats after feeding choline deficient diets. Germfree rats, however, are not completely refractory to acute dietary choline deficiency as indicated by an increase in liver fat (2) and a decrease in intestinal acetylcholine levels.

As explanations for the lessened development of the stigmata of acute choline deficiency by germfree rats fed choline deficient diets we have stated that the exogenous choline requirement of the germfree is less because of : (a) absent bacterial conversion of choline to trimethylamine in germfree rats, (b) absent bacterial utilization of methionine in the germfree animals, (c) the possible lesser level of certain pressor amines in germfree rats (derived in open animal room rats from the bacterial decarboxylation of amino acids), (d) the lower metabolic rate of the germfree rats (5), and (e) the greater endogenous synthesis of choline by the liver of germfree rats fed choline deficient diets (6).

We postulated in an earlier publication (3) that the nephropathy of acute choline deficiency was mediated via a decrease in kidney acetylcholine resulting in an imbalance of vasoactive neurohumoral substances, leading to renal ischemia, tubular necrosis and hemorrhage. This view is supported by our observations of the circulation of the meso-appendixes of choline deficient rats. We found (7) that in addition to a generalized ischemic capillary bed in the choline deficient rats, a striking alteration in the characteristic reaction of the microvessels to topical epinephrine. In the controls (choline supplemented rats), an average of 0.48 μg/ml epinephrine (0.1 to 1.0 μg/ml) resulted in maximal responses of the precapillary sphincters, lesser responses of the precapillary arterioles but no reaction of the supplying small artery. In marked contrast, 0.03 μg/ml epinephrine (0.01 to 0.10 μg/ml) resulted in pronounced vasoconstriction of the supplying small artery with minimal or no reactions of precapillary arterioles and precapillary sphincters in the rats fed choline deficient diet. These data are consistent with our report showing a decrease in intestinal acetylcholine concentractions during acute choline deficiency with no changes in the activity of acetylcholinesterase to account for these changes.

We also had speculated that one of the reasons that germfree rats develop lesser nephropathy than open animal room or conventionalized rats under the condition of nutritional choline deficiency might be due to differences in their responsiveness to vasoactive materials. In this regard, Baez and Gordon (8) found the microcirculation of the meso-appendix of germfree rats hyporesponsive to epinephrine when compared to that of open animal room rats.

In our recent experiments, we have shown that the rapid decrease in the concentration of acetylcholine in the kidney, characteristically found in the open animal room rat after 5 days of ingesting a choline deficient diet, which we believe to be basic to the nephropathy of acute choline deficiency, does not occur in germfree rats ingesting the same choline deficient diet.

These data support our view that the requirement for choline from an exogenous source by germfree rats is less than by conventionalized or open animal room rats. These data also are consistent with our view that the nephropathy of acute choline deficiency is mediated via a neurohumoral vascular mechanism.

Supported in part by grants 5 P01 AM05664 AMP (Germfree Research Program) and 5K5-GM-14,208 (Career Award, S.M. Levenson) from the NIH to the Albert Einstein College of Medicine, and grants NB-03304 and 1K3-NB-Q5,133 (Career Award, W-D. Dettbarn) from the NIH and grant GB4844 from the NSF to the College of Physicians and Surgeons, Columbia University, New York, N.Y.

REFERENCES

1. Salmon, W., and P. Newberne 1962 Effect of antibiotics, sulfonamides and a nitrofuran on development of hepatic cirrhosis in choline deficient rats. J. Nutrition 76:483.
2. Levenson, S. M., A. L. Nagler, E. F. Geever, and E. Seifter 1968 Acute choline deficiency in germfree, conventionalized and open animal room rats: Effects of neomycin, chlortetracycline, vitamin B_{12} and coprophagy prevention. J. Nutrition 95: 1.
3. Nagler, A. L., W-D. Dettbarn, E. Seifter and S. M. Levenson 1968 Tissue levels of acetylcholine and acetylcholinesterase in weanling rats subjected to acute choline deficiency. J. Nutrition 94: 13.
4. Nagler, A. L., W-D. Dettbarn and S. M. Levenson 1968 Tissue levels of acetylcholine and acetylcholinesterase in weanling germfree rats subjected to acute choline deficiency. J. Nutrition (In press).
5. Levenson, S.M., D. Kan, M. Lev and F. S. Daft 1968 Influence of microörganisms on mammalian metabolism and nutrition with specific reference to oxygen consumption, carbondioxide production and colonic temperatures. J. Nutrition (In press).
6. Levenson, S. M., D. Kan, E. Kwong, and R. H. Barnes 1968 Choline biosynthesis in germfree rats. J. Nutrition (In press).
7. Nagler, A. L., S. M. Levenson and S. Baez 1968 Microcirculation in acute choline deficiency. Fed. Proc. 27
8. Baez, S., and H. Gordon 1957 Microvascular refractoriness to epinephrine in germfree rats. Symposium on Gnotobiotic Research June 11-13, 1967, Park Motor Inn, Madison, Wisconsin.
9. Levenson, S. M. and B. Tennant 1963 Some metabolic and nutritional studies with germfree animals. Fed. Proc. 22:109.

COMPARISON OF THE 1-^{14}C-GLUCOSE METABOLISM BETWEEN RED BLOOD CELLS OF GERMFREE AND PATHOGEN-FREE RATS ON RIBOFLAVIN AND THIAMINE DEFICIENT DIETS

Y. F. Herman, H. E. Sauberlich and R. H. Herman

Chemistry Division and Metabolic Division, U. S. Army Medical Research and Nutrition Laboratory, Fitzsimons General Hospital, Denver, Colorado

We have shown that various riboflavin compounds stimulate the pentose phosphate pathway of intact and hemolysed human red blood cells (1). This was demonstrated by measuring the amount of radioactive carbon dioxide ($^{14}CO_2$) produced from glucose with the radioactivity on the first carbon (1-^{14}C-glucose). In addition, data obtained from red blood cells of normal human volunteers on a thiamine-deficient diet showed a marked decrease in $^{14}CO_2$ production during the deficiency period. Thus, during the course of studies on riboflavin (2) and thiamine nutrition in male germfree (GF) and male pathogen-free (PF) rats, red blood cells (RBCs) of these animals were obtained at various periods of the dietary regimen and the 1-^{14}C-glucose metabolism of these rodent RBCs was compared.

METHODS

The rats employed were GF and PF, strain CDF, obtained from the Charles River Breeding Laboratories, Wilmington, Mass. The basal diet was a heat-sterilized ration containing 20% "vitamin-free" casein, 67% cornstarch, 4% Crisco, 3% Alphacel non-nutritive bulk (ground cellulose), 4% salt mixture (U.S.P. XIV, 1950) and 1% corn oil. The complete diet (supplemented with both riboflavin and thiamine) was prepared by mixing 6.201 grams of vitamin mix (Table I) with 3.799 grams of cornstarch and adding this to each kilogram of the basal diet. The riboflavin-deficient diet was prepared by omitting riboflavin from the vitamin mix; similarly, the thiamine-deficient diet was prepared by omitting thiamine from the vitamin mix. In the thiamine nutrition study an additional 3 gms of dl-methionine was added to each kilogram of diet. All types of diets were assayed microbiologically before heat-sterilization for

TABLE I

VITAMIN MIX

	g/Kg Basal Diet
RIBOFLAVIN	0.03
THIAMINE HCl	0.06
Biotin	0.001
Folic Acid	0.010
Nicotinamide	0.05
Nicotinic Acid	0.05
Vitamin B_{12} (0.1% trit.)	0.003
DL CaPantothenate	0.3
Inositol	1.0
Ascorbic Acid	2.0
Choline Chloride	2.0
Pyridoxine HCl	0.02
Pyridoxamine $(HCl)_2$	0.005
PABA	0.050
Menadione (K)	0.10
Vitamin E (alpha-tocopherol)	0.5 (1 mg=1.1 IU)
Vitamin A	0.02 (10,000 IU)
Vitamin D	0.002 (1,000 IU)
TOTAL	6.201 (add per Kg Basal Diet)

either riboflavin (B_2) or thiamine (B_1) and again after heat-sterilization to determine the degree of destruction, if any. Lactobacillus casei, ATTC #7469, was used as the test organism for the B_2 assay; Lactobacillus viridescens, ATTC #12706, for B_1.

The production of $^{14}CO_2$ from 1-^{14}C-glucose was measured with a few modifications by the method previously described (3). After the animals were anesthesized with ether, blood was obtained by cardiac puncture and placed into heparinized tubes. The plasma was removed by centrifugation at 2400 rpm for 10 minutes and the cells were washed 3 times in Krebs-Ringer-phosphate buffer, pH 7.4. The buffy coat and top layer of the packed red blood cells (RBCs) were discarded. A 1:1 suspension of washed packed RBCs, 0.4 or 0.2 ml, from a pool of blood of several animals was incubated, in duplicate, in 25 ml Ehrlenmeyer flasks, with approximately 22,200 dpm of 1-^{14}C-glucose (S.A. 1.79 mc/mmole) in Krebs-Ringer-phosphate buffer, pH 7.4, total volume 3.0 ml, for one hour at 37° C in room air. Hemolysates of the RBCs were prepared by substituting water for the buffer in the incubation flasks. Each flask, which was equipped with a plastic center well attached to the rubber vaccine stopper, was shaken at 40-50 cycles/min. At the end of the incubation period, 0.2 ml of hyamine hydroxide (1 M in methanol) was injected through the stopper into the center well and then 0.3 ml of 0.4 N sulfuric acid was injected through the stopper into the RBC suspension. Incubation was continued for one hour to effect $^{14}CO_2$ trapping in the hyamine. The plastic center well containing the hyamine was placed into a vial containing 15 ml of scintillation fluid (2,5-diphenyloxazole, 4 g and 1,4-bis-2(5-diphenyloxazolyl) benzene, 0.05 g/l of toluene) and counted in a Nuclear Chicago liquid scintillation counter. The results were calculated as the mean per cent of the initial radioactivity of the 1-^{14}C-glucose recovered as $^{14}CO_2$. The effect of various cofactors on the RBCs was measured by addition of each as follows: nicotinamide adenine dinucleotide phosphate (NADP), 2 mM; riboflavin (B_2), 30 µg; thiamine (B_1), 2 mM per incubation flask.

In the riboflavin study, PF and GF rats were divided into 4 groups. Groups I, II, and III received the B_2 deficient diet (-B_2) for 56 days. Each group was injected intraperitoneally, daily, with 60, 30, and 10 µgs of B_2, respectively. Group IV received the complete B_2 diet (+B_2). At the end of the 56 days groups I, II, and III were continued on the -B_2 diet without any further B_2 injections for 45, 45 and 46 days, respectively, until the RBC studies were done. Group IV was maintained on the +B_2 diet until the RBC studies were done.

In the thiamine study, PF and GF rats were divided into 6 groups. Groups 1, 2 and 3 were kept on a B_1 deficient diet (-B_1) for 28 days. Groups 4, 5 and 6 were maintained on a B_1 supplemented diet (+B_1) for the same length of time. Groups 2 and 5 had 10 mg of penicillin added to every 100 g of their diets, while groups 3 and 6 had 500 mg of succinyl sulfathiazole added to every 100 g of their diets.

RESULTS

Table II summarizes the results obtained from the RBCs of PF and GF rats on the -B_2 diet as compared to those on the +B_2 diet. The intact RBCs of the PF rats on the +B_2 diet (group IV) showed more $^{14}CO_2$ production than the intact RBCs of the corresponding GF animals. The hemolysate of the PF animals was more active than that of the GF animals when no cofactors were used. The hemolysates of both PF and GF animals were equally as active with the cofactors, B_2 and NADP. The intact RBCs of the PF rats of group I had somewhat lower $^{14}CO_2$ production than the corresponding intact RBCs of the GF rats. The hemolysates of the PF rat RBCs, without any cofactor, had more than twice the activity as that of the intact cells, while the corresponding GF rat RBC hemolysate only increased

TABLE II. The production of $^{14}CO_2$ from 1-^{14}C-glucose by intact and hemolyzed red blood cells of PF and GF rats depleted of vitamin B_2. Values are given as the per cent of initial radioactivity recovered as $^{14}CO_2$. The initial amount of 1-^{14}C-glucose used was 22,200 dpm. Volume of RBCs was 0.4 ml.

Amount of B_2 injected (µg/day)	No.of Days Rats injected	Type of diet given to rats	Days on Diet	Group	Additions	PF Rats Intact RBCs	PF Rats Hemolysate	PF Rats No.of Rats	GF Rats Intact RBCs	GF Rats Hemolysate	GF Rats No.of Rats
					None	13.9	28.4		15.2	20.2	
60	56	$-B_2$	45	I	Vit. B_2	14.8	44.7	4	17.6	62.7	2
					NADP	17.6	----		19.5	----	
					None	15.6	11.0		16.7	22.0	
30	56	$-B_2$	45	II	Vit. B_2	15.8	54.2	4	18.8	62.5	4
					NADP	18.0	----		20.9	----	
10	56	$-B_2$	46	III	None	16.3	13.1		15.3	17.7	
					Vit. B_2	16.5	62.3	4	16.1	64.8	4
					NADP	18.5	37.0		16.1	48.6	
					None	17.6	19.7		10.4	8.7	
--	--	$+B_2$	48	IV	Vit. B_2	19.4	57.3	4	12.9	55.4	3
					NADP	19.2	35.6		11.3	38.5	

TABLE III. The production of $^{14}CO_2$ from 1-^{14}C-glucose by intact and hemolysed red blood cells of PF and GF rats depleted of vitamin B_1. Values are given as the per cent of initial radioactivity recovered as $^{14}CO_2$. The initial amount of 1-^{14}C-glucose used was 22,200 dpm. Volume of RBCs was 0.2 ml.

Diet	Antibiotic	Group	Days on Diet	Additions	PF Rats Intact RBCs	PF Rats Hemo-lysate	PF Rats No.of Rats	GF Rats Intact RBCs	GF Rats Hemo-lysate	GF Rats No.of Rats
				None	7.7	9.4		10.4	35.3	
$-B_1$	None	1	28	Vit. B_1	8.8	28.9	2	11.6	27.8	2
				Vit. B_2	9.7	36.5		9.7	37.5	
				None	11.9	21.8		12.9	35.6	
$-B_1$	Penicillin	2	28	Vit. B_1	13.1	14.4	2	14.5	23.6	2
				Vit. B_2	11.2	22.2		12.2	41.7	
	Succinyl			None	10.8	28.7		10.5	34.0	
$-B_1$	sulfa-	3	28	Vit. B_1	12.5	28.8	2	13.0	28.0	2
	thiazole			Vit. B_2	11.7	39.9		10.0	37.1	
				None	10.3	18.2		10.5	13.8	
$+B_1$	None	4	28	Vit. B_1	9.2	26.2	6	10.7	13.2	6
				Vit. B_2	12.1	40.6		11.6	37.1	
				None	11.8	19.8		13.3	13.4	
$+B_1$	Penicillin	5	28	Vit. B_1	13.8	20.8	2	13.5	21.4	2
				Vit. B_2	14.2	34.8		12.9	46.2	
	Succinyl			None	10.9	25.8		12.9	36.8	
$+B_1$	sulfa-	6	28	Vit. B_1	12.3	27.2	2	13.2	34.2	2
	thiazole			Vit. B_2	11.8	39.2		13.5	45.7	

by a third. The hemolysate of the PF animals responded less to B_2 than did the hemolysate of the GF animals.

The intact RBCs of the PF animals of group III had about the same activity as the corresponding intact RBCs of the GF animals. The hemolysate of the PF animals without cofactors did not have an increased $^{14}CO_2$ output, while the corresponding hemolysate of the GF animals had only a slightly increased $^{14}CO_2$ output. Both PF and GF hemolysates responded equally to B_2 and NADP addition but the GF hemolysates had greater stimulation with NADP than did the PF hemolysate.

The thiamine studies (Table III) showed that in the B_1 deficient PF animals (group 1) little stimulation of $^{14}CO_2$ production occurred in the intact RBCs with B_1 and B_2 as compared to the RBCs with no cofactors added. Similar results occurred with the intact RBCs of the GF animals. The hemolysates of the RBCs from the PF animals showed about the same $^{14}CO_2$ production as the intact cell and good stimulation with added B_1 and B_2. In contrast, the hemolysate of the RBCs from the GF animals showed increased $^{14}CO_2$ without cofactors but no stimulatory effect of added B_1 and B_2. The RBCs of the penicillin-fed PF animals on the $-B_1$ diet (group 2) showed slightly higher $^{14}CO_2$ output in the intact state and higher $^{14}CO_2$ output in the hemolysate with no stimulation from added B_1 and B_2. Similar results were seen in the hemolysate of the $-B_1$ succinyl sulfathiazole fed PF rats (group 3), except added B_2 in the hemolysate did stimulate $^{14}CO_2$ production. The corresponding GF animals showed no particular change in the intact RBCs and the hemolysates showed increased $^{14}CO_2$ regardless of cofactors added. However, the hemolysate of both $-B_1$ antibiotic-fed GF animals seemed to have lesser $^{14}CO_2$ production with added B_1.

In the intact RBCs of the $+B_1$ diet-fed PF animals (groups 4, 5 and 6) no great increase in $^{14}CO_2$ output occurred with added cofactors whether or not antibiotics were used in the diet. The hemolysates of the PF animals showed an increase of $^{14}CO_2$ production over the intact RBCs and further stimulation with added B_1 and B_2 as in the corresponding group of $-B_1$ diet-fed PF rats (groups 1, 2 and 3). The hemolysates of the PF animals on the $+B_1$ penicillin diet could be stimulated by B_2 addition, but not by B_1; while the $+B_1$ sulfathiazole diet-fed PF hemolysates had higher $^{14}CO_2$ production and could be stimulated also only by added B_2 but not by B_1.

The intact RBCs of the $+B_1$ diet-fed GF animals showed similar results as that obtained for the PF rats (groups 4, 5 and 6). In contrast to the PF rats, however, the hemolysate of the GF animals of group 4 showed stimulation only with added B_2. In hemolysates of the $+B_1$ penicillin diet-fed GF animals of group 5, stimulation occurred with added B_1 and even more so with B_2. In hemolysates of $+B_1$ sulfathiazole diet-fed GF animals of group 6 the $^{14}CO_2$ output was quite high, though not altered with added B_1, and was stimulated to a greater degree with B_2.

DISCUSSION

These studies show that the pentose phosphate pathway activity as measured by $^{14}CO_2$ output from 1-^{14}C-glucose did not seem to be much different in the intact GF rat RBCs as compared to the intact PF rat RBCs, whether or not a B_2 deficiency existed. Only minor changes were observed in the hemolysates between the GF and PF animals in the B_2 study.

Similarly the intact RBCs of the GF and PF rats showed little change between the groups in the B_1 study. These RBCs seemed to be little influenced by the lack or presence of dietary B_1, antibiotics or the cofactors, B_1 and B_2. The main differences occurred in the hemolysates. In group 1 the pentose phosphate pathway was stimulated by both B_1 and B_2. This stimulation was abolished by antibiotics in groups 2 and 3 except that added B_2 stimulated the hemolysate of the $-B_1$ penicillin diet-fed GF animals of group 2 and B_2 stimulated, the $-B_1$ sulfathiazole diet-fed PF animals of group 3. The hemolysate of the $+B_1$ diet-fed PF rats (group 4) behaved similarly to that of the $-B_1$ diet-fed PF rats of group 1. The feeding of antibiotics to the PF rats (groups 5 and 6) blocked the stimulatory effect of B_1 but not B_2 in the hemolysates.

These results are complex. What is demonstrated is the great resistance of intact rat RBCs to dietary changes and to exogenous cofactors. Whereas intact human RBCs are stimulated by B_2 and NADP (1), intact rat RBCs are stimulated poorly, if at all. The presence or absence of dietary B_2 and dietary B_1 and the antibiotics, penicillin and succinyl sulfathiazole, had no great effect on the $^{14}CO_2$ production by intact rat RBCs. The intact RBCs of PF and GF rats gave comparable results. The germfree state did not lead to any demonstrable difference in activity of the pentose phosphate pathway. In the B_2 study only some few differences were seen in hemolysates of PF and GF rat RBCs. Hemolysates of the B_1 deficient animals demonstrated various degrees of activity of the pentose phosphate pathway being dependent on diet, the presence of antibiotics in the diet, and added cofactors.

SUMMARY

The germfree state can lead to subtle differences in rat red blood cell metabolism of 1-^{14}C-glucose as compared to the pathogen-free state and a variety of maneuvers are necessary to demonstrate these differences. Riboflavin deficiency is resisted well by both PF and GF rat RBCs and does not demonstrate appreciable differences between pathogen-free and germfree states. A deficiency of B_1 together with antibiotic feeding can cause alterations in the pentose phosphate pathway activity in rat RBC hemolysates leading to complex changes. The mechanism whereby these changes occur involves all of the factors capable of affecting the pentose phosphate pathway.

The Principles of Laboratory Animal Care as promulgated by the National Society for Medical Research were observed.

REFERENCES

1. Herman, R. H., Herman, Y. F., Sauberlich, H. E. and Plunket, D. Fed. Proc. 26:305, 1967.

2. Herman, Y. F., Sauberlich, H. E. and Raica, N. Bact. Proc., 68, 1967.

3. Herman, R. H., Berkowitz, J., Clayton, L. E. and Sherman, J. L. Nature, 195:500, 1962.

NUTRITIONAL CONTROL OF NEMATODE DEVELOPMENT

George J. Jackson

Rockefeller University, New York

The cultivation of worms in species isolation has been tried since the beginning of our century. Much of this effort was motivated by the hope that important parasitic species might be more adequately studied, understood and, therefore, controlled if they could be separated from their host without being physiologically damaged. First results, however, were with roundworms (nematodes) living as predators of microorganisms in the soil or as saprophytes on decaying organic matter. There is no telling when, originally, samples of soil or detritus were brought into the laboratory and purposely enriched to yield greater numbers of worms. The late Ellsworth C. Dougherty, historian and neologist of the culturing of invertebrate metazoa (1959), mentioned Perez (1866) as being the first to grow worms, probably a mixture of rhabditid nematode species, in chicken ovalbumin, a medium differing significantly from their natural environment (Dougherty & Calhoun, 1948). However, no rigorous attempt was made to dissociate the worms from other microbes. This was tried by Metcalf (1903) with Rhabditis brevispina, "a nematode associated with plant decay", which he grew on an asparagus juice agar that was partially digested by microbial action and then heat sterilized. Modern information that nematodes require heat labile substances makes it likely but not certain that the worm eggs inoculated onto the medium were not free of contaminants. Zimmermann (1921), with cleaned vinegar worms, Turbatrix aceti, in a peptone medium containing an alcoholic extract of yeast, is commonly thought to have been the first to cultivate an invertebrate metazoan for continuous generations in the absence of other species.

A decade later, Glaser (1931) reared a nematode parasite of Japanese beetles, Popillia japonica, in association with yeast and an unknown population of microorganisms (xenically) and then (1940) in species isolation (axenically) on pieces of presumably non-metabolising avian or mammalian tissues.

A variety of nematodes are now being grown routinely in species isolation. These include free living, saprophytic and insect parasitic species, a plant parasite (Myers, 1967) and a mammalian wound parasite (Yarwood & Hansen, 1968). Despite considerable work, the important roundworm parasites of warm blooded animals have not been established in continuous culture. The hang-up seems to be sexual reproduction. With parasitic flatworms, the tapeworms (cestodes) and flukes (trematodes) whose life cycles include sexually and asexually reproducing stages, only the asexual reproduction of a few species has been obtained *in vitro*.

The sometimes fascinating, often frustrating details of culturing invertebrate animals, including worms, can be traced from a few key references. The classical cook book for nurturing the invertebrates (Needham *et al.*, 1937) is available in a popular reprint (1959). Newer work on metazoa can be found in three volumes: Dougherty (1959), Tiner (1966), Taylor (1967), and in Silverman's review (1965). Also, there are now a variety of immunological (Douvres, 1960; Jackson, 1961, 1963, 1964; Silverman *et al.*, 1962; Soulsby, 1963, 1966) and physiological studies (Pramer & Stoll, 1959; Weinstein, 1966; Rothstein & Mayoh, 1966 and previous) with worms in cultures.

CULTIVATION AND NUTRITIONAL PHYSIOLOGY OF *Neoaplectana glaseri*

The insect parasite cultivated by Glaser (1940) has been the subject of numerous physiological studies. It was described and named *Neoaplectana glaseri* by Steiner (1929) and is a not untypical small nematode with four larval stages that are separated from each other and from the adult male and female by a molt. Embryonated eggs develop and hatch in the uterus of the mother. Normally the first stage larvae are vigorously ejected through the birth pore. The survival form of *N. glaseri* between insects in the moist soil or in low levels of water between laboratory cultures is the ensheathed third stage larva, the sheath being the second stage's cuticle that is not shed under "adverse" conditions but instead retained as additional covering over a new cuticle. Although biographies of isolated individuals show differences in time of attainment and duration of some morphological characteristics, there are population norms for the nematodes' maturation in a given medium.

Stock supplies of *N. glaseri* and related species are still most conveniently maintained in Glaser's type of culture environment: a wide, glass test tube, plugged with cotton to allow gas exchange, that contains an agar slant enriched with meat infusion broth, peptone, dextrose and saline; a moist, sterile piece of mammalian (rabbit kidney) or avian (chick embryo) tissues is placed at the bottom of the slant. Individuals in a water suspension of ensheathed N. glaseri, inoculated into such a tube, develop and reproduce at the bottom of the slant, eventually consuming the tissue slice. As food becomes scarcer and the nest crowded and soiled, new individuals accumulate in the thin films of moisture on upper reaches of the agar or on the inside glass walls. These accumulations are visible macroscopically as fine white networks. Examined microscopically, they are almost pure populations of third stage ensheathed larvae. The larvae are harvested with a pipette or wire loop for storage in low levels of refrigerated water, to be used in starting new cultures almost immediately or after many months. In this way N. glaseri has been grown continuously apart from its natural insect host(s) since 1944, yet axenic populations retain their infectivity (Stoll, 1953a; Jackson & Moore, 1968).

During the 1950s, Stoll (1953b, 1959) developed a liquid medium for N. glaseri that consists of a heat labile extract of raw liver (RLE), a heat stable broth (PF) of Pfanstiehl peptone and beef heart infusion, and dextrose. Dextrose (1-3%) supplemented PF, RLE or commercially dehydrated liver digests suffice for small population increments (10x) but the total Stoll medium is required for a more complete expression (100-200x) of the population potential (Jackson, 1962a). Even with this rich medium it is difficult to cultivate worms continuously by passing them from culture to culture. Presumably, after a few generations, a rest period for ensheathed larvae must be interposed between periods of population growth. Continuous culture columns that provide both rest and growth opportunities are currently being explored for Neoaplectana species (Hansen, Yarwood & Jackson,1967).

During the 1960s, a chemically defined medium for N. glaseri has been developed (Jackson 1962a, b) and used to study nutritional requirements and physiological products (Sherman & Jackson, 1963; Jackson & Siddiqui, 1965; Jackson, 1966). In addition to dextrose, trace metals, salts, purines and pyrimidines, the recipe calls for high concentrations of several amino acids and B-vitamins compared to commonly used media for mammalian cell culture (McQuilkin et al., 1957). In these respects, the medium for N. glaseri is closely related to Trager's (1957) for the culture form or presumed insect-inhabiting leptomonad stage of the lizard-parasitizing protozoan, Leishmania tarentolae, as well as to the defined part of a complex medium (Dougherty et al., 1959) for the hermaphroditic soil nematode Caenorhabditis briggsae. Studies to improve the medium for N. glaseri and extend its use for related species (Jackson, 1965) are continuing. Here, the intent is to discuss how chemical quality and quantity of a nutritional medium and the length of time it is made available, can affect and, once this is known, can be used to control the rate of development and reproduction and certain morphological features of worms in a population.

The first clear cut example of a single component in the defined medium affecting cultures of N. glaseri was with a B-vitamin. In the absence of folic acid or the related biologically active folinic acid, the worms develop apparently normally from ensheathed third stage larvae into morphological adults but there is no reproduction. Total inhibition of reproduction but not development also occurs in complete medium with folic acid antagonists such as aminopterin and amethopterin at 50x the folic acid concentration. Bacterial assays of reproducing cultures showed that folic acid disappears from the medium only with considerable reproduction. These studies have been detailed previously (Jackson & Siddiqui, 1965).

Results with media deficient in single amino acids are not as clear cut as with a vitamin like folic acid. However, if instead of testing amino acids singly, the nutritional value of amino acid groups is to be tried, it then becomes important to compensate for loss of bulk. Overall concentrations must be maintained because N. glaseri adjusts to dilute media only within limits.

For instance (Table 1), if all but the sugar and salt parts of a PF medium are diluted, the first effect is on the worms' reproduction. It takes longer for a new generation to appear although the parental generation has become adult on schedule. In broths diluted below a concentration represented by 1 mg N per culture, third stage larvae develop into adults more and more slowly but there is no reproduction. This sequence can be duplicated by saline dilutions of only the amino acid portion of the defined medium (Table 2).

Keeping, therefore, the total level of amino acids and salts and the relative ratios of retained compounds close to those of the optimal defined medium, it was of interest to test the nutritional value for these worms of "mammalian essential and nonessential" amino acids. It is possible to make a medium that includes the essential amino acids proportionally at high enough concentrations to compensate for the missing bulk of the nonessentials. Because of insolubilities, this is not possible with a recipe that calls for certain nonessentials (glycine, arginine, serine, tyrosine, proline, alanine, glutamic and aspartic acids, taurine, (urea)), as the sole amino acids; in these experiments I settled for less bulk than the desired 2.67 g amino acids per 100 ml medium. Third stage N. glaseri develop into 4th stage larvae and into stunted adults in medium with mammalian essentials

Table 1
Axenic Cultures of the Insect
Parasitic Nematode, Neoaplectana glaseri:
NITROGEN LEVELS AND GROWTH RATE

mg N/1.5 ml culture	Days to: Adulthood	Reproduction
2.466	4	7
2.192	4	11
1.918	4	11
1.644	4	18
1.370	4	18
1.096	5	18
.822	7	--
.548	11	--
.274	(25)	--
0	--	--

Cultures were started with ca. 30 ensheathed larvae from the same stock and contained similar concentrations of saline (0.85%) and dextrose (1.38%) irrespective of nitrogen (N) concentrations. The nitrogen source was PF, a meat infusion broth and its dilutions. The time recorded is that for development of adults and appearance of newly shed larvae in all of three triplicate cultures.

Table 2
Axenic Cultures of the Insect
Parasitic Nematode, Neoaplectana glaseri:
AMINO ACID LEVELS AND GROWTH RATE

mg AA/1.5 ml culture	Days to: Adulthood	Reproduction
21.8	5, 5, 6	8, 9, 6-9
16.4	5-8, 5, 6	8-12, 9, 9
10.9	12, 9, 9	20-23, 17, 14
5.5	20, 13, 14	--, --, --
2.2	S, S, S	--, --, --
1.1	--, --, --	--, --, --
0	--, --, --	--, --, --

All cultures were started with ca. 30 ensheathed larvae. Results with three different worm stocks are shown for each amino acid (AA) dilution; each number or range represents five replicate cultures. Amino acid dilutions were made with saline of comparable tonicity. All other components of the chemically defined medium were kept constant. S signifies an occasional, stunted adult.

Table 3
Axenic Cultures of the Insect Parasitic Nematode, *Neoaplectana glaseri*:
FEEDING PULSE AND GROWTH

days in complete medium	multiple washings	saline-dextrose	replaced in: NCTC-109	complete medium
$\frac{1}{4}$	"	survive.	survive; some exsheath.	reproduction by day 7.
1	"	survive; some exsheath; some molt once.	survive; some exsheath; some molt once; some molt twice: (stunted adults).	"
2	"	some stunted adults.	some adults.	"
3	"	stunted adults; some adults.	reproduction by day 17.	"
4	"	stunted adults; some adults; reproduction by day 7.	reproduction by day 7.	"

Cultures were started with ca. 150 ensheathed larvae from the same stock. The complete medium was PF (peptone-meat infusion broth) with dextrose (1.38%) and saline (0.85%).

(lysine, tryptophane, histidine, phenylalanine, leucine, isoleucine, threonine, methionine, valine) as the only amino acids. The stunting is externally visible and in females also internally because of the failure to produce eggs. In medium without the essentials there is no larval development. High total concentrations of nonessential amino acids are even lethal. Although these results do not include the possible effects of nutritional balance (differing ratios of essentials or nonessentials), one suspects that a greater variety of amino acids may be required by these worms than by mammals.

The ability of N. glaseri to accumulate nutrients from greatly diluted media suggested pulse studies with complete medium (Table 3). In an "all or none" situation, worms must remain in complete medium for at least 4 days in order then to reproduce in saline-dextrose. The appearance of a new generation is not retarded, compared to controls, as it is in an "all or some" situation when worms from complete medium are placed in NCTC-109 on day 3.

DISCUSSION

Dilution and pulse studies show limits in the ability of Neoaplectana glaseri to accumulate and concentrate nutrients. From tracer and oral ligation experiments with a few nematode species (Zam et al., 1963; Lee, 1965) we know that the uptake of nutrients is mostly into and across the digestive tract, not across the outer surface as in the case of even those flatworms such as flukes that have digestive tubes. Observations with N. glaseri suggest that the nematode intestine may serve as a site for concentrating and even storing substances ingested from the medium. The intestine of the large female worm can be seen into easily at low magnification with the light microscope. Natural pigments dissolved in proteinaceous media or dyes such as neutral red and bromcresol green added to media are concentrated in the anterior intestinal lagoon; scattered particles such as liver cell precipitate in Stoll's extract (RLE) or L. tarentolae leptomonads added to the defined medium are packed in this site. Moreover, that many of the added protozoa are active in the anterior intestine further suggests that digestion takes place posteriorly. Much remains to be learned about the physiology of nematode digestion, especially absorption of substances across the intestinal wall. Using the chemical quality or quantity of a nutritional medium to control the rate of development and reproduction and, as in the case of certain amino acid deficiencies, even the morphology of N. glaseri, is of advantage in a variety of physiological studies, for example: in telling when, during the nematode's life cycle, nutrients are required. If also applicable to nematodes that produce serious diseases in man and domestic animals, properly controlled cultures could supply attenuated organisms for vaccination. Meanwhile, optimally reproducing cultures of Neoaplectana species have (McCoy & Glaser, 1936; Niklas, 1967) and are being used to control insect pests.

REFERENCES

Dougherty, E.C. (including Dougherty *et al.*) 1959. Ann. N.Y. Acad. Sci. 77 (Art. 2): 27-406.

Dougherty, E.C. & Calhoun, H.G. 1948. Proc. Helminthol. Soc. Washington 15: 55-68.

Douvres, F.W. 1960. J. Parasitol. 46 (5, Sect. 2): 25-26.

Glaser, R.W. 1931. Science 73: 614-615.

1940. Proc. Soc. Exptl. Biol. & Med. 43: 512-514.

Hansen, E.L.; Yarwood, E.A. & Jackson, G.J. 1967. Prog. 42nd Ann. Meeting Am. Soc. Parasitol.: 57.

Jackson, G.J. 1961. Exper. Parasitol. 11: 241-247.

1962a. Exper. Parasitol. 12: 25-32.

1962b. Trans. N.Y. Acad. Sci.(Series II) 24: 954-965.

1963. Am. J. Hyg. Monogr. 22: 109-112.

1965. Parasitol. 55: 571-578.

1966. Ann. N.Y. Acad. Sci. 139 (Art. 1): 91-97.

1966. Proc. 1st Internat. Congr. Parasitol. (Pergamon Press, Oxford, G.B. & Tamburini Ed., Milano) 1: 578-579.

Jackson, G.J. & Moore, G.E. 1968. In preparation.

Jackson, G.J. & Siddiqui, W.A. 1965. J. Parasitol. 51: 727-730.

Lee, D.L. 1965. "The Physiology of Nematodes" (Oliver & Boyd, Edinburgh & London, G.B.).

McCoy, E.E. & Glaser, R.W. 1936. New Jersey Dept. of Agr., Bureau Plant Ind. Circular 265: 1-10.

McQuilkin, W.T.; Evans, V.J. & Earl, W.R. 1957. J. Nat. Cancer Inst. 19: 885-907.

Metcalf, H. 1903. Trans. American Micr. Soc. 24: 89-102.

Myers, R.F. 1967. Proc. Helminthol. Soc. Washington 34: 251-255.

Needham, J.G. *et al.* (1937) 1959. "Culture Methods for Invertebrate Animals" (Dover Publications, New York).

Niklas, O.F. 1967. Mitteilungen Biol. Bundesanst. Land.- & Forst-wirtsch. Berlin-Dahlem 124: 1-40.

Perez, (J.) 1866. Ann. Sci. Nat. 6: 152-307.

Pramer, D. & Stoll, N.R. 1959. Science 129: 966-967.

Rothstein, M. & Mayoh, H. 1966. Comp. Biochem. Physiol. 17: 1181-1188.

Sherman, I.W. & Jackson, G.J. 1963. J. Parasitol. 49: 392-397.

Silverman, P.H. 1965. Adv. in Parasitol. 3: 159-222.

Silverman, P.H.; Poynter, D. & Podger, K.R. 1962. J. Parasitol. 48: 562-571.

Soulsby, E.J.L. 1963. Ann. N.Y. Acad. Sci. 113 (Art. 1): 492-509.

Soulsby, E.J.L. 1966. "Biology of Parasites" (Academic Press, New York): 255-276.
Steiner, G. 1929. J. Washington Acad. Sci. 19: 436-440.
Stoll, N.R. 1953a. "Thapar Commemoration Volume", J. Dayal & K.S. Singh, eds. (U. of Lucknow, India): 259-268.
Stoll, N.R. 1953b. J. Parasitol. 39: 422-444.
Stoll, N.R. 1959. Ann. N.Y. Acad. Sci. 77 (Art. 2): 126-136.
Taylor, A.E.R. 1967. "Problems of in vitro Culture" (Blackwell Sci. Publications, Oxford and Edinburgh, G.B.).
Tiner, J.D. 1966. Ann. N.Y. Acad. Sci. 139 (Art. 1): 1-272.
Trager, W. 1957. J. Protozool. 4: 269-276.
Weinstein, P.P. 1966. "Biology of Parasites" E.J.L. Soulsby, ed. (Academic Press, New York): 143-154.
Yarwood, E.A. & Hansen, E.L. 1968. J. Parasitol. 54: 133-136.
Zam, S.G.; Martin, W.E. & Thomas Jr., L.J. 1963. J. Parasitol. 49: 190-196.
Zimmermann, A. 1921. Rev. Suisse Zool. 28: 357-380.

GNOTOBIOTICS AND HUMAN CELL CULTURE

G.E. Moore, J.W. Kullen, H.A. Franklin, & N. Kinsley

Cell Laboratories, Roswell Park Memorial Institute,

Buffalo, New York

Some gnotobiotic techniques have been adopted for certain types of cell culturing. Related techniques are required for the maintenance of "cell plants" and complex attempts to provide an optimal environment for normal cells. This brief dissertation will review the interface of gnotobiotics and some of the practical problems encountered in our cell laboratory.

I. CONTAMINATION

Human cell cultures are initially set up from fetal tissues; placenta and membranes; biopsy specimens at the time of surgery; bone marrow aspirations; fluid from collections in the body cavities; and the blood. The most frequent cultures are temporary ones of the skin and leukocytes for chromosomes studies and immunological reactions. Maximal attempts are made to insure sterility and antibiotics, penicillin, streptomycin and Nystatin, are added to the cultures. More intensive care is required for establishing long-term cultures. As a surgeon I was shocked at the frequency with which biopsy specimens obtained from the operating room were contaminated. There are several reasons for this: often the biopsy site includes tumor tissue which may be partially necrotic and such tissues contain a variety of bacteria which may not be clinically evident. It is difficult to immediately place a specimen into a sterile bottle. Further, I am personally convinced that the combination of cell culture medium and a serum suppplement plus cells is far more sensitive for microorganisms than the traditional ones used by the bacteriologists. These problems can be reduced; laminar flow units will provide cleaner operating

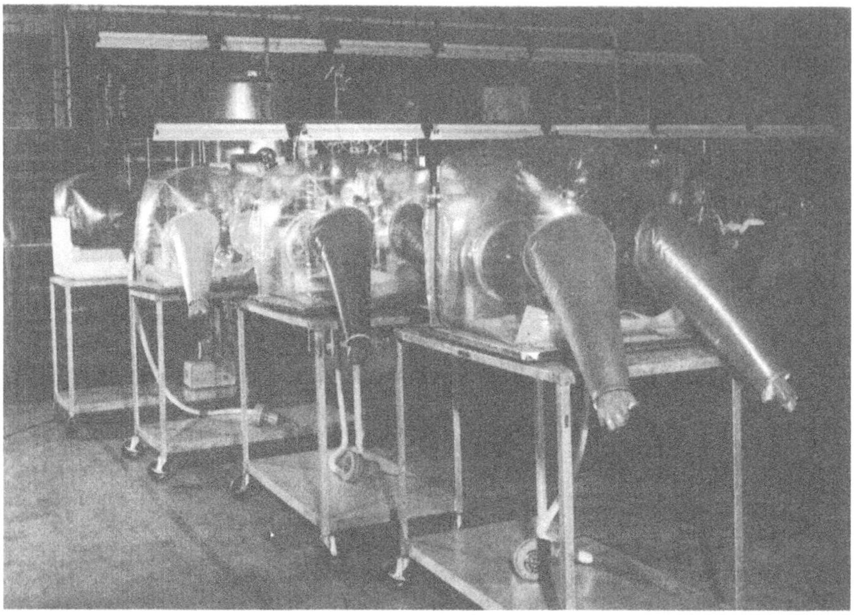

Figure 1

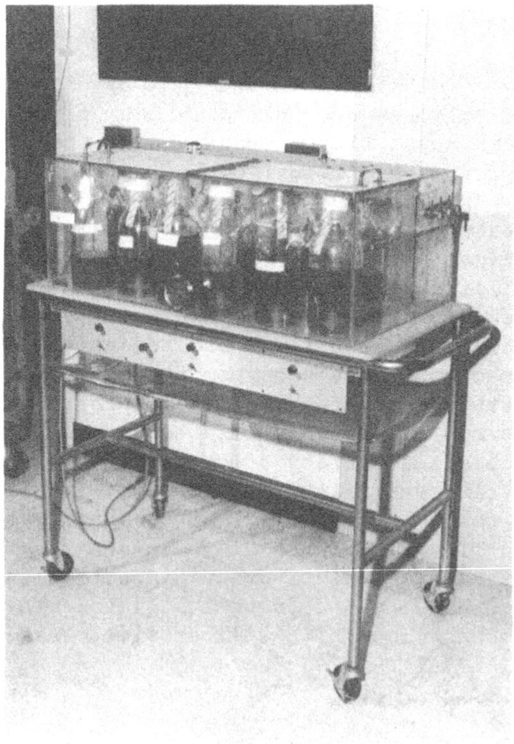

Figure 2

rooms as soon as better and more reliable designs are available. Better antibiotics will also be helpful.

The collection and maintenance of initial cultures free from extraneous viruses is more difficult. Incinerated and filtered air would be desirable. The use of tissues from patients who have been immunized selectively should be of some value. There is a need for closed systems of harvesting cells from solid tissues and better enzymes for tissue disassociation.

II. MAINTENANCE OF CELL CULTURES

An ingress or exchange of air into culture flasks is a continuing source of possible contamination with microorganisms. The substitution of remote or "bulb" pipetting is mandatory. Manipulations of several cell cultures in the same hood or "sterile room" and their maintenance in incubators with other kinds of cell cultures provide the additional hazard of cell contamination. This danger does exist but it has been exaggerated (1) in my opinion.

We have modified a standard gnotobiotic unit* with two heat sources balanced with a thermocouple. An additional layer of plastic over the entire unit with a 2" layer of air space provided the necessary insulation (Figure 1). Original specimens and all the media, pipettes, scrapers, etc., for uninterrupted care of the cell cultures were passed into the chamber after sterilization with 4% peracetic acid*. These cell cultures were maintained in the gnotobiotic incubator until they become established as cell lines.

No alteration in the frequency of leukovirus (2,3) in the human lymphoblast cell lines was noted.

Hematopoietic Cell Lines

We have established over 400 human hematopoietic cell lines in the last several years from a variety of patients and normal persons (2,4,5). All new cell lines are assayed for the leukovirus by antigenic reactions and electron microscopy. Routine smear preparations for morphological studies and chromosome analyses are completed as soon as possible. A minimum of 6 to 10 ampules of cells are preserved in a liquid Nitrogen cell bank after preparation with 12.5% dimethylsulfoxide and slow freezing. Interesting cell lines are maintained in inexpensive Brockway 16 oz. GEM oval bottles** as static or so-called "hover" cultures since the cells

---

* Snyder Manufacturing Co., New Philadelphia, Ohio

** Brockway Glass Co., Pittsburgh, Pa.

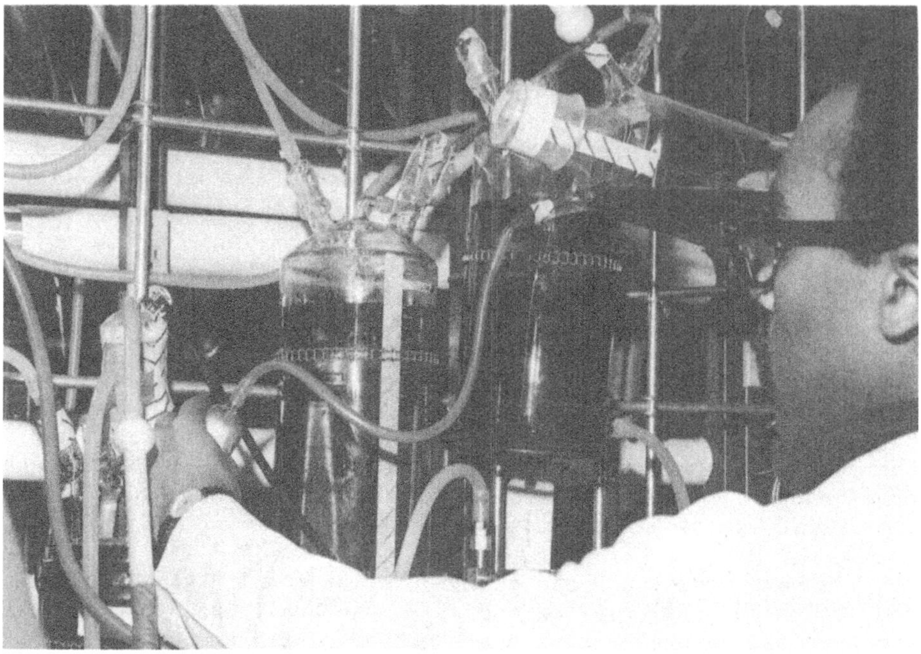

Figure 3

Figure 4

do not attach to the glass. No antibiotics are used except for the initial treatment of cultures from biopsy specimens or, in rare instances, for the treatment of mycoplasma. The extent of contamination of cell cultures containing antibiotics with saprophytic organisms was not really appreciated until we examined the material recovered by continuous flow ultracentrifugation. Unexpected bacteria and mycoplasma were detected by electron microscopy even though routine cultures of the media had been sterile.

It must be emphasized that contrary to published statements from many laboratories, the use of antibiotic free media does not insure rapid detection of all infections with bacteria, mycoplasma, and L-forms. A sudden development of scattered instances of unexplained bacterial contamination or a decrease in cell viability and growth rate is sufficient evidence that subclinical contamination may be present.

We have not identified any epidemics of infection with extraneous viruses and therefore cannot comment on the frequency of such incidents--which probably occur.

Ortho and Magna Suspension Cultures

It would be advantageous to modify the incubators which are used for "spinner" cultures (Fig. 2) or enclose them in gnotobiotic chambers. This would be an excellent technique when handling cells which would be used for clinical purposes.

When transferring one culture to another culture, e.g., a spinner culture to a trophocell culture, a simple transfer unit is used (Fig. 3). This unit consists of an air filter with a stopper and a filling attachment*, 1 3/4" OD, connected to a piece of tubing with a stopper. The air filter is inserted into one port of the spinner and the tubing unit into the other port. The port of the vessel that is to receive the culture is flamed in order eliminate any moisture surrounding the inside of the port. A 4 x 4 gauze sponge, saturated with Cidex**, is wrapped around the "lip" of the port of the receiving vessel. The filling attachment is placed over the gauze and the culture is poured into the vessel (Fig. 3). The gauze acts as a filter for any reflux of air into the receiving vessel. We are aware of the faults of this arrangement but it has provided a practical solution for transferring cells in an unsterile atmosphere. Routine feeding and sampling of spinner cultures in laminar flow hoods is recommended. The effectiveness of the air filters of all laminar flow hoods should be checked with a particle counter.

---

* Corning Glass Co., Corning, N.Y.

** Arbrook, Somerville, N.J.

Figure 5

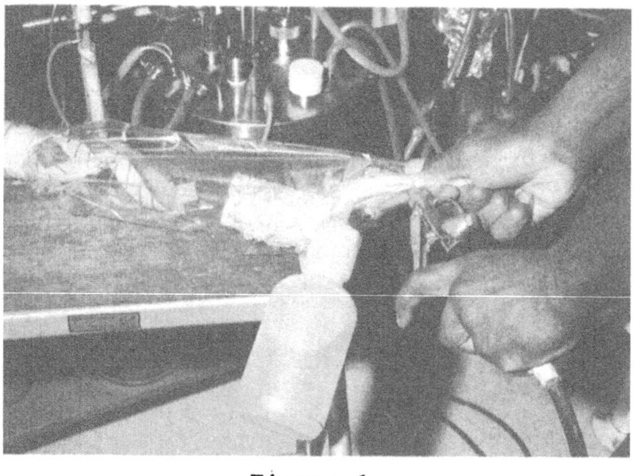

Figure 6

Cell Plant

Figure 4 provides a general view of one level of the "cell plant" which we use for growing large amounts of human lymphoblasts for virus studies (3), immunological and genetic studies (6), and the production of antilymphocyte serum (globulins) (7). The intermediary suspension culture units or trophocell have an initial stage with a capacity of 1.5 liters and a secondary unit which has a capacity of approximately 25 liters. The trophocell units are equipped with various sensors and servo-mechanisms depending on the nature of the experiment. Technical details of these culture units have been published (8). Similarly, the 200-liter stainless steel culture units (Fig. 5) have been described (8).

Pumps with metal bellows* have provided an excellent system for the continuous feeding of the trophocell units. Roller pumps, such as the one manufactered by Travenol for hemodialysis units, are used for feeding and harvesting the 200-liter culture vessels.

The modification of a gnotobiotic technique for making sterile connections between tubing (9) has solved a major source of contamination. Figure 6 illustrates the filling of a plastic sleeve with peracetic acid vapor in preparation for such a connection. One end of the sleeve is closed over a tube utilizing a plastic tie by Panduit**, the sterile wrappings, if any, on the tube ends are loosened, the other tube is inserted into the sleeve, peracetic acid vapor is sprayed into the sleeve, and the second end of the sleeve closed about the second tube (Fig. 6a).

After approximately 30 mintues the connection is made. All connections are tested to insure sterility. For example, when a culture vessel is initially set up (after sterilization) and several connections have been made, approximately 1 liter of culture and 2-3 liters of media are pumped into the vessel. Twenty-four hours later a sample of the culture is taken to determine sterility. If the various connections were not sterile, additional amounts of culture are added. In the case of a bad connection, the entire culture is not lost, only the small portion that was used for testing the connection.

When 2 tubes are disconnected, each having the possibility of being connected again to other units, special precautions are taken to guard against bacteria getting into the tubing while it is not connected. The tubing is double clamped, using a Hoffman*** clamp,

---

* Research Appliance Co., Allison Park, Pa.
** Panduit Corporation, Tinley Park, Ill.
*** Humboldt Manufacturing Corp.,

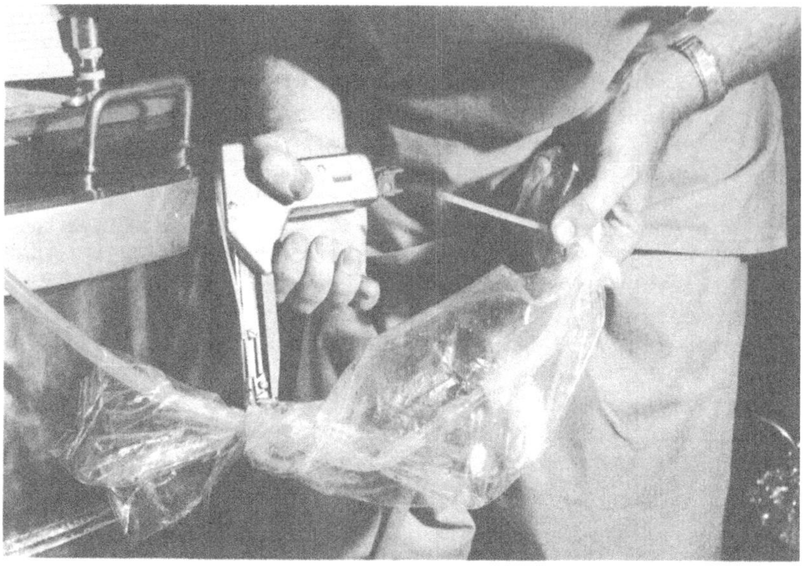

Figure 6a

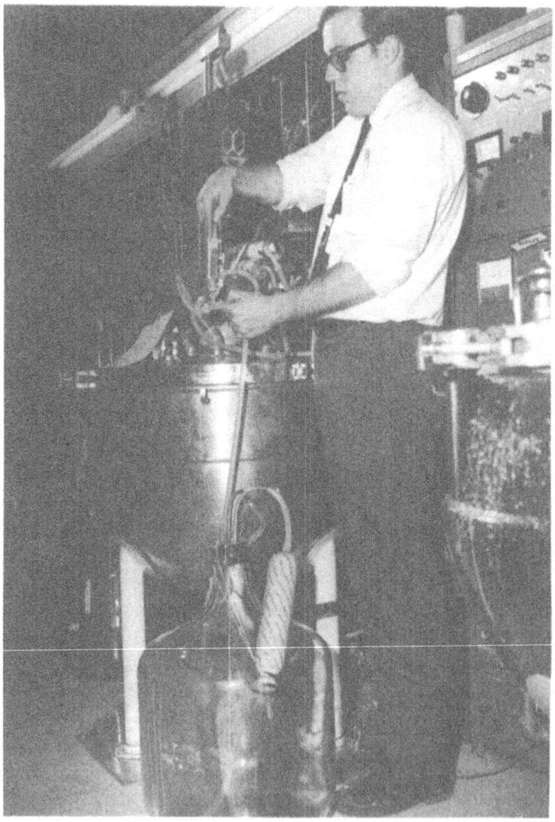

Figure 7

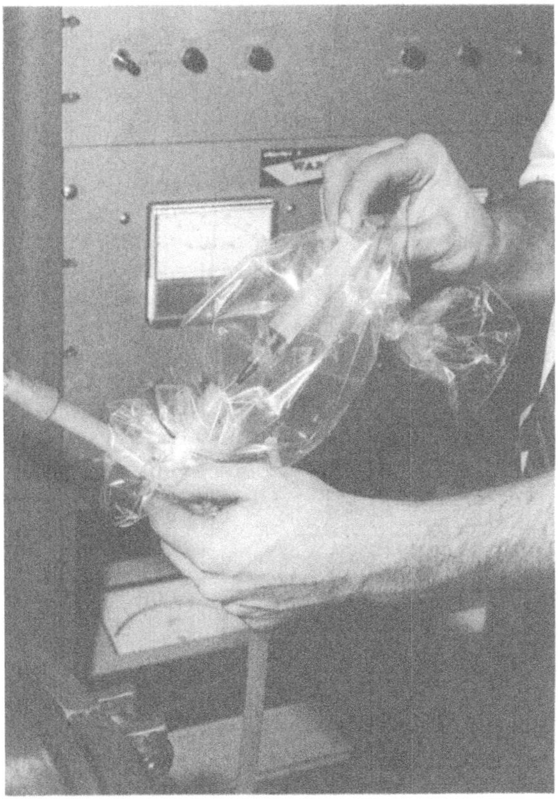

Figure 8

Figure 9

with an open side, approximately 2" from the end. The two ends are inserted into separate plastic sleeves, tied off at one end, filled with Cidex and fastened with a plastic tie to the tubing. When the tubing end is replaced in a gnotobiotic sleeve for resterilization, the distal clamp is removed so the clamped area will be exposed to the peracetic acid.

We have not found a perfect solution to the problem of sampling large culture units. In most instances a flanged rubber stopper of the type used for bottles with injectable fluids is attached to a plastic tee which is inserted into a tubing that is connected to a port that has tubing attached (inside vessel) in order to reach the bottom of the vessel. The tubing is clamped off until a sample is to be drawn, at which time the clamp is removed allowing a small amount of culture to flow through the tubing into a sterile bottle. The inverted flange of the stopper is filled with Cidex; after approximately 60 seconds a needle is inserted and a sample is withdrawn from the tee (Figure 7). The clamp is again loosened to allow a small portion of the culture to flow through the tubing in order to flush any bacteria that might have been introduced, by the needle, into the sterile bottle. This guards against the possibility of any of the solution from being "sucked" back into the culture vessel. If the bottle becomes contaminated, it is simply disconnected and a sterile bottle is connected. It does not affect the culture in the vessel.

A modification of the sampling procedure was to enclose the sampling unit and syringe in a plastic sleeve (Figure 8), but this was discontinued due to the amount of time involved.

We tested the use of peracetic acid vapor for the sterilization *in situ* of the 200-liter culture vessels with attached air filters and short lengths of tubing. Approximately 200 to 500 ml of the acid was pumped into the sealed vessel and the vessel gently warmed. This method was successful occasionally but not as effective as steam sterilization. It was ineffective, as might be expected, in the presence of media containing serum.

Maintenance of a steady state of feeding and harvesting is extremely difficult because of variation in the multiplication of the leukovirus and other unknown factors affecting the growth of these particular cells. For example, the accumulation of lactic acid over 200 mgs. % seemed to have a deleterious effect on the culture despite neutralization with sodium hydroxide. Dialysis units were helpful in removing lactic acid and other unidentifiable cell products plus partially correcting deficits in glucose and amino acid levels (10). A preliminary trial of a pre-sterilized hemodialysis unit designed for clinical use was successfully applied to one of the 200-liter culture units (Figure 9).

Figure 10

Harvesting a large volume of intact human cells is difficult without interrupting sterile conditions. Most continuous flow centrifuges* (Figure 10) have a certain capacity and beyond that, cells may be lost. Consequently, harvesting must be interrupted, the cells removed, and the centrifuge resterilized before continuing to harvest a culture. It would be desirable to enclose the centrifuges in a gnotobiotic chamber.

Tissue Culture Media

All tissue culture medium, sera, combinations of the two, balanced salt solutions, and Trypsin solution are sterile filtered through a millipore high volume sterilizing filter holder utilizing millipore GS (0.22 micron) pore size, 293 mm diameter filter discs under maximum pressure of 16 pounds.

The sera used is heat inactivated at 56° C for 35 minutes in an agitating water bath.

Routine sterility checks on the above are performed. After an initial sample has been filtered into a bottle containing sterility broth (Brain Heart Infusion Broth, yeast extract, and glucose) the filter disc is checked by increasing the pressure to approximately 45-47 lbs. to insure that the disc is intact. Once this is determined, another sample is taken for sterility check, the filtering and bottling procedure then begins. Upon completion, another sterility sample is taken and tested in BHI broth. The filter disc is checked again as previously described. At least six (6) bottles each containing approximately 333 ml are selected at random and also placed on sterility check. These samples are kept at 37° C for a period of at least seven (7) days, during which time the entire bottled batch is isolated in a cold room at 4° C. If the batch of media is sterile, it is distributed to the various laboratories. Once a bottle is opened for culture feeding, it is kept at 37° C so that contamination will show up in 24-48 hours.

III. DISCUSSION

The purpose of this brief exposition was to illustrate some of the ways in which gnotobiotic techniques have been used in our cell laboratories. Admittedly, our experiments and cell lines are of a special nature and these methods may require modification for other purposes. Further, there are many practical problems which have

---

* Sharples, Division, Pennsalt Chemical Corp., Warminster, Pa.

not been solved satisfactorily. The "sterile" rooms for handling cultures should be redesigned so as to provide a laminar flow of air which is free of virus, bacteria and toxic gases. I do not advocate a return to the era in which the mechanics of cell culture stifled its development as an important biological device. However, the level of precautions should match the eventual use of the cultured cells. If human cells and cell products are to be used for clinical purposes, the techniques to avoid contamination of the cells with toxins, micro-organisms, and other cells should be continuously improved.

It was noted that the use of antibiotic free media does not insure immediate recognition of contamination with micro-organisms. Therefore, better techniques of preparing media, serum supplements, and initial cell cultures are needed.

The culture of lymphoblastic cells may provide an additional complication since they produce interferons and immunoglobulins which may suppress but not destroy some microorganisms. We have completed preliminary studies which demonstrated that these lymphoblastic cells produced specific antibodies to a phage-antigen (11). Some cells of these human lymphoblastic cell lines are also capable of phagocytosis which may further complicate the problem.

There is great need for the application of creative bioengineering to cell culture methods. There is little doubt but that some erroneous scientific conclusions will be made because of contamination and that unnecessary dangers will attend clinical applications of cell cultures until better cell culture systems are devised.

CONCLUSIONS

1) Gnotobiotic techniques have been adapted for the improvement of cell culture methods.

2) There is a great need for providing microorganisms--free environment for cultured cells which will be used directly or indirectly for clinical applications.

REFERENCES

1. Gartler, S. M., Apparent HeLa Cell Contamination of Human Heteroploid Cell Lines. Nature 217: 750-751, 1968.

2. Moore, G. E., Gerner, R. E., and Minowada, J., Studies of Normal and Neoplastic Human Hematopoietic Cells In Vitro. The 21st Annual Symp. Found. Cancer Res. Houston, Texas, in press.

3. Moore, G. E., Minowada, J. and Kitamura, H., Incidence of "Leukovirus" in Cultured Human Hematopoietic Cells, in press.

4. Moore, G. E., Grace, J. T., Citron, P., Gerner, R. E. and Burns, A. A.: Leukocyte cultures of patients with leukemia and lymphomas. New York State Journal of Medicine. 66: 2757-2764, 1966.

5. Moore, G. E., Gerner, R. E., Kitamura, H., Minowada, J. and Fjelde, A.: Lymphocytic cell lines derived from normal individuals. The Third Annual Leukocyte Conference, Univ. of Iowa, Iowa, November 9-11, 1967. In press.

6. Tanigaki, N., Yagi, Y., Moore, G. E., Pressman, D. Immunoglobulin Production in Human Leukemic Cell Lines, Journal of Immunology, 97, 5:634-646, 1966.

7. Merkel, F., Moore, G. E., Najarian, J.: Preparation, Testing and Use of Human lymphoblast serum. Federation Proceedings, April 1968.

8. Moore, G. E., Hasenpush, P., Gerner, R. E., Burns, A. A.: A pilot plant for Mammalian Cell Culture, in press, September, 1967, Symposium of Biotechnology and Bioengineering.

9. Moore, G. E., Technique for Sterile Connections of Cell Culture Apparatus, Applied Microbiology, 16: 163, January, 1968.

10. McLimans, W. F., Mount, D. T., Bogitch, S., Crouse, E. J., Harris, G. and Moore, G. E.: A Controlled Environment System for Study of Mammalian Cell Physiology; Annals of the N. Y. Acad. of Sciences. 139, 1:190-213, 1966.

11. Kamei, H., Moore, G. E. Antibody Production Stimulated in Vitro in Burkitt Lymphoma Cells, submitted for publication. March 1968.

ETHYLENE-OXIDE STERILIZATION OF PLASTIC-FILM ISOLATORS: A SIX-YEAR STUDY OF OPERATIONAL EFFICIENCY

Donald M. Robie and Harry E. Walburg, Jr.

Biology Division, Oak Ridge National Laboratory

Oak Ridge, Tennessee

INTRODUCTION

Ethylene oxide (ETO), an alkylene oxide, has been used for many years to kill microorganisms and insects (1,2). The principle use in the past has been the fumigation of foodstuffs and spices; ethylene oxide is still used extensively in agricultural industries for this purpose (3). During the last 20 years the effect of time, temperature, relative humidity, and concentration of ETO on sterility has been repeatedly discussed (4,5,6,7,8).

Pure ETO is dangerous because of its flammability, but as early as 1959 the use of nonflammable mixtures of ETO and other gases was reported (9).

The use of ETO for sterilization of non-agricultural products has increased in recent years with the increase in need for sterile medical equipment; i.e., blood transfusion sets, heart-lung machines, disposable syringes and needles, and other plastic goods (10,11,12). As early as 1960, ETO was recommended as a sterilant for interplanetary space vehicles (13), and at least one report on ETO sterilization of equipment used in rearing germfree animals has appeared in the literature (14).

This paper reports observations on the successful routine ETO sterilization of equipment which has been used for rearing and maintaining germfree mice by the Gnotobiotics Group of the Biology Division, Oak Ridge National Laboratory, over the past six years.

MATERIALS AND METHODS

In our facility, isolators, husbandry equipment, and research equipment required in the gnotobiotics program are routinely sterilized with ETO. Food, water and bedding are not sterilized with ETO but are sterilized with high pressure steam and passed into isolators with the standard peracetic acid transfer technique. Polyvinyl chloride (PVC) plastic film isolators (15), either 26", 48", or 60" long, 24" wide and 24" high, purchased primarily from the Snyder Mfg. Co. (New Philadelphia, Ohio), are constructed of 20 mil laminated press-polished PVC film, with metal inlet filters and fiberglass outlet traps and door ports. Twenty-five mil thick neoprene gloves are attached to the isolator with compression-type aluminum glove rings, and the entire unit is supported in a plywood or fiberglass base. Air inlet and outlet connector tubes are of high quality "Tygon". These tubes, as well as the door port and outer cap, are attached to the isolator with 471 pressure sensitive tape (Minnesota Mining and Mfg. Co., St. Paul, Minnesota).

Husbandry equipment which is routinely placed inside the isolator for sterilization includes acetate-butyrate disposable plastic cages, glass pint water bottles with latex or neoprene rubber stoppers and stainless steel (S-S) drinking tubes, S-S cage covers, S-S forceps with latex rubber tips, thin polyethylene and paper bags, swabs (cotton on wood), glass or plastic bottles for fecal samples, leather and cloth work gloves, PVC food tray covers, cloth diapers, and a metal ear punch. Additional equipment frequently sterilized with the isolators include S-S scissors, wooden pencils, plexiglass X-ray turntable (including electric motor), incandescent lamp and bulbs, electric hair clippers, mouse holders, syringes, needles, pipettes and automatic pipettors.

Prior to sterilization, all isolators are carefully washed, inspected and electronically "freon-checked" for holes, which are patched if found. Equipment is arranged in the isolators to permit adequate circulation of the sterilant during the sterilizing cycle. The door port inner cap and its rubber retaining ring are placed inside the isolator, but they are not installed on the door port so that the door port air lock may be sterilized. The door port outer cap is taped in place, and the outer plastic sleeve on the inlet filter is pulled down to permit unobstructed entry and exit of gases during the cycle.

The equipped isolator units (see figure 1) are covered with plastic sheeting to protect against excessive moisture and the small droplets of liquid ETO which often enter the sterilizer chamber despite proper gas conditioning. Up to two 60" isolators or four 26" isolators are positioned on a double-decked cart and

Fig. 1. An isolator equipped and ready for gas sterilization.

placed in a 62" x 84" x 84" American Sterilizer Co. (Erie, Pa.) double-walled pressure vessel (sterilizer), which is designed to operate with either high-pressure steam or ETO as sterilant. Vacuum is attained with a Siemen & Hirsch Co. (Germany) water seal pump.

A variety of conditions have been used in the sterilizing cycle, but the most common, and those presently in use, are shown in figure 2. An initial vacuum of 23" Hg is produced and maintained while humidity is increased to 35% relative humidity (R.H.). This "dwell" time is approximately 30 minutes. ETO is introduced into the chamber to 18 psig, which represents a concentration of 650 mg/liter of ETO when 20% ETO in carbon dioxide is the sterilant. This concentration of ETO is maintained for 6 hours; then the sterilizer chamber is exhausted to atmospheric pressure and further evacuated to a vacuum of 23" Hg. The sterilizer chamber is then backfilled with carbon dioxide to atmospheric pressure. Another vacuum is produced and the chamber is backfilled with filtered air. This much of the cycle is automatically controlled by the sterilizer and takes approximately 8 hours. Early in our operation, ETO concentrations in excess of 500 ppm in the surrounding air were measured after opening the

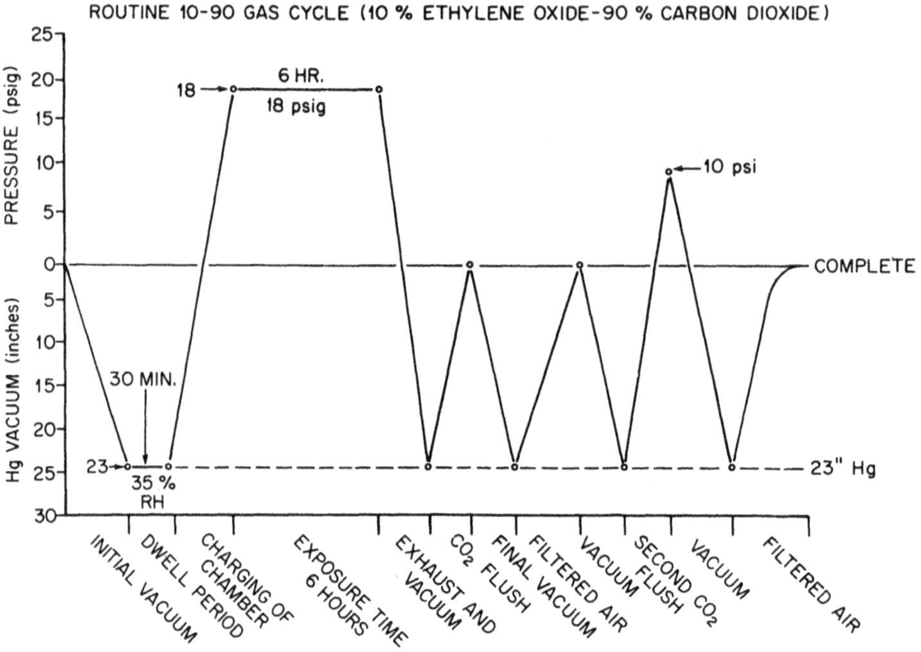

Fig. 2. Conditions obtained during routine ethylene oxide sterilizing cycle.

the door of the sterilizer at the completion of the cycle. Because residual ETO in the air and in the neoprene gloves posed health problems (16), the following manual operations were added to the automatic cycle. A fourth vacuum, to 23" Hg, is produced; the chamber is charged with carbon dioxide to a sterilizer chamber pressure of 10 psig and then exhausted to atmospheric pressure. A final vacuum followed by a filtered air backfill completes the standard cycle. Total cycle time is nine hours.

The isolators are removed from the sterilizer within 12 hours after completion of the sterilizing cycle. The protective plastic sheets are removed and the isolators inspected visually for defects; i.e., stoppers blown off, tape slippage, excessive moisture. The sterilizer chart is examined to verify that adequate temperature, sterilant pressure and exposure time were used. The inlet filter sleeve is replaced, the inner door port cap is installed, and the isolators are attached to an air supply. The isolators are ready for use within an hour after completion of these procedures, but they are routinely aerated for a minimum of 24 hours prior to introduction of mice, food, water, or bedding.

Sterility of isolators was determined by the culture of bacterial spore strips (Bacillus subtilis spores on paper strips ["Spordex"], American Sterilizer Co., Erie, Pa.) placed at various locations in the isolators.

RESULTS AND DISCUSSION

Although the most common method of sterilizing isolators and the equipment contained in them is fogging with a 2% peracetic acid aerosol (17,18), sterilization of isolators and equipment with ETO gas mixtures has many advantages when compared to the peracetic acid aerosol method. Fogging with peracetic acid requires a considerable amount of labor and presents technical difficulties; i.e., equipment becomes slippery and visibility is poor in the isolator during the procedure. The aerosol is an irritant to the respiratory tract and eyes, resulting in extreme discomfort to the operator or requiring protective measures. By contrast, ETO sterilization of isolators is completed with little labor and no discomfort to the operator. When isolators are sterilized with peracetic acid, the unit must be disassembled, the inlet filter sterilized separately, and the isolator reassembled. This procedure involves considerable labor and poses an additional risk of contamination to the sterilized isolator. With the technique described in this report, the intact unit is sterilized by exposure to ETO, and sterility need not be compromised by reassembly. Peracetic acid leaves the isolator and equipment coated with a film of this extremely corrosive sterilizing solution. The great amount of moisture in the chamber and the corrosiveness of the peracetic acid vapor require that the unit be dried prior to use. These conditions also prevent the sterilization of sensitive instruments, electric motors, paper goods, and other important research equipment. By contrast, ETO sterilized isolators are dry and usable at the end of the sterilizing cycle. In addition, almost any required piece of equipment can be sterilized along with the isolator.

Both ETO and peracetic acid are economical, readily available through commercial sources, and equally convenient to store and handle. Although ETO causes some deterioration of neoprene and therefore necessitates careful inspection of gloves which have been repeatedly sterilized, a scheduled change of isolator gloves after 12 months of use or exposure to 6 ETO sterilizing cycles has prevented loss of isolators due to glove deterioration in our facility. Moreover, the initial cost of sterilizer installation for use with ETO is sizable; hence wherever possible such sterilizers should have multiple uses; e.g., sterilization of large amounts of material for use in a barrier facility or for other purposes in addition to sterilization of isolators and their equipment.

TABLE 1. ROUTINE USE OF VARIOUS ETHYLENE OXIDE STERILANTS IN THE GNOTOBIOTIC FACILITY

ETO %	Dilutant	Time Period in use	Number of Cycles	Number of Isol.	Cylinder Pressure (psig)	Commercial Name
20	Carbon Dioxide	1961 to Nov 5, 1967	548	908	640	Oxyfume 20
12	Freon*	Aug 9,1966 to Nov 3,1966	11	17	60	Penn Gas Oxyfume 12
10	Carbon Dioxide	March 8, 1966 to April 1968	41	75	750	Carboxide

*Dichlorodifluoromethane CCL_2F_2 (UCON-12)

Over the past six years, approximately 600 sterilizing loads containing over 1000 isolators have been successfully sterilized with ETO using the techniques described above. Several mixtures of ETO in other gases have been used, including 10% ETO in carbon dioxide, 12% ETO in "freon" and 20% ETO in carbon dioxide, the latter receiving the most extensive use (Table 1). There have been only two failures of routine ETO sterilizing procedures during the 6-year study period. One was due to a failure to maintain proper temperature during the cycle; the other, which occurred only recently, apparently was due to improper spacing of equipment in the isolators, preventing proper circulation of the sterilant.

After approximately four years of experience with 20% ETO carbon dioxide mixture, the effectiveness of 12% ETO in "freon" was studied. Seventeen isolators were satisfactorily sterilized with this gas mixture, which provided ample proof of its sterilizing capability (Table 1). Additional test cycles were performed with this gas mixture. Only one spore strip was not sterilized in the test cycles and that one was placed in the center of a thick book inside an isolator (Table 2). It is noteworthy that sterility was achieved with concentrations of ETO and sterilization times considerably below the manufacturer's recommendations of 650 mg/liter ETO (19), but the results are consistent with studies which show that a concentration of 450 mg/liter of ETO will kill microorganisms under proper conditions (20,21).

It should also be noted that sterility of isolators can be achieved in the absence of moisture, despite considerable evidence that humidification of certain other materials is an important

TABLE 2. STERILIZING EFFICIENCY OF 12% ETHYLENE OXIDE IN FREON

ETO Conc. mg/liter*	Expos. Time (hrs.) 130°F	Chamber Pressure (psig)	Relative Humidity (%)	No. of Test Cycles	No. spore strips Pos./Tested
700	6	10	35	3	0/18
680	4	9	35	1	0/6
680	6	9	35	1	0/5
680	3	9	35	2	0/10
680	2	9	35	1	0/5
680	4	9	0	2	1/10**
600	2	7	0	1	0/5
500	2	4	0	5	0/25

*Determined by use of graph on page 12, "Linde Gas Sterilants" Union Carbide Corporation.

**The one positive spore strip was placed in the center of a thick book inside isolator.

factor in obtaining sterility with ETO (7,20).

Because halogenated hydrocarbons ("freon") are solvents for the plasticizers in many flexible plastics, including the PVC used in the isolator and the plastic used for backing 471 pressure-sensitive tape, problems arose with the use of ETO-freon gas mixtures. Studies were then begun with another ETO gas mixture, 10% ETO in carbon dioxide. After initial testing provided evidence that sterility could be achieved with practical pressure, exposure time, and temperature requirements, routine sterilization of isolators with this sterilant was begun. Since March 1966, 75 isolators have been successfully thus sterilized (see Table 1). Additional data (Table 3) suggest that not only is this sterilant effective below the manufacturer's recommended concentration and time, but also below the minimum effective concentration reported in the literature (20,21). The concentration of ETO was determined by mass spectrometric analysis of a sample obtained from within the isolator during a sterilizing cycle.

These data are consistent with other reports which demonstrate the importance of temperature, exposure time, and concentration of ETO in achieving sterility (20-24). Manufacturer's recommended conditions provide a sizable safety factor for achieving sterility. All of the ETO gas mixtures used were highly effective in achieving sterility of the isolators and the equipment which they contain. These techniques are safe and efficient; they are also economical

TABLE 3. STERILIZING EFFICIENCY OF 10% ETHYLENE OXIDE IN CARBON DIOXIDE

ETO Conc. (mg/liter)*	Expos. Time (Hrs.) 130°F	Cham. Press. (psig)	Relative Humidity (%)	No. Test Cycles	Spore Strips Pos/Tested
<350	6	18	35	3	0/10
<350	2	18	35	4	0/13
<350	1	18	35	2	1/10
<350	1/2	18	35	2	3/8
<200	2	3-5	35	2	1/6
<200	1	4	35	3	0/12

*Determined by Mass Spectrometry and the Ideal Gas Law

if the initial sterilizer installation costs are amortized over a period of years.

ACKNOWLEDGEMENT

We would like to express our appreciation to the Union Carbide Corporation, Linde Division, and to the Pennsylvania Engineering Company (Philadelphia, Pa.) for providing some of the sterilants used in the testing program.

REFERENCES

1. Cotton, R. T., and Roark, R. C.: Indust. and Engin. Chem. 20: 805 (1928).
2. "Fumigating with 'Carboxide Gas', pamphlet, Carbide and Carbon Chemicals Division, Union Carbide and Carbon Corp.
3. Yesair, J., and Williams, O. B.: Spice contamination and its control. Food Res. 7: 118 (1942).
4. Phillips, C. R., and Kaye, S.: Am. J. Hyg. 50: 280 (1949).
5. Perkins, J. J.: "Principles and Methods of Sterilization." Charles C. Thomas, Springfield, Ill. 1956.
6. Gilbert, G. L., Gambill, V. M., Spiner, D. R., Hoffman, R. K., and Phillips, C. R. Appl. Microbiology 12: 496 (1964).
7. Ernst, R. R., and Shull, J. J.: Appl. Microbiology 10: 337, (1962).
8. Ernst, R. R., and Shull, J. J.: Appl. Microbiology 10: 342 (1962).
9. Haenni, E. O., Affens, W. A., Lento, H. G., Yeomans, A. M., and Fulton, R. A.: Ind. Eng. Chem. 51: 685 (1959).

10. Kaye, S.: J. Lab. and Clin. Med. 35: 823 (1950).
11. Znamirowski, R., McDonald, S., and Roy, T.: Canad. M. A. J. 83: 1004 (1960).
12. Snow, J. C., and Anderson, M. L.: Mod. Hospital 105: 128 (1965).
13. Phillips, C. R., and Hoffman, R. K.: Science 132: 991 (1960).
14. Reed, L. L., Gidner, R. S., and Dinger, H. S.: Apparatus for Sterilizing Components of Interplanetary Craft. American Sterilizer Co., Erie, Penna. (1962).
15. Trexler, P. C. and Reynolds, L. I.: Appl. Microbiology 5: 406 (1957).
16. Royce, A., and Moore, W. K.: Brit. J. Ind. Med. 12: 169 (1955).
17. Barrett, J. P., Jr.: Proc. Animal Care Panel 9: 127 (1959).
18. Wescott, R. B., and Gardner, J. A.: "Apparatus and Method for the Steam Sterilization of Food for Germfree Laboratory Animals." U. S. Army Chemical Corps Biological Laboratories, Fort Detrick, Maryland. April 1962.
19. Personal communication, American Sterilizer Company, Erie, Penna. (1961).
20. Lloyd, R. S., and Krahe, J. B.: "Gaseous Sterilization with Ethylene Oxide (A supplement to Chapter XXI), "Principles and Methods of Sterilization" by Perkins, J. J. American Sterilizer Co., Erie, Penna. (1958).
21. Claghorn, A.: J. Am. Assn. for Inhalation 11: 76 (1966).
22. Phillips, C. R.: "The Sterilizing Properties of Ethylene Oxide." Paper read at Symposium on "Recent Developments in the Sterilization of Surgical Materials," organized by the Pharmaceutical Society of Great Britain and Smith and Nephew Research Ltd. and held at the School of Pharmacy, University of London on April 11-13, 1961, printed in England by Eyre and Spottiswoode Limited, Her Magesty's Printers at the Thanet Press, Margate.
23. Perkins, J. J., and Lloyd, R. S.: "Applications and Equipment for Ethylene Oxide Sterilization." Paper read at above symposium (Ref. 22).
24. Stierli, H., Reed, L. L., and Billick, I. H.: Public Health Monograph 68 pages 1-15, (1962).

(Research sponsored by the U. S. Atomic Energy Commission under contract with the Union Carbide Corporation.)

REPRODUCTION IN GERMFREE BEAGLES

James B. Heneghan, Ph.D., Salvador G. Longoria, D.V.M. and Isidore Cohn, Jr., M.D.

Department of Surgery, School of Medicine, Louisiana State University Medical Center, New Orleans, Louisiana

At the time the abstract for this meeting was submitted, we had observed a successful mating in a breeding pair of germfree beagles; hence, the title "Reproduction in Germfree Beagles." However, now this title could be expanded to "Successful Reproduction in Germfree Beagles." On April 7, 1968, second generation germfree dogs were whelped in our laboratory. The following is a description of the events leading up to this accomplishment.

INTRODUCTION

I would like to briefly review the history of second generation germfree animals in other species as seen in Table I.

Table I

Second Generation Germfree Animals

Year	Species	Laboratory
1948	Chickens	LOBUND
1951	Rats	LOBUND
1958	Mice	LOBUND
1959	Guinea Pigs	LOBUND
1960	Japanese Quail	G.F. Center, Tampa
1961	Rabbits	LOBUND
1968	Dogs	L.S.U.

In 1948, second generation germfree chickens were reared at LOBUND. In 1951 the LOBUND group reared second generation germfree rats. Second generation germfree mice, however, were not obtained until 1958 at LOBUND. One year later in 1959, LOBUND obtained second generation germfree guinea pigs. Reyniers in Tampa, Florida obtained second generation Japanese quail in 1960. Germfree rabbits failed to reproduce until 1961 when LOBUND obtained the first litter of germfree rabbits which did not nurse well and soon died. It was not until 1963, after various corrections in hand feeding techniques and diet, that second generation rabbits were weaned and reared to adulthood. Thus, successful reproduction had been obtained only in rodents and in birds. However, now canines have been added to this list by our laboratory.

MATERIALS AND METHODS

In order to obtain successful reproduction, two breeding pairs of dogs were maintained for approximately two years according to the techniques previously reported by our laboratory (1,2). Three of the dogs, two females and one male, were littermates born on April 7, 1966. The fourth dog, a male, was born on May 14, 1966. After 6 months of age each breeding pair was maintained together in the same isolator. Up to 18 months of age each female had gone through two normal heat periods with no mating observed. At the time of the first heat period, approximately 11 months of age, viable sperm could not be demonstrated in either male. However, 6 months later, at the time of the second heat period, viable sperm were observed in both males and the failure to breed was attributed to a lack of experience on the part of the male dogs. At this time artificial insemination was attempted. Although we were able to successfully collect semen from conventional dogs which were experienced studs, all efforts to collect semen from the germfree males failed. It was then decided that, when the third heat period was indicated by vaginal smears, the female in heat would be transferred into the isolator with the other male. Fortunately the first female to come into heat was the one which was already housed with her littermate, thus we moved her into the isolator with the male which was not her littermate. Three days after the transfer a mating was observed; followed by another mating one day later.

RESULTS

A summary of the effort required to maintain these 4 dogs for 2 years is found in Table 2.

Table 2

Second Generation Germfree Dogs
Maintenance Required
4 dogs 4-7-66 to 4-7-68

Isolator sterilizations	18	Milk	30 Liters
Glove entries	4,093	Water	1,671 Liters
Sterile lock entries	256	Feed	1,354 Pounds

Eighteen different isolator sterilizations were necessary; there were over 4,000 glove entries into these isolators and there were over 256 sterile entries with food and water. This amounted to the sterilization of over 1,700 liters of water and milk plus 1,350 pounds of dog food, which is over a half ton of feed. All these manipulations were accomplished without a contamination. The results of the experiment can be seen in Figure 1.

Figure 1. Second Generation Germfree Beagles.

Four male puppies were born on April 7, 1968 which, coincidentally, was the bitch's two year old birthday. Figure I illustrates that the puppies were healthy and viable. The delivery was normal and no assistance was given or required. The puppies were not weighed immediately for fear of disturbing the mother. However, we extrapolated the growth curve backwards (Figure 2) and we feel that they must have weighed approximately 300 grams at birth, since they weighed an average 375 grams at 3 days. They weighed 560 grams at 1 week; 920 grams at two weeks; 1290 grams at 3 weeks; and 1560 grams at 4 weeks of age. Not only did we obtain good viable second generation germfree puppies but the resultant maternal care promoted a rate of growth which was comparable to conventional beagles and was twice as fast as our hand reared Cesarean derived first generation germfree puppies.

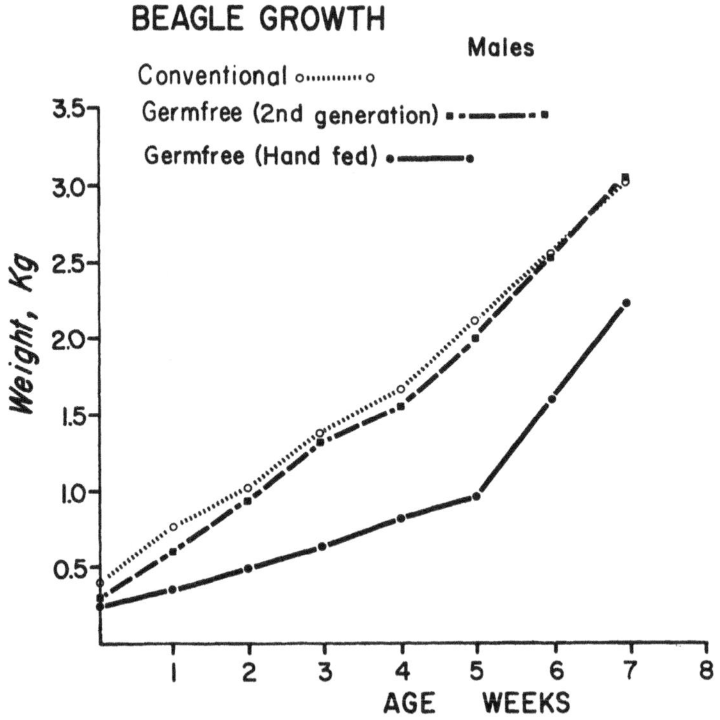

Figure 2. Growth of Germfree and Conventional Beagles.

DISCUSSION

Obtaining second generation germfree dogs proved that the techniques employed in sterilizing diets, the vitamin fortification and an amino acid supplementation were adequate to support reproduction and growth. As you know, past experience in most of the other species of germfree animals, has shown that reproduction is one of the last things which has become normal in the overall process of technical development. Also, we have been searching for a way to improve the growth of our puppies during this early stage of life; hand feeding is definitely a poor substitute for maternal care. Animals which grow normally during early development make better experimental animals than hand fed animals. This brings up the question of whether it would be practical to breed and maintain a germfree dog colony. An important item to consider is that it may not be necessary to wait for each bitch to become two year's old before she reproduces. Indeed, if it does, it might not be feasible to maintain a germfree colony. However, I feel that experienced males will mate with the females when they come into heat the first time, that is, between 9 to 12 months of age. Each female would come into heat approximately twice a year and four to six puppies could be expected from each litter. An increase in the number of viable offspring weaned per female maintained is probable; this has happened in S.P.F. cats, which have a higher percentage of pregnancies and higher weaning rate than conventional cats (3).

CONCLUSIONS

Second generation pure bred germfree beagles have been obtained in the Germfree Laboratory at L.S.U. These four beagle puppies have grown at a rate approximately twice that of hand fed first generation germfree dogs. In the future, colonies of pure bred germfree dogs may be maintained.

REFERENCES

1. Heneghan, J.B. Floyd, C.F., and Cohn, I., Jr.: Gnotobiotic dogs for surgical research. J. Surg. Res. 6:23-30, 1966.

2. Heneghan, J.B., Longoria, S.G., and Cohn, I., Jr.: Maintenance and growth of gnotobiotic beagle dogs. "Advances in Germfree Research and Gnotobiology." Ed. M. Miyakawa and T.D. Luckey, Chemical Rubber Co. Press, pp. 63-67, 1968.

3. Griesemer, R.A. (personal communication).

THE ANIMAL TECHNICIAN: RESIDENT AND TRANSIENT MICROORGANISMS

Howard W. Clapp

Research Laboratories, The Upjohn Company

INTRODUCTION

The requirements for the production of healthy prime quality controlled laboratory animals include isolation facilities that protect and maintain these animals from organisms which cause disease. Important requisites to accomplish this are: adequately trained personnel, particular design and engineering of buildings and equipment, exacting animal husbandry practices, and a definite awareness of the microbes in the animal environment.

Duguid and Wallace (4) and Hare and Thomas (7) have shown that human beings disseminate microorganisms into their surrounding environment even in the absence of fecal, genito-urinary, respiratory and purulent discharges. Eichenwalt _et al_ (5) and Hare and Ridley (8) demonstrated that people can shed pathogenic organisms while revealing no lesions, signs or symptoms of illness. These people have been termed "carriers."

A comprehensive sanitizing program that uses heat, gases, filtration, irradiation and chemicals makes it practical to sterilize or disinfect the majority of materials that come into contact with laboratory animals. However, the animal technician who frequently handles these animals can not be sterilized.

The purpose of this study was twofold: 1) to enumerate the microorganism present on the external skin surface of the animal technician and 2) to test procedures that would qualitatively and quantitatively alter these microorganisms.

MATERIALS AND METHODS

The animal technicians who participated in this study worked in similar environments. The technicians were divided equally in each experiment so that one-half had thick and one-half had thin body hair coverage.

Rinse vs. Shower

Two animal technicians participated in this experiment. Three areas of skin surface were tested. 1) A 2 x 8 inch surface on the right forearm exposed to the external environment. 2) A 4 x 6 inch surface under the left arm over the rib cage not exposed to the external environment. 3) A 4 x 6 inch surface on the left thigh over the rectus muscle not exposed to the external environment.

The subjects rinsed with water or bathed with bar soap in a shower for 5 minutes. The rinsing, washing and drying procedures were from head to foot with the genital and anal areas dried last. A sterile bath towel was used for drying.

The skin surfaces were tested for microorganisms before and after the rinse or shower as follows: a Caligswab was first wetted in sterile brain-heart infusion broth and then moved forward and backward 10 times to cover each of the marked out skin areas. A new swab was used for each area. The culture media were then inoculated with these swabs.

Lotion vs. Bar

Four animal technicians participated in this experiment. A 2 x 8 inch skin surface on the right forearm was the test area. The subjects lathered with soap for 20 seconds, rinsed and dried with a sterile towel. After this procedure, for the arms covered experiment, the technicians put on a sterile long-sleeved shirt. The prescribed skin surface was tested as described above; before, immediately after, and 3 1/2 hours after treatment.

Bacteria on the Hands

Two animal technicians participated in this experiment. A shower was taken with lotion soap. The procedure was the same as previously described. Immediately after the shower, the subjects depressed the palm of their right hand onto the surface of the culture media.

Media and Microbiological Procedures

Brain-heart infusion agar and brain-heart infusion agar supplemented with 5 percent rabbit blood were the test media. Incubation was aerobic at 37°C for 24 hours. No anaerobic methods were used. Colonies were counted and identified using standard microbiological procedures. *Corynebacterium* sp. colony counts were not included in the quantitative microbiology. Accurate counts of these minute colonies were not possible because of the opaqueness of the culture medium.

The media was dispensed in sterile plastic petri plates and/or bioassay trays.

RESULTS

Rinse vs. Shower

Table 1 shows the species of microorganisms detected on the skin surface and the percent of recovery in decreasing order before and after a rinse or shower. Assays were done to compare the efficacy of a rinse (water only) with a shower (wash with hexaclorophenated bar soap and water) for reduction of these skin microorganisms. Eleven species of microorganisms were discovered on the skin surface of the subjects before a rinse or shower. *Corynebacterium* sp. (diphtheriods) and *Staphylococcus aureus* (mostly coagulase negative) were also consistently present after a rinse or shower. Although not shown, *Corynebacterium* sp. was noted in greatest numbers before and *S. aureus* was found in greatest numbers after the rinse or shower. For this reason *Corynebacterium* sp. and *S. aureus* were considered to be resident skin bacteria, and all other strains were grouped as transient microorganisms. A reduction in the recovery for 7 of the 9 species of transient microorganisms occurred after a rinse. Exceptions to this reduction were the increase of *Escherichia coli* and *Aerobacter aerogenes*. After a shower, the incidence of recovery for 8 of the 9 transient species was reduced. The detection of *E. coli*, the one exception, was increased. Greater reduction occurred in the recovery of total microorganisms after a shower than after a rinse.

The averaged microbial colony counts for 3 successive days appears in Table 2. These counts represent the microbes present on 16 square inches of skin surface. The areas tested were the arm, rib and leg. Combined total counts from all three areas were highest before a rinse or shower. After either a rinse or shower,

TABLE 1.

SPECIES OF MICROORGANISMS DETECTED AND THEIR PERCENT RECOVERY*

ORGANISMS ISOLATED	BEFORE RINSE OR SHOWER[1]	AFTER RINSE[2]	AFTER SHOWER[3]
CORYNEBACTERIUM SP.	100	100	100
S. AUREUS	100	100	100
BACILLUS SP.	62.4	33.3	10.2
STREPTOCOCCUS SP.[4]	16.6	11.2	0.0
FUNGI	12.5	2.5	0.0
E. COLI	4.2	11.2	7.6
SARCINA SP.	4.2	0.0	2.5
BREVIBACTERIUM SP.	2.1	0.0	0.0
A. AEROGENES	2.0	2.5	0.0
PSEUDOMONAS SP.	2.0	0.0	0.0
PROTEUS SP.	1.2	0.0	0.0

*BRAIN-HEART INFUSION AGAR + 5% RABBIT BLOOD WAS THE MEDIUM. INCUBATION WAS 37° C FOR 24 HOURS. PERCENTAGES WERE THE TOTAL INCIDENCE OF RECOVERY FOR 2 SUBJECTS.

[1]TOTAL FOR 48 ASSAYS

[2]TOTAL FOR 36 ASSAYS

[3]TOTAL FOR 78 ASSAYS: BAR SOAP CONTAINED 2 PERCENT HEXACHLOROPHENE.

[4]ALPHA AND GAMMA HEMOLYSIS

these total counts were reduced. The lowest levels occurred after a shower. The number of colonies after a rinse in the arm and rib areas increased to greater quantities than before rinse counts, but a decrease was seen in the colony numbers from the leg area. Fifty percent of these after rinse organisms were S. aureus and 50 percent were transient microorganisms. After a shower, an increase in the number of colonies from the rib area was discovered, while a decrease was seen in the arm and leg areas. Eighty percent of these after shower organisms were S. aureus and 20 percent were transient microorganisms.

TABLE 2.

THE AVERAGED MICROBIAL COLONY COUNTS FOR 3 DAYS/16 SQUARE INCHES OF SKIN SURFACE*

AREA	BEFORE RINSE OR SHOWER	AFTER RINSE[1]	AFTER SHOWER[2]
ARM	451	789	416
RIB	98	535	275
LEG	1177	198	240
TOTAL	1726	1522	931

*BRAIN-HEART INFUSION AGAR + 5% RABBIT BLOOD WAS THE MEDIUM. INCUBATION WAS AT 37° C FOR 24 HOURS. COUNTS WERE THE AVERAGE NUMBER OF COLONIES DETECTED FOR 2 SUBJECTS.

[1]50% *S. AUREUS* POST-RINSE. TRANSIENTS 50%.

[2]80% *S. AUREUS* POST-SHOWER. TRANSIENTS 20%.

Figure 1 represents a series of tests that depict the bacteriology of the external skin surface over the rib cage under the left arm. Bacterial counts were taken before and after a shower on 3 successive days. The counts increased immediately after a shower to greater numbers than before shower counts, but with successive daily showers, this increase was substantially lowered. These counts continued to decrease with additional showers. A variation in this trend occurred with Subject #1 on the third day when the increase in after shower counts was greater than the previous days after shower counts. It should be noted that these after shower counts consisted almost entirely of *S. aureus*, while before shower counts showed both resident and transient microorganisms.

FIGURE 1.

THE BACTERIOLOGY OF RIB BEFORE AND AFTER A SHOWER FOR THREE SUCCESSIVE DAYS.

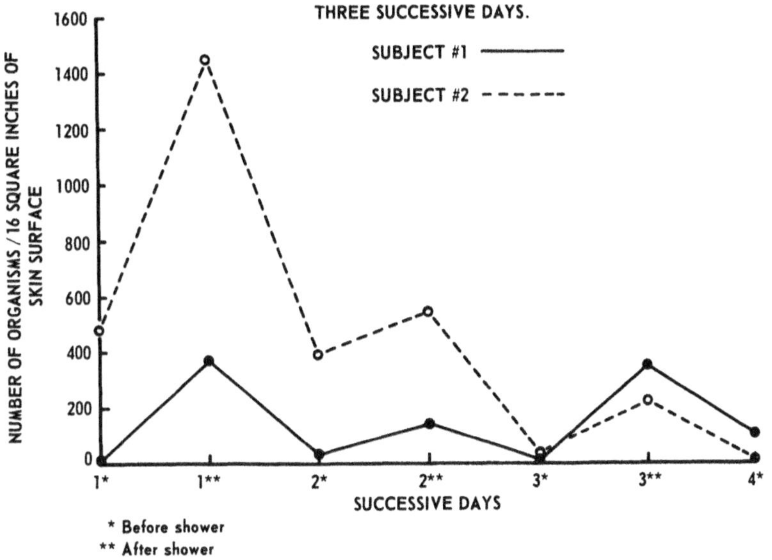

Lotion vs. Bar

Farquharson *et al* (6) found that a wash solution containing hexachlorophene when used in a bathing regimen reduced pyogenic infections due to staphylococci in newborn infants. Table 3 presents a summary of studies concerned with the disinfection of arm-surface while the arms were exposed to the work environment. A comparison of the effect between a lotion and a bar soap is shown. Hexachlorophene was present in both preparations. Microbial colony counts were taken before, immediately after and 3 1/2 hours after the washing procedure. Total microorganisms and *S. aureus* counts were observed. Figures represent the total number of colonies detected in a 6 day test period.

Significant reduction (Table 3) in the total and S. aureus counts occurred immediately after the washing procedure for both lotion and bar soaps. Three and one-half hours later the total and S. aureus counts had increased 3 to 6-fold greater than before wash counts. It was hypothesized that a major portion of these organisms were a contamination from the work environment and not strictly a result of increased counts of resident skin bacteria.

TABLE 3.

SUMMARY OF STUDIES WITH ARM-SURFACE DISINFECTION WHILE ARMS WERE EXPOSED TO THE WORK ENVIRONMENT*[1]

SOAP	BEFORE WASH COUNTS		IMMEDIATELY AFTER WASH COUNTS		3½ HOURS AFTER WASH COUNTS	
	TOTAL	SA[4]	TOTAL	SA	TOTAL	SA
LOTION[2]	2895	2598	276	274	9643	8624
BAR[3]	519	454	52	36	2835	2785

*BRAIN-HEART INFUSION AGAR + 5% RABBIT BLOOD WAS THE MEDIUM. INCUBATION WAS AT 37° C FOR 24 HOURS.

[1]FIGURES REPRESENT TOTAL NUMBER OF COLONIES DETECTED FOR 6 DAYS

[2]CONTAINS 3% HEXACHLOROPHENE ON A TOTAL WEIGHT BASIS

[3]CONTAINS 2% HEXACHLOROPHENE ON A TOTAL WEIGHT BASIS

[4]*S. AUREUS*

Table 4 shows a summary of experiments which were done to test this hypothesis. The subjects wore sterile long sleeved shirts to shield the arm from the external environment. As noted in Table 3 and now shown in Table 4 a similar reduction of total and S. aureus microorganisms occurred immediately after the washing regimen. This reduction of total microorganisms was 91.4 percent for the lotion and 88.5 for the bar soap. The S. aureus reduction was 89.4 and 88.0 percent, respectively. The counts taken 3 1/2 hours after the wash treatment had increased to greater numbers than the immediately after wash counts, but a continual sustained reduction was maintained at lower numbers than the before wash counts. This phenomenon is opposite to that noted in Table 3.

TABLE 4.

SUMMARY OF ARM SURFACE DISINFECTION STUDIES*[1]

SOAP	BEFORE WASH COUNTS		IMMEDIATELY AFTER WASH COUNTS		% REDUCTION IMMEDIATELY AFTER TREATMENT		3½ HOUR COUNTS AFTER TREATMENT	
	TOTAL	SA[4]	TOTAL	SA	TOTAL	SA	TOTAL	SA
LOTION[2]	2865	2312	247	245	91.4	89.4	376	371
BAR[3]	190	159	22	19	88.5	88.0	55	52

*BRAIN-HEART INFUSION AGAR + 5% RABBIT BLOOD WAS THE TEST MEDIUM. INCUBATION WAS AT 37° C FOR 24 HOURS.

[1]ARMS WERE COVERED TO WORK ENVIRONMENT. FIGURES REPRESENT TOTAL NUMBER OF COLONIES DETECTED FOR 2 DAYS.

[2]CONTAINS 3% HEXACHLOROPHENE ON A TOTAL WEIGHT BASIS.

[3]CONTAINS 2% HEXACHLOROPHENE ON A TOTAL WEIGHT BASIS.

[4]*S. AUREUS*

A comparison of the relative bacteriostatic action for the lotion and bar soap is shown in Table 5. An increase in the number of colonies was noted 3 1/2 hours after the wash compared with the immediately after wash counts. This increase of total microorganisms was 1.5-fold greater for the lotion and 2.5-fold for the bar soap. The S. aureus increase was 1.5-fold greater for the lotion and 2.7-fold for the bar soap.

TABLE 5.

THE RELATIVE BACTERIOSTATIC ACTION OF LOTION AND BAR SOAP*[1]

SOAP	IMMEDIATELY AFTER WASH COUNTS		3½ HOUR AFTER WASH COUNTS		BACTERIOSTATIC ACTION FOR 3½ HOUR PERIOD AFTER WASH COUNTS FOLD-INCREASE	
	TOTAL	SA[4]	TOTAL	SA	TOTAL	SA
LOTION[2]	247	245	376	371	1.5	1.5
BAR[3]	22	19	55	52	2.5	2.7

*BRAIN-HEART INFUSION AGAR + 5% RABBIT BLOOD WAS THE TEST MEDIUM. INCUBATION WAS AT 37° C FOR 24 HOURS.

[1]ARMS WERE COVERED TO WORK ENVIRONMENT. FIGURES REPRESENT TOTAL NUMBER OF COLONIES DETECTED FOR 2 DAYS.

[2]CONTAINS 3% HEXACHLOROPHENE ON A TOTAL WEIGHT BASIS.

[3]CONTAINS 2% HEXACHLOROPHENE ON A TOTAL WEIGHT BASIS.

[4]*S. AUREUS*

Bacteria on the Hands

The data given in Table 6 was obtained after a standard showering procedure had been developed for the animal technicians before entering into the work areas. Human contact with the animals occurs most often with the palm of the hand. The species of organisms that remained on this surface after a shower and the incidence of recovery is shown in this table. Six species of bacteria were detected. S. aureus was recovered most often and Paracolobacterium sp. was least frequently detected.

TABLE 6.

BACTERIA DETECTED ON THE PALM OF THE HAND POST-SHOWER*

BACTERIAL SPECIES	INCIDENCE OF RECOVERY[1]
S. AUREUS	49/60
BACILLUS SP.	33/60
COLIFORMS[2]	22/60
STREPTOCOCCUS FECALIS	6/60
PARACOLOBACTERIUM SP.	1/60

*BRAIN-HEART INFUSION AGAR WAS THE MEDIUM. INCUBATION WAS AT 37° C FOR 24 HOURS.

[1]INCIDENCE OF RECOVERY/TOTAL OF 60 ASSAYS.

[2]*E. COLI* AND *A. AEROGENES*.

DISCUSSION

The bactericidal and bacteriostatic efficacy of soaps is variable against a wide spectrum of microorganisms. The chief effect of soap under practical conditions is its ability to suspend and wash away bacteria and particulate matter. Since Koch first reported the bactericidal activities of soap in 1881, considerable work has been done to enhance these antibacterial properties. The addition of chemicals such as hexachlorophene and 3,4,4'-trichlorocarbanilide in recent years has extended its action against microorganisms.

This study demonstrates that a showering regimen with hexaclorophenated soap is superior to a rinsing regimen in suspending and washing away bacteria on the skin surface (Table 2). It also shows that bacteriostatic action on the skin bacteria can be maintained for at least 3 1/2 hours after a shower if the body is shielded from its external environment (Tables 4 and 5). However, these antibacterial effects were variable from one area of skin to another (Table 2). In the rib area an increase in total counts was noted after the showering regimen while a decrease was seen in the arm and leg areas. The reason for the increase in the rib area is not known, but may be related to the location of glands and hair follicles that are randomly distributed over the body surface (12). Most of these organisms were resident bacteria (*S. aureus*) but some transients were also detected. Lovell (10) demonstrated resident bacteria in sebaceous glands, hair follicles and stratum corneum and suggested that they could not be completely removed without injurying the skin. Barbeito *et al* (1) showed that the beards of men retained transient microorganisms, when inoculated experimentally, despite washing with hexachlorophenated soap. Washing did, however, reduce the number of infectious particles.

Figure 1 shows that bacterial counts in the rib area increased immediately after a shower to greater numbers than before shower counts, but successive daily showers reduced this increase from one day to the next and this reduction continued with additional daily showers. An exception is noted for Subject #1 on the third successive day of the test when an increase in after shower counts was greater than the previous days after shower counts. Bernard *et al* (3) and Heldman *et al* (9) reported that the rate at which human beings shed microorganisms increased after the skin had been washed. Riemensnider (11) showed that a daily regimen of bathing with a germicidal detergent reduced shedding of microorganisms. As previously noted, the increase in bacterial counts (Figure 1) consisted mostly of *S. aureus*. Transient microorganisms appeared infrequently and in low numbers. It would seem that the washing regimen allowed the release of resident bacteria located and held within the deeper layers of the skin.

Baxby and Woodroffe (2) studied the location of bacteria in pig and human skin. It was found in pig skin that vigorous washing with toilet soap removed most of the transient surface bacteria. The majority of remaining bacteria were detected in the outer 0.3 mm of the stratum corneum with decreasing numbers in progressively deeper sections. Results obtained from closely adjacent areas from the same animal were comparable but occasional high counts were noted indicating that the distribution of bacteria along the skin surface was not constant. Williams (13) stated that it is common to find clusters of S. aureus colonies on impression plates taken from the interior of surgeons' gloves. This might be expected if these bacteria came from single skin glands.

Sebaceous glands and hair follicles could act as reservoirs for bacteria in human and pig skin thus providing a continuous supply of bacteria to the skin surface. It would seem helpful to define these microorganisms in terms of resident and transient bacteria, and to determine potential pathogenicity by appropriate tests. This could be of epidemiological significance.

SUMMARY

Several experiments were completed to define the microbial skin flora of the animal technician and to test methods which would alter this flora. These tests were designed to detect changes in the resident and transient microorganisms found on the skin of the body surface. Four salient trends emerged and are summarized below:

1) The microbial skin flora of the animal technician can be divided into two categories: a) organisms which apparently colonize on the skin and become residents, b) organisms which do not colonize on the skin but are present as contaminates from the work environment (transients).

2) A systematic rinsing or showering regimen can reduce the number of resident and transient skin microorganisms. Greater reduction of these skin microorganisms occurred after the showering regimen.

3) The microorganisms detected after a rinsing regimen consisted of 50 percent S. aureus and 50 percent transient microorganisms. The microorganisms detected after a showering regimen consisted of 80 percent S. aureus and 20 percent transient microorganisms.

4) The reduction of microorganisms which occurred after a washing regimen was sustained for at least 3 1/2 hours when the skin surface was shielded from the work environment.

REFERENCES

1. Barbeito, M. S., Mathews, C. T. and Taylor, L. A. Microbiological laboratory hazard of bearded men. Appl. Microbiol. 15: 899-906, 1967.

2. Baxby, D. and Woodroffe, R. C. S. The location of bacteria in skin. J. Appl. Bact. 28: 316-321, 1965.

3. Bernard, H. R., Speers, R., O'Grady, F. W. and Shooter, R. A. Airborne bacterial contamination - investigation of human source. Arch. Surg. 91: 530-533, 1965.

4. Duguid, J. P. and Wallace, A. T. Air infection with dust liberated from clothing. Lancet 2: 845-849, 1948.

5. Eichenwald, H. F., Kolsevalov, D. and Fassco, L. A. The "cloud baby": an example of bacterial-viral interaction. Am. J. Diseases Children 100: 161-173, 1960.

6. Farquharson, C. D., Penny, S. F., Edwards, H. E. and Barr, E. The control of staphylococcal skin infections in the nursery. Can. Med. Ass. J. 67: 247-249, 1952.

7. Hare, R. and Thomas, C. G. A. The transmission of *Staphylococcus aureus*. Brit. Med. J. 2: 840-844, 1956.

8. Hare, R. and Ridley, M. Further studies on the transmission of *Staphylococcus aureus*. Brit. Med. J. 1: 69-73, 1958.

9. Heldman, D. R., Sunga, F. E. C. A. and Hedrick, T. I. Microorganism shedding by human beings. Contamination Control 5: 28-31, 45, 1967.

10. Lovell, D. L. Skin bacteria: their location with reference to skin sterilization. Surg. Gynec. Obstet. 80: 174-177, 1945.

11. Riemensnider, D. K. Quantitative aspects of shedding of microorganisms by humans. Proceedings of a Conference on Spacecraft Sterilization Technology, National Aeronautics and Space Administration. 97-103. Contamination Control 6:45, 1966.

12. Sarkany, I. A method for studying the microtopography of the skin. Brit. J. Derm. 74: 254-259, 1962.

13. Williams R. E. O. Healthy carriage of *Staphylococcus aureus*: its prevalence and importance. Bact. Rev. 27: 56-71, 1963.

IN VITRO COLONY-FORMING CELLS IN THE BONE MARROW OF GERMFREE MICE

Donald Metcalf, M. D. and Roger S. Foster, Jr., M. D.

Roswell Park Memorial Institute, Buffalo, New York

Using the semi-solid agar culture system (Bradley and Metcalf, 1966; Robinson et al, 1967; Metcalf and Foster, 1967) it has been shown that certain cells in the bone marrow are capable of proliferation in vitro with the formation of colonies of granulocytic and mononuclear cells (Metcalf et al, 1967). Indirect evidence suggests that the in vitro colony-forming cells are actively dividing in vivo (Fox, M. and Lajtha, L., personal communication) and are probably primitive members of the granulocytic series, sharing a common ancestor with red cell precursors (Bradley et al, 1967; Metcalf 1968).

Colony formation in vitro by normal bone marrow cells is dependent on the continuous presence in the culture system of a factor termed the "colony stimulating factor" (Robinson et al, 1967; Metcalf and Foster, 1967; Metcalf and Foster, 1967a). Detectable levels of this factor are present in the sera of many normal conventional mice and serum factor levels are elevated in mice with leukemia and reticulum cell sarcoma. Chemical studies on the factor in mouse serum suggest that it is probably protein in nature with a molecular weight of approximately 50,000 (Stanley et al, 1968).

In an earlier study (Metcalf, Foster and Pollard, 1967) a survey was made of colony stimulating factor levels in germfree and conventional young adult Swiss mice. This survey showed that serum colony stimulating factor levels were exceptionally low in germfree mice, although a similar analysis of sera from germfree mice with lymphoid leukemia showed that high levels of colony stimulating factor were present as in conventional leukemic AKR mice.

These data suggested that although factor levels were elevated in leukemia, another cause for elevation of serum factor levels might well be infections of various types. This was confirmed by studies in Swiss mice in which it was shown that infection by the LDH-elevating virus of Riley led to a sharp rise in serum factor levels in the acute stages of the infection (Foster et al, 1968). Further, surveys of factor levels in human sera revealed that serum levels were elevated in patients with acute infections (Foster et al 1968a) and mononucleosis (Metcalf and Wahren, 1968).

In the present study, assays were performed on the bone marrow cells from conventional and germfree Swiss mice to determine the incidence of in vitro colony-forming cells.

MATERIALS AND METHODS

Mice. Mice used were 12 week old conventional and germfree mice obtained from the Charles River Breeding Laboratory and maintained in the germfree facility of the Roswell Park Memorial Institute.

In Vitro Culture System. The culture system used and the methods for assaying for colony-forming cells have been described at length elsewhere (Robinson et al, 1967; Metcalf and Foster, 1967; Bennett et al, 1968). In brief, a plug of bone marrow cells was removed from a single femur from each animal under study and pipetted up and down in 2 ml. of bone marrow collecting fluid (modified Eagle's medium containing 10% fetal calf serum and 10% trypticase soy broth) until a single cell suspension resulted. Nucleated cells were counted with a hemocytometer. Double strength modified Eagle's medium was mixed with an equal volume of 0.6% agar in distilled water (previously boiled for 2 minutes and held at 40°C). This mixture was held at 37°C and sufficient bone marrow cells were added to aliquots of the mixture to give final cell concentrations of 5×10^4 and 2.5×10^4 nucleated cells per 1 ml. of culture medium. 1 ml. of each cell suspension was pipetted into 35 mm plastic petri dishes containing .025 ml. of pooled mouse serum of known colony stimulating activity. Four replicate plates were prepared for each cell concentration. The plates were incubated for 7 days without media change in humidified incubators at 37°C with a continuous flow of 5% CO^2 in air. Colonies were scored using a X 30 dissecting microscope. Samples of day 3, 5 and 7 colonies were removed using a fine Pasteur pipette, placed on microscope slides and stained with 0.6% orcein in acetic acid. The number of cells in each colony was determined by direct cell counts and the colony cells classified according to morphological types (Metcalf et al, 1967).

RESULTS

A total of 12 germfree and 10 conventional Swiss mice were individually assayed. Germfree mice were killed immediately after removal from the isolator and the bone marrow cell suspensions prepared without delay.

The colonies formed in vitro by bone marrow cells from germfree mice were similar in gross morphology to the loose globular aggregates of cells developing from conventional bone marrow cells. Cytological examination revealed that early colonies from germfree cells were granulocytic in nature whereas at day 7 of incubation, most colonies contained pure populations of phagocytic mononuclear cells. This is the same sequence of changes as is shown by colonies developing from conventional bone marrow cells.

The incidence of colony-forming cells per 10^5 bone marrow cells is shown in Table 1. It is evident from the results that the incidence of colony-forming is extremely similar in both germfree and conventional mice. Although the incidence of such cells was slightly lower in germfree mice, the difference was not statistically significant. The numbers of colony-forming cells per 10^5 bone marrow cells in the present experiments are lower than those reported previously (Bennett et al, 1968) because of the use of a low dose of serum to stimulate colony formation, and the present results should only be used for comparative purposes between the two groups. Earlier results indicated that linear relationships between the numbers of colonies developing and the dose of colony stimulating factor used (Metcalf and Foster, 1967) and this was found to be true also for germfree bone marrow cells. It should be noted that there was a tendency for a greater uniformity in the incidence of colony-forming cells in germfree mice than in conventional mice.

Table 1

Incidence of In Vitro Colony-Forming Cells in the Bone Marrow of Germfree and Conventional Swiss Mice.

Type	No. of Mice Assayed.	Mean No. of In Vitro Colony-Forming Cells per 10^5 Bone Marrow Cells.
Germfree Swiss	12	36 ± 5 *
Conventional Swiss	10	45 ±13

* ± Standard error.

DISCUSSION

Since in vitro colony-forming cells are probably early members of the granulocytic cell series, the present results suggest that despite the lack of bacterial and protozoal infections in germfree mice, the level of granulopoiesis in the bone marrow is similar to that in conventional mice. This conclusion is in agreement with other observations indicating that blood polymorph levels are normal in germfree mice (Metcalf, Foster and Pollard, 1967) and that total bone marrow cell counts and differential cell counts are similar in germfree and conventional mice (Boggs et al, 1967).

The function of granulocytes is not clearly understood, but the level of granulopoiesis clearly alters during acute infections. Whilst germ-free mice have no bacterial or protozoal infections, they are known to be infected by vertically transmitted viruses, e.g. the Gross leukemia virus, and possibly such virus infections may affect the level of granulopoiesis in germfree mice.

Examination of mice injected with colony-stimulating factor have shown that such mice develop a polymorph leucocytosis and elevated numbers of in vitro colony-forming cells in the bone marrow, spleen and blood (Bradley et al, 1968). However the present data indicate that in vitro colony-forming cells in the bone marrow are present in normal numbers in germfree mice despite the remarkably low serum levels of colony stimulating factor in such mice. These results suggest strongly that whilst the colony-stimulating factor may be one of the factors regulating granulopoiesis in the body it cannot be the only regulatory factor.

SUMMARY

Assays for in vitro colony-forming cells in the bone marrow of 12 germfree Swiss mice and 10 conventional Swiss mice revealed a similar incidence of such cells despite the abnormally low serum levels of colony-stimulating factor in germfree mice. The colony-stimulating factor seems unlikely to be the only factor regulating granulopoiesis in the body.

This investigation was supported by National Institutes of Health Research Grants CA-08847-01, CA-07745-03 and FR-OU294, The John A. Hartford Foundation and the Carden Fund of the Anti-Cancer Council of Victoria.

REFERENCES

BENNETT, M., G. CUDKOWICZ, R. S. FOSTER and D. METCALF (1968). Hemopoietic progenitor cells of W anemic mice studied in vivo and in vitro. J. Cell. Physiol. 71, 211-226.

BOGGS, D. R., P. A. CHERVENICK, J. C. MARSH, H. I. PILGRIM, G. E. CARTWRIGHT and M. M. WINTROBE (1967). Granulocytopoiesis in germfree mice. Proc. Soc. Exp. Biol. Med. 125, 325-330.

BRADLEY, T. R. and D. METCALF (1966). The growth of mouse bone marrow cells in vitro. Aust. J. Exp. Biol. Med. 44, 287-300.

BRADLEY, T. R., D. METCALF, M. SUMNER and R. STANLEY (1968). Characteristics of in vitro colony formation by cells from haemopoietic tissues. In Vitro 2, (in press).

BRADLEY, T. R., W. ROBINSON and D. METCALF (1967). Colony production in vitro by normal, polycythaemic and anaemic bone marrow. Nature, 213, 511.

FOSTER, R., D. METCALF and R. KIRCHMYER (1968). Induction of bone marrow colony stimulating activity by a filterable agent in leukemic and normal mouse serum. J. exp. Med. 127, 853-866.

FOSTER, R., D. METCALF, W. A. ROBINSON and T. R. BRADLEY (1968a). Bone marrow colony stimulating activity in human sera. Results of two independent surveys in Buffalo and Melbourne. Brit. J. Haematol. 15, 147-159.

METCALF, D. (1968). The effect of bleeding on the number of in vitro colony-forming cells in the bone marrow. Brit. J. Haematol. (in press).

METCALF, D., T. R. BRADLEY and W. ROBINSON (1967). Analysis of colonies developing in vitro from mouse bone marrow cells stimulated by kidney feeder layers or leukemic serum. J. Cell. Physiol. 69, 93-108.

METCALF, D. and R. FOSTER (1967). Bone marrow colony-stimulating activity of serum from mice with virus-induced leukemia. J. Nat. Cancer Inst. 39, 1235-1245.

METCALF, D. and R. FOSTER (1967a). Behavior on transfer of serum stimulated bone marrow colonies. Proc. Soc. Exp. Biol. Med. 126, 758-762.

METCALF, D. R. FOSTER and M. POLLARD (1967). Colony stimulating activity of serum from germfree normal and leukemic mice. J. Cell. Physiol. 70, 131-132.

METCALF, D. and B. WAHREN (1968). Bone marrow colony stimulating activity of sera in mononucleosis. Brit. Med. J. 3, 99-101.

ROBINSON, W., D. METCALF and T. R. BRADLEY (1967). Stimulation by normal and leukemic mouse sera of colony formation in vitro by mouse bone marrow cells. J. Cell. Physiol. 69, 83-92.

STANLEY, E. R., W. A. ROBINSON and G. L. ADA (1968). Properties of the colony stimulating factor in leukaemic and normal mouse serum. Aust. J. Exp. Biol. Med. (in press).

MICROBIC INTERACTIONS IN THE GNOTOBIOTIC MOUSE

Morris Wagner and Theodore J. Starr

Department of Microbiology, University of Notre Dame

Notre Dame, Indiana

It is axiomatic that the microflora associated with animal life, including man, may be vital to the well-being of the host and that disturbances in the intermicrobial equilibrium could become detrimental to health. Long term confinement of man or animals in a sealed-off environment could cause a disturbance in the microbial equilibrium and possibly result in the elimination of beneficial organisms, the build up of pathogens, or both. Microbial exchanges among individuals in a closed environment could transfer potential pathogens from a nonsymptomatic carrier host to a susceptible recipient. The study of the above relationships in man or in conventionally-reared animals is quite complex because of the diversity and largely unknown nature of the microbial flora. Therefore, attempts have been made to establish experimental models whereby gnotobiotic mice, bearing only known, purposely introduced microorganisms can be studied in terms of interorganism compatibilities and incompatibilities. Attempts have also been made to manipulate these interflora relationships via dietary means.

This report contains findings on the fate and numerical establishment of certain pure cultures of bacteria in the intestinal tract of previously germfree mice after oral inoculation. The effect of feeding high carbohydrate, high protein or high fat diets on the microbial population of the intestinal contents in mono-associated mice and in mice bearing combinations of different microbial species has also been observed.

METHODS

Bacteria-free mice of the Lobund CFW colony were housed in

Trexler-type plastic isolators (1) and associated with pure cultures of specific bacteria to form a number of separate bacterially-mono-associated mouse colonies. Progeny of these animals were studied in the mono-associated state. When animals bearing two bacterial species (di-associated) were used, mice from the respective mono-associated colonies were transferred and pooled in a separate isolator so that exchange of their original respective monofloras could take place. The quantitation of bacteria in the feces was done by viable plate counts using appropriate media for the quantitative recovery of each species in the mixed flora. Lactobacilli were enumerated on Rogosa SL Agar (2); streptococci on Azide Blood Agar Base with 2 ppm crystal violet added (2); salmonella on Brilliant Green Bile Agar (2).

Diets used in these studies were: Teklad T-1, a general purpose diet adapted for normal rearing of germfree animals (3); Diet 128 (4) fed with 5% sucrose water ad lib. as a high carbohydrate diet; Hi-Protein Diet (GBI) containing casein, 65%, rice starch, 22%, vegetable oil, 8%, salt mix U.S.P. XIV, 4%, and vitamin mix GBI, 1% (5); Hi-Fat Diet (GBI) containing corn oil, 20%, protein as casein and gelatin, 22%, carbohydrate as corn starch and sucrose, 41%, cellulose, 10.6%, salt mix and fortified vitamin mix (5).

RESULTS

Figure 1 illustrates the levels of Pseudomonas aeruginosa cultured from the feces of gnotobiotic mice which were mono-associated with a pure culture of the above organism and fed the following diets sequentially: (a) the general purpose commercial type diet T-1; (b) high carbohydrate diet; (c) high protein diet; (d) high fat diet. Each point on the curve represents the mean of P. aeruginosa counts per gram of dry feces taken on eight individual mice.

It was anticipated that P. aeruginosa might establish itself in higher numbers in mice fed the high protein and high fat diets and in lower numbers when fed high carbohydrate diet since this bacterium is known to be strongly proteolytic and strongly lipolytic in vitro but only weakly saccharolytic. However, these in vitro characteristics of P. aeruginosa did not reflect a particular advantage or disadvantage to the organism as far as its numerical population in the feces was concerned. There was some fluctuation of the count to the high side when the high protein diet was fed but dips in the count during the same dietary period did not indicate a real advantage of the high protein diet to the Pseudomonas population. The high fat diet also did not indicate that the strong lipolytic activity of this bacterium was advantageous in vivo. In fact, the counts during the high fat diet period tended to be lower.

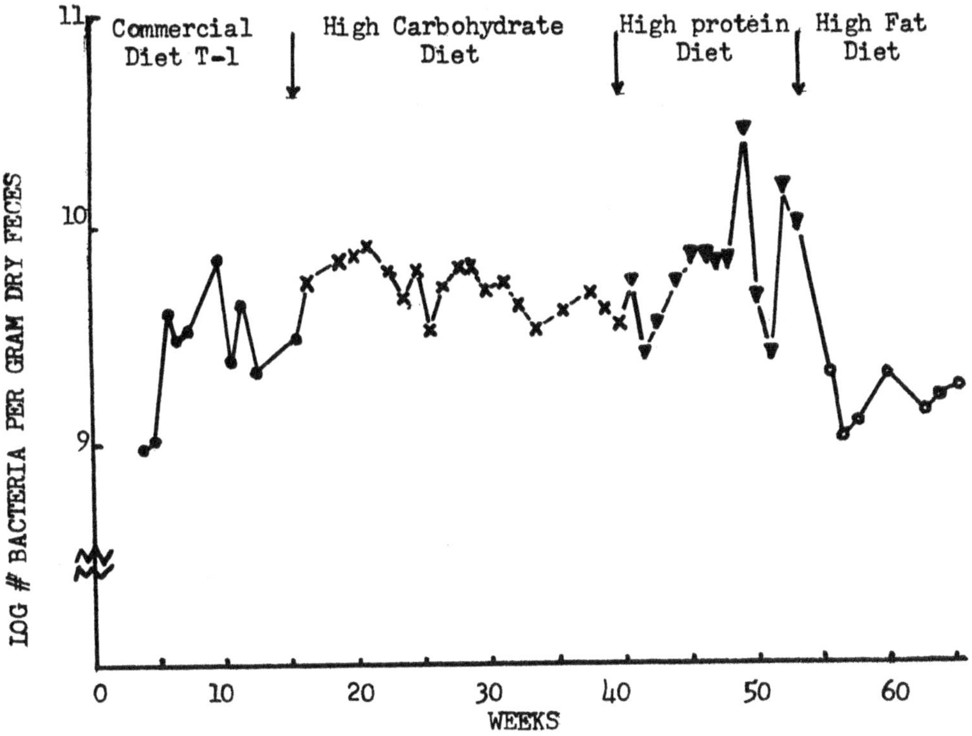

Figure 1. Bacterial counts in feces of mice mono-associated with *Pseudomonas aeruginosa* and fed various diets.

Fecal bacterial counts in mice mono-associated with *Lactobacillus brevis* were quite variable, particularly during the time that the commercial general maintenance diet was being fed (Fig. 2). The number of *L. brevis*, an organism with saccharolytic but no proteolytic or lipolytic activity, was not enhanced by feeding a high carbohydrate diet to the host. As a matter of fact, some of the lowest per gram of dry feces were observed during the early weeks of the high carbohydrate dietary period. However, there was a gradual buildup in numbers to some of the highest counts seen as time progressed through this same dietary period. It seems questionable that this buildup was due to high carbohydrate alone since discontinuation of this diet and the initiation of of the high protein regimen did not alter the numerical population of *L. brevis* appreciably.

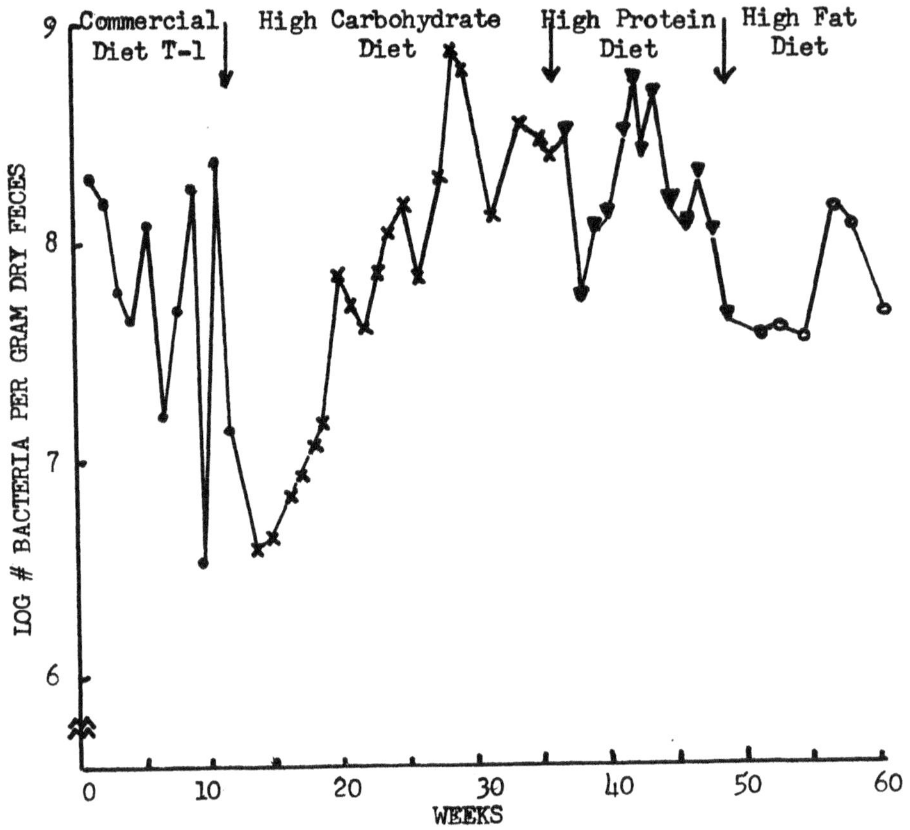

Figure 2. Bacterial counts in feces of mice mono-associated with *Lactobacillus brevis* and fed various diets.

The quantitative level attained by *Salmonella typhimurium* in the feces of gnotobiotic mono-associated mice is presented in Figure 3. The count approached 10^{11} cells per gram dry feces and remained relatively constant during the periods that the commercial general maintenance type diet and the high carbohydrate diet were fed. Another group of mice fed the high protein diet (shown in figure 3 to the right of the vertical zigzag line) indicated that the high protein content of the diet also failed to alter the *S. typhimurium* population appreciably.

Other bacteria established as mono-associates in gnotobiotic CFW mice attained the following fecal populations: (log of numbers)

Escherichia coli	10.40	Streptococcus faecalis	10.39
Proteus vulgaris	10.19	Pseudomonas aeruginosa	9.51
Erwinia carotovora	10.41	Lactobacillus bifidus	8.79
Salmonella typhimurium	10.81	Lactobacillus brevis	7.97

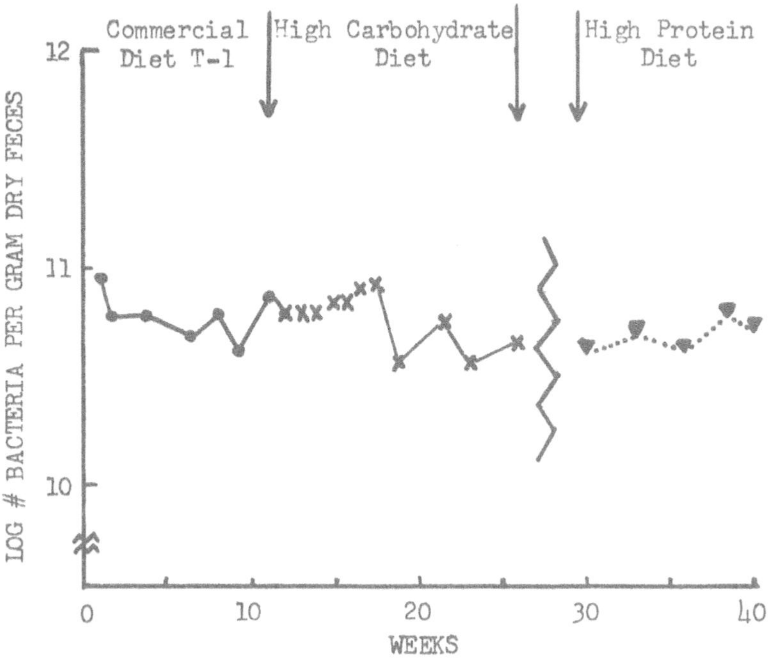

Figure 3. Bacterial counts in feces of mice mono-associated with Salmonella typhimurium and fed various diets.

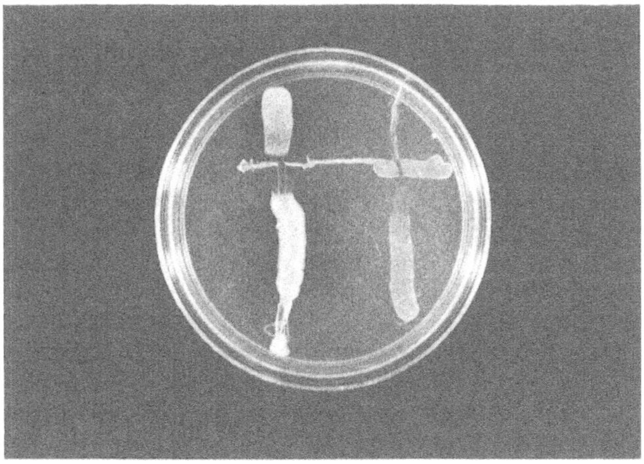

Figure 4. Inhibition of Salmonella typhimurium (left) and Pseudomonas aeruginosa by Streptococcus faecalis (horizontal).

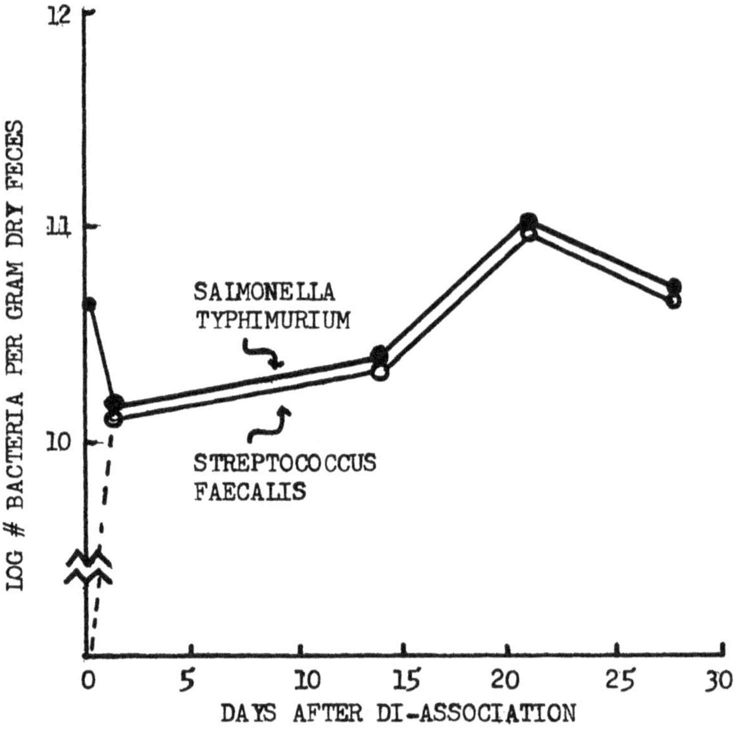

Figure 5. Interbacterial compatibility between *Salmonella typhimurium* and *Streptococcus faecalis* as di-associates in gnotobiotic mice. (Mice originated from *S. typhimurium* mono-associated colony)

Interbacterial compatibility between two organisms when established *in vivo* in the intestinal tract of the gnotobiotic mouse is demonstrated in figure 5. Two groups of mice previously mono-associated with *Streptococcus faecalis* and *Salmonella typhimurium* respectively, were pooled in an isolator and allowed to exchange floras by contact. The resultant di-association showed the establishment of approximately equal numbers of *S. faecalis* and *S. typhimurium*. The levels attained were essentially the same as those observed when each microorganism was present as a pure culture in the mono-associated mice. There was no evidence for antibiosis nor did there appear to be any competition for so-called "biological space". Each species seemed to occupy its own space independently of the other.

The *in vivo* compatibility exhibited by these two bacteria

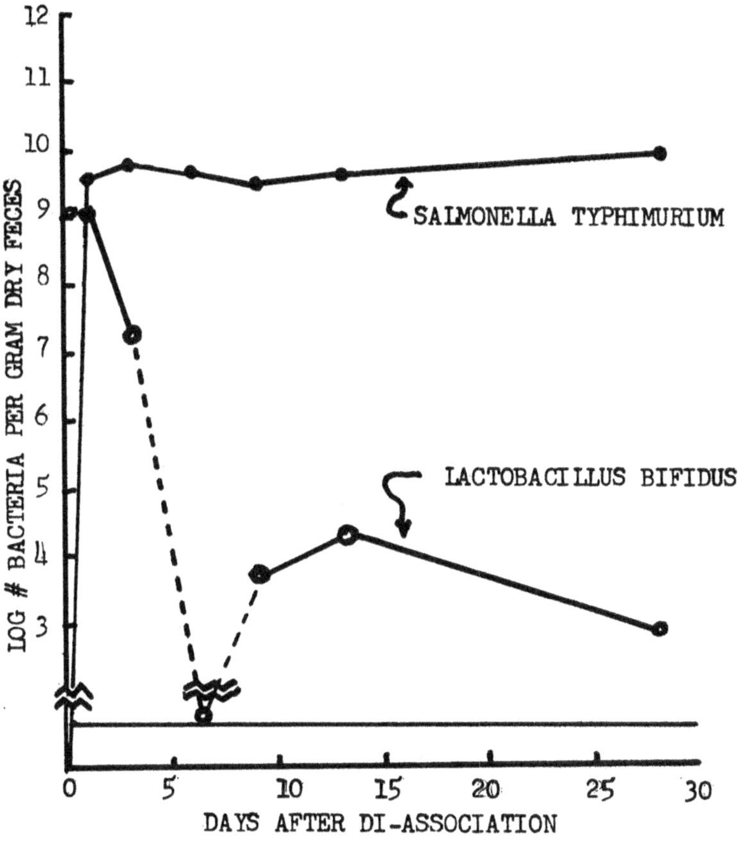

Figure 6. Interbacterial incompatibility between *Salmonella typhimurium* and *Lactobacillus bifidus* as di-associates in gnotobiotic mice. (Mice originated from *L. bifidus* mono-associated colony).

was interesting in that the combination was originally chosen in an attempt to suppress the pathogenic *S. typhimurium* by the introduction of *Streptococcus faecalis* as a representative of the normal intestinal flora of the mouse. *In vitro* tests such as the one illustrated in figure 4 had shown that the growth of pathogenic strains of *Salmonella* and *Pseudomonas* was inhibited by a cross-streak of *S. faecalis* on the nutrient agar surface. Thus, the *in vitro* inhibition was not confirmed in the *in vivo* experiment.

Bacterial incompatibility was observed in mice di-associated with *Lactobacillus bifidus* and *Salmonella typhimurium*. Mice which originated from the mono-associated *L. bifidus* colony are shown in figure 6. When these mice were pooled with the

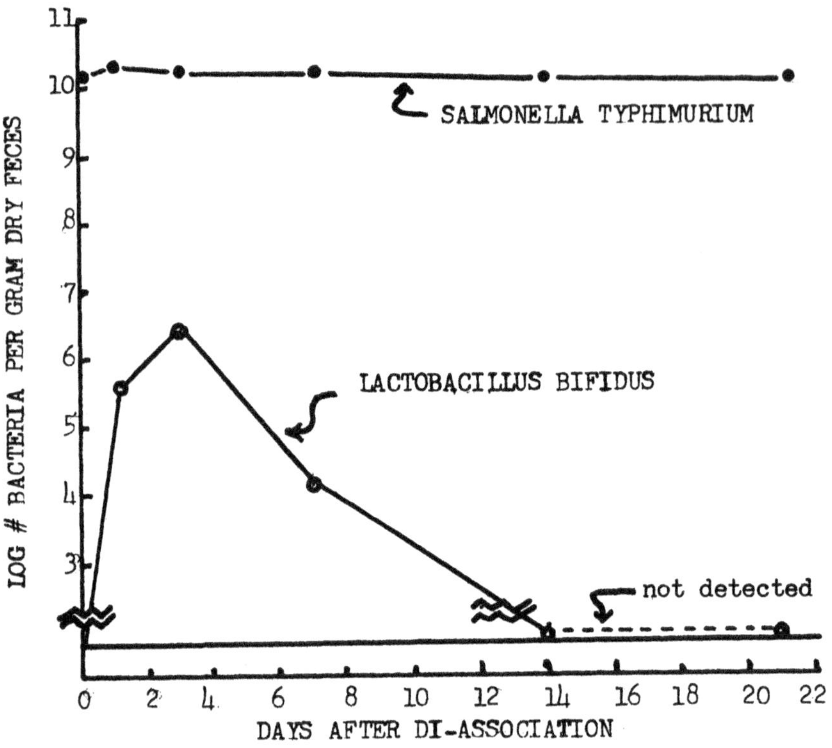

Figure 7. Interbacterial incompatibility between Salmonella typhimurium and Lactobacillus bifidus as di-associates in gnotobiotic mice. (Mice originated from S. typhimurium mono-associated colony)

Salmonella derived group, they were newly exposed to the Salmonella. The heterologous organism was readily picked up and established but a precipitous drop in the homologous Lactobacillus population was observed. In the reciprocal mice, those originating from the Salmonella mono-associated colony (see figure 7) there was a transient low level uptake of the heterologous lactobacilli but later the lactobacilli could not be recovered. There was no effect on the original resident Salmonella population. The latter two experiments are examples of microbial suppression and microbial exclusion but unfortunately, the exclusion was in the opposite direction to what was intended since the in vitro tests had indicated that L. bifidus might suppress Salmonella .

The effect of various diets on the di-association of S. typhimurium and Lactobacillus brevis in mice is shown in figure 8. There was some indication that the high carbohydrate may have had a transient effect on the relative populations of bacteria as seen

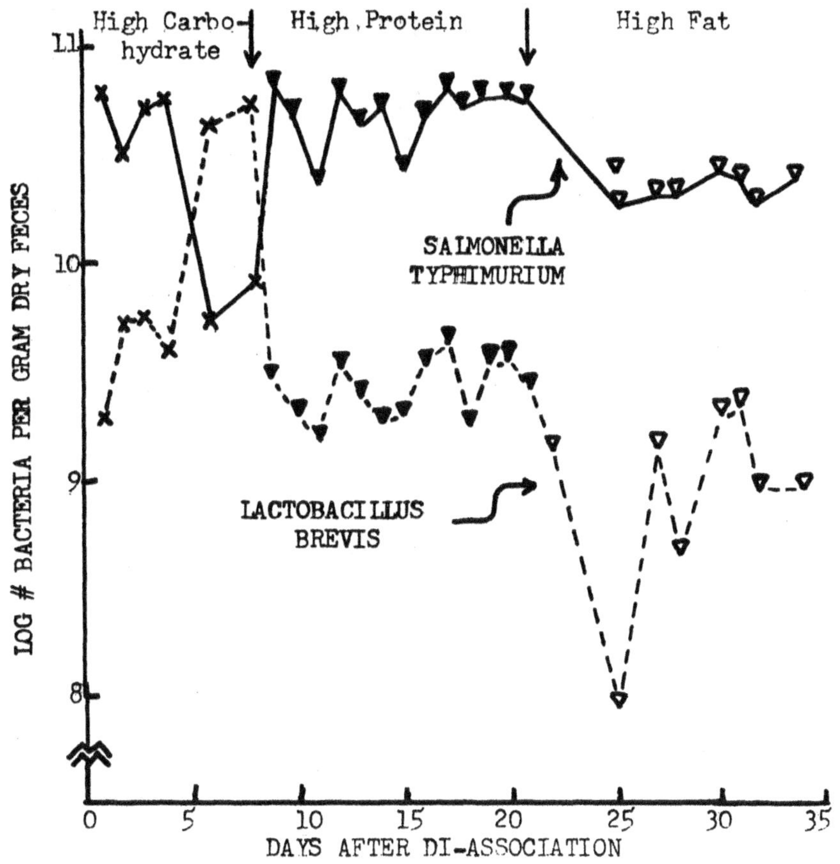

Figure 8. Bacterial counts in mice di-associated with Lactobacillus brevis and S. typhimurium and fed various diets.

by the inversion of the curves for several days but the Salmonella sustained a substantial level, even at its lowest point and did not represent the extreme suppression seen in the previous L. bifidus experiment

In summary, (1) the numbers that a variety of bacteria attain in feces when grown as pure cultures in the intestinal tract of gnotobiotic mice has been reported. (2) The saccharolytic, proteolytic or lipolytic activities of specific bacterial species do not necessarily enhance their numerical population in the intestinal contents when high carbohydrate, high protein or high fat diets are fed to the host. (3) In vitro inhibition of one bacterium by another is not necessarily reflected in a similar inhibition in the same direction in vivo. Indeed, the reverse may be true.

Other bacteria are currently under investigation for their ability to suppress or exclude enteric pathogens.

REFERENCES

1. Trexler, P.C. and Reynolds, L.I.: Flexible film apparatus for the rearing and use of germfree animals. Appl. Microbiology, 5:406, (1957)
2. Difco Supplementary Literature and Difco Manual, Difco Laboratories, Detroit, Michigan. (1966)
3. Kellogg, T.F. and Wostmann, B.S.: Rat and mouse stock diet L-485. Gnotobiotics Newsletter 3(3):1, Sept.(1967). This diet is available from Teklad, Inc., Monmouth, Illinois as L-485 Autoclavable Rat and Mouse Diet.
4. Orland, F.J. et al.: Use of the germfree animal technic in the study of experimental dental caries. I. Basic observations on rats reared free of all microorganisms. J. Dent. Res. 33: 147, (1954). Available from General Biochemicals, Chagrin Falls, Ohio. See Price List 1967, pg. 70.
5. Special High Protein and High Fat (20%) Rat Diets were adapted by General Biochemicals, Inc., Chagrin Falls, Ohio for pelleting and autoclaving for use in germfree experimentation.

This work was supported by NASA Grant NGR 15-004-017.

GERM-FREE MICE AND LYMPHOCYTE DIFFERENTIATION

Ben W. Papermaster

Roswell Park Memorial Institute, New York State

Department of Health, Buffalo, New York

Lymphocytes have been the object of intensive study for the last fifty years, (1) yet we know surprisingly little about them. Our ignorance is based in part on difficulties in locating individual members of a circulating cell population *in vivo* and attempting to ascribe functions to heterogeneous cells in the complex organs which provide a matrix for their differentiation. The word, lymphocyte, is a term loosely applied to a family of cells which differ in size, morphology, anatomical location and function. Small, medium and large lymphocytes are found ubiquitously throughout the body, but in particular in bone marrow, thymus, lymph nodes, and spleen. To compound difficulties further, the morphological resemblance of one group of lymphocytes to another has been corroborated by physical studies, such as size distribution in electronic cell counters and distribution in density gradients. Thus lymphocytes comprise a group of cells which vary from 6 to 25 microns in diameter (1) and between 1.06 and 1.07 in buoyant density (2).

The early morphological studies on lymphocytogenesis (3) suggested a pattern of differentiation beginning with a larger type of precursor cell and progressing to a smaller cell which, eventually, enlarges again. Cells do not have arrows on them in tissue sections, and all the conclusions were drawn purely by inference. We have become more sophisticated recently and now realize that the different sizes of blood cells represent different states in differentiation during the course of development. Lymphocytes have been examined by electron microscopy during various stages of organ and cell stimulation by such exogenous factors as antigen, endotoxin, phytohemagglutinin and other reticuloendothelial stimulants or direct cell to cell contact at the surface (1). Changes in size correlate with different stages of function during one type of reaction or another (1). Thus morphology has provided a guide to our understanding of lymphocyte differentiation, but has given us no unequivocal answers. Direct studies of function at the single cell level are preferable wherever possible (4,5). When single cell studies are not feasible, cell populations can be evaluated by inference using statistical methods and other approaches to formulate a more accurate picture of the direction of differentiation (6,7,8).

I. Stages of Lymphocyte Development

For the purposes of discussion, I would like to divide the problem of lymphocyte differentiation arbitrarily into three stages of cell development without categorizing them morphologically, since at this point, we cannot be certain of the morphology of a heterogeneous population which can be defined at a functional level only. The stages discussed must be understood to represent windows in a continuum of differentiation and, as such, form a very tentative basis for discussion.

Lymphocytes progress through developmental steps involving cell division and differentiation from bone marrow stem cells to antibody producing lymphocytes and plasma cells. The concepts and experimental approaches involved in establishing these presumptive relationships have been documented in recent symposia, (6) and I shall discuss them only briefly here. Three compartments of differentiation which are presently useful for functional analysis in the hematopoietic system are the following: Stem cells, Transitional cells, and Effector cells. This is a personal terminology which has been evolving and on which I will elaborate specifically with reference to lymphocytes.

Stem Cells

One group of cells which is related to lymphocytes, and about which there is an increasing general agreement fulfills the function of a stem cell, i.e. an original source of the cells in other compartments. In the adult animal, hematopoietic stem cells are found in the bone marrow. The function of this group of cells is to supply other organs with sources of cells. One particular attribute of stem cells is the capacity for self-renewal (7). Bone marrow cells are the source of red cells, granulocytes, liver macrophages, and circulating monocytes in the peripheral blood (9,10). On the basis of studies by Ford and others (9), we know that thymus lymphocytes are derived at one point from a source of cells in the bone marrow. Thus it seems that all the hematopoietic cells in the spleen, peripheral blood, and throacic duct (the circulating pool of lymphocytes) are derived from stem cell populations in the bone marrow (9). Whether these cells morphologically resemble or in fact are the lymphocyte-like cells found in marrow cannot be determined at present.

Transitional cells

A second group of cells, which at this point I prefer to call transition cells, are on their way to becoming end stage cells. They are derived from a source such as the bone marrow; and they exist, for example, in the thymus, the Bursa in the chicken, and the analogous organs in mammals such as Peyer's patches in the gut-tract. Lymphocytes in the thymus and Bursa do not react to most of the various activating agents, such as antigens, endotoxins and phytohemagglutinin (11,12). Much evidence now implicates the thymus in the differentiation of lymphocytes. While the thymus derives its source cells from bone marrow, lymphocytes undergo further differentiation within the thymus and other lymphoid organs. The indigenous population cannot be separated from the transient population, and these cells engage in cooperative interactions with other cells to complete the final steps of differentiation (13,14, 15).

Effector cells

A third group of cells in the spectrum of lymphocyte differentiation, I call effector cells. Effector lymphocytes react to a variety of surface active substances by enlargement, mitosis, and protein synthesis. Effector cells are the easiest to characterize functionally and thus we know most about them. Antibody-producing cells, as an example, comprise one important group of effector lymphocytes which have been studied extensively in the last few years (16). End stage antibody-producing cells are restricted in function. They secrete only one type of antibody molecule from among the thousands found in the animals serum (17), and are, therefore, functionally restricted at the level of gene-expression.

II. Experimental Approaches

The most useful experiments in laboratories working on lymphocyte differentiation are presently directed along four main lines of effort: 1) Quantitative assays of function *in vivo*, 2) Cell separation experiments aimed at purifying homogenous fractions from heterogenous cell populations, 3) *In vitro* approximations of differentiating cell systems, and 4) Biochemical studies of immunoglobulin formation and secretion and attendant aspects of regulation at the molecular level.

Because of the potential contribution of germ-free mice to the study of lymphocyte differentiation, I would like to discuss the *in vivo* studies first and in the greatest detail.

Mice have been used to great advantage in immunological research in tumor and tissue transplantation (18). Only during the last ten years has the mouse become one of the important animals contributing to knowledge of antibody formation and lymphocytogenesis (6). The use of mice for quantitative studies in antibody formation and in particular the lethally irradiated mouse as an *in vivo* tissue culture environment for donor cells was demonstrated for the most part in Makinodan's laboratory (19). Makinodan's group demonstrated a linear relationship between hemagglutinin titers in recipients sera and the number of donor spleen cells used to transfer the secondary response. These studies coupled with the use of diffusion chambers for donor cells proved the recipient's antibody response was derived strictly from donor cells. Such an assay system involving transfer of competent donor cells to syngeneic lethally irradiated recipients became a very useful model for future experiments involving donor cell populations (6). Another impetus to use of the syngeneic mouse cell transfer system was provided by Till and McCulloch who described a colony forming stem cell for hematopoietic functions which could be cloned in the spleens of irradiated recipients (20).

Since these earlier investigations use of the mouse cell transfer system has gained wide acceptance and has been useful in unraveling a number of events in lymphocytogenesis and antibody formation. It has become the *in vivo* experimental system of choice in studying the early phase of effector cell differentiation in antibody formation involving the precursor and cellular interactions in the development of antibody producing cells (15). Evidence that bone marrow cells behave as a stem cell source population was obtained by combining cell transfer techniques with the use of donor cell populations containing obvious marker chromosomes which could readily be distinguished from the host cells of the recipient. Consequently, the three stages of lymphocyte development discussed under Section I of this paper have been delinated almost entirely with the aid of cell transfer experiments in irradiated mice.

Evidence has accumulated from recent experiments by Claman (13), Miller's group (15) and Davies (14) which suggest that cells from bone marrow undergo a cooperative interaction with thymus cells and then differentiate into antibody producing cells. Since the acquisition of recognition specificity for diverse antigens is a key attribute of lymphocytes involved in immune responses, one of the most immediate problems involves identifying the population or populations which develop antigen recognition capability. Once antigen recognition has been acquired the effector population undergoes clonal expansion by division (6) to increase the frequency of antibody producing cells. While the cells producing antibody are numerous and exist at a frequency of approximately 10^{-3} after antigenic stimulation, cells reacting to antigen are infrequent before antigenic stimulation (frequency of 10^{-5} or less in conventional mice) and therefore, difficult to ascribe to any particular class of cells or site of origin.

Recently it has become obvious that a baseline for the frequency an antigen reactive population of effector cells in conventional mice cannot be established with validity (21). Numerous experiments have indicated the background plaque forming cells in mice existing before antigenic stimulation with sheep red cells are not a precursor pool, but in fact are end stage, differentiated, antibody producing cells. Does a background population of plaque-forming cells exist in germ free mice? The experiments of Bosma and Schearer, Walburg and Cudkowicz reported at this meeting show that there are no significant differences in the background frequency or the kinetics of plaque-forming cell increase between germ-free and conventional mice. However, the experiments of Sterzl (21) and Kim and Watson (22) in the antigen-free, colostrum deprived piglet reared on a non-antigenic liquid diet have cast doubt on the immunological studies in germ-free mice raised on solid food diets which may contain antigens and endotoxins. Sterzl found that piglets raised under an antigen-free environment had no background of plaque-forming cells, and Sterzl and Kim and Watson have both shown that such animals are agammaglobulinemic. At this meeting Wostmann and his associates have demonstrated that a similar antigen-free liquid diet will render germ-free mice agammaglobulinemic and free of background plaque-forming cells for long periods of time.

Now that it is possible to utilize germ-free antigen-free, inbred mice for cell transfer studies, the study of primary reactivity to antigenic stimulation at the cellular level must be repeated in these low-background mice in order to get an accurate picture of the contribution by various cell populations to the development of immune recognition. Cell interactions involved in key differentiative steps can be evaluated more accurately without concern for contamination by circulating cells which have been antigenically stimulated. Much further work will be needed on interacting cell populations in all transfer systems before those differentiative steps which can be studied by cell-transfer techniques will be delineated completely.

Next to the immunological status of the donor and recipient, the single most pressing problem is to produce homogeneous populations of cells from suspensions of lymphoid and hematopoietic organs. After a shaky beginning, separation techniques for mammalian hemopoietic cells were given great impetus by the report of Leif and Vanograd (23) on the separation of erythrocytes by differences in buoyant density. Applications of buoyant density separations to lymphoid cell populations have been most thoroughly studied by Shortman (2). Shortman was able to separate a discontinuous population of lymphocytes in albumin gradients and peak separation was reproducible to a level of ≠ 2.5% minimum to ≠ 25% maximum. While differences in biological activity were separable and various fractions could be enriched there was a high degree of overlap in functional activity. Functions which could be separated to some extent included antigen-sensitive cells from plaque-forming cells, and colony forming cells from spleen (2). The studies by Shortman represent a beginning in the application of more rigorous biophysical techniques to cell separation. Other separation approaches based on velocity sedimentation (Personal communication of Dr. Robert Phillips) also seem promising. These approaches when combined with appropriate animal assay systems such as the antigen-free irradiated mouse will yield more reliable analysis of lymphocyte functions.

In vivo assay systems are but a stop gap to more carefully controlled studies in vitro. The recent successes of Dulton and Mishell (24,25) in perfecting the first reproducible system for induction of antibody formation in vitro in suspension cultures of mouse spleen cells have overcome a 50 year obstacle in the long sought goal of studying lymphocyte differentiation in culture. It has not yet been possible to clone a precursor population of antigen-sensitive cells as yet in their studies, however two types of cells separable in albumin density gradients must undergo interaction in order to induce antibody formation (Dr. R. Dutton - personal communication). Successful long term culture of peripheral blood lymphocytes from normal humans which produce immunoglobulins has been accomplished by Dr. George Moore (26) and his associates and their experience with these cells is reported at this symposium. Cellular interactions between bone marrow thymus and spleen have so far only been explored in organ cultures by Globerson and Auerbach (27) and reproduction of thymus-marrow interaction effects in suspension cultures has not been achieved at this writing. In cell culture experiments utilizing cells of mouse origin, it will be important ultimately to obtain starting cell suspensions from antigen-free mice for the same reasons outlined above.

The biosynthesis of immunoglobulin and its genetic regulation is presently being studied most productively in myeloma tumors maintained in mice by transplantation or from humans and mice in permanent cell culture (28,29). These studies have shown that immunoglobulins are assembled on polyribosomes of different size distribution for the H and L chains (30) and that kinetics of immunoglobulin assembly involve production of $L \rightarrow LH \longrightarrow L_2H_2$ for a complete 7S molecule (31). This rapidly expanding field of study must include further work on polyribosome purification, messenger RNA isolation and characterization, transfer RNA purification and identification before a useful cell free protein synthesizing system can be developed. Further studies on polysome purification and MRNA isolation (Dr. M. Saifer unpublished) will soon allow mapping of immunoglobulin cistrons on DNA from myeloma cells and other hemopoietic cells. A complete characterization of macromolecular machinery will necessarily antedate similar approaches with normal, inducible lymphoid cell populations.

I have attempted to point out some of the more current excuses for our lack of knowledge on lymphocyte differentiation in the last half century. The last twenty years in biology has witnessed the overwhelming contribution of bacterial cells and their viruses to the study of gene function and regulation at the molecular level. The prospects are bright for lymphocytes contributing in a similar fashion to the study of mammalian cell differentiation in the ensuing twenty years. Instead of culture broth tubes, agar, and E. coli our sources and tools are antigen-free mice, humans and cell culture systems. The increase in difficulty is matched only by the increased expense in facilities, animals and cell culture supplies.

The studies reported in this paper were supported by U.S.P.H.S. Grant No. AI-06501, American Cancer Society Grant #E-338 and a fund from U.S.P.H.S. General Research Support granted to Roswell Park Memorial Institute No. FR-05648-02.

REFERENCES

1. Elves, M.W. (1967). The lymphocytes. Year Book Med. Publ. In. 35 East Wacker Drive, Chicago.
2. Shortman, K. (1968). The separation of different cell classes from lymphoid organs II. The purification and analysis of lymphocyte populations by equilibrium density gradient centrifugation. Aust. J. Exp. Biol. Med. 46, 375.
3. Maximow, A. (1969). Der lyphozyt als gemeinsome stammzelle der verschiedenen Blutelemente in der embryonalen Entwicklung und im pastfetalen Leben der Saugetiere. Folia haemat. (Frankfurt) 8, 125.
4. Nossal, G.J.V., K. Shortman, J. Miller, G. Mitchell, and Haskill, (1967). The target cell in the induction of immunity and tolerance. Cold Spring Harbor Symp. on Quant. Biol. 32, 369.
5. Makela, O. (1967). The specificities of antibodies produced by single cells. Ibid p. 423.
6. Papermaster, B. (1967). The clonal differentiation of antibody producing cells. Ibid. p. 447.

7. Till, J.E., R.A. Phillips and Siminovitch (1967). Analysis of differentiating clones derived from marrow. Ibid. p. 461.
8. Schearer, G.M. and G. Cudkowicz, S . James Connel and R. Priore, (1968). Cellular differentiation of the immune system in mice I. Separate splenic antigen-sensitive units for different types of anti-sheep antibody forming cells. J. Exp. Med. 128, 437.
9. Ford, C.E., (1966). Traffic of lymphoid cells in the body. p. 131. In G.E.W. Wolstenhome and R. Porter (ed.) Ciba Foundation Symposium on The Thymus. Little Brown and Co., Boston.
10. Volkman, A., (1966). The origin and turnover of mononuclear cells in peritoneal exudates in rats. J. Exp. Med. 124, 241.
11. Cooper, M.D., A.E. Gabrielsen, M.A. and R.A. Good (1967). Role of the thymus and other central lymphoid tissue in immunological disease. Ann. Rev. Med. 18, 113.
12. Winkelstem, A. and Craddock, C. (1967). Comparative response of normal human thymus and lymph node cells to phytohemagglutinin in culture. Blood 29, 594.
13. Claman, H.N., E. Chaperon and R.F. Triplett. (1966). Thymus marrow cellcombinations - synergism in antibody production. Proc. Soc. Exp. Biol. Med. 122, 1167.
14. Davies, A.J., E. Leuchars, V. Wallis, R. Marchant and E. Elliot. (1967). The failure of thymus-derived cells to produce antibody. Transplantation. 5, 22.
15. Mitchell, G.F. and J.F. Miller (1968). Cell to cell interaction in the immune response II. The source of hemolysin-forming cells in irradiated mice given bone marrow and thymus or thoracic duct lymphocytes. J. Exp. Med. 128, 821.
16. Frisch, L. (ed.) Cold Spring Harbor Symposium on Quantitative Biology. (1967) Antibodies vd. 32.
17. Pernis, B. (1967). Relationships between the heterogeneity of immunoglobulins and the differentiation of plasma cells. Cold Spring Symp. op. cit. 32, 333.
18. Snell, G.D. and J.H. Stimpfling. (1966). Genetics of tissue transplantation in Biology of the Laboratory Mouse. Ed. by Earl L. Green, McGraw-Hill Book Co., New York, p. 457.
19. Makinodon, T., and Albright, J.F., (1967). Proliferative and differentiative manifestations of cellular immune potential. Progr. Allergy 10, 1.
20. Till, J.E., and E. A. McCulloch. (1961). A direct measurement of the radiation sensitivity of normal mouse bone marrow. Rad. Res. 14, 213.
21. Sterzl, J. (1967). Factors determining the differentiation pathways of immunocompetent cells. Cold Spring Harbor Symp. op. cit. 32, 493.
22. Kim, Y.B., S.G. Bradley and D.W. Watson (1968). 19S G and 7SγG antibody synthesis in germfree colostrum-deprived piglets. in Advances in Germfree Research and Gnotobiology, M. Miyakawa & T.D. Luckey (ed.) CRC Press.

23. Leif, R.C. and J. Vinograd. (1964). The distribution of buoyant density of human erythrocytes in bovine albumin solutions. P.N.A.S. 51, 520.
24. Mishell, R.I., and R.W. Dutton. (1967). Immunization of dissociated spleen cell cultures from normal mice. J. Exp. Med. 126, 423.
25. Raidt, D.J., R.E. Mishell, and R.W. Dutton. (1968). Cellular events in the immune response. Analysis and *in vitro* response of mouse spleen cell populations separated by differential flotation in albumin gradients. J. Exp. Med. 128, 681.
26. Moore, G.E., R.E. Gerner, and H.A. Franklin. (1967). Culture of normal human leukocytes. J. Am. Med. Assn. 199, 519.
27. Globerson, A. and R. Auerbach. (1967). Reactivation *in vitro* of immuno-competence in irradiated mouse spleen. J. Exp. Med. 126, 223.
28. Cohn, M. (1967). Natural history of the myeloma. Cold Spring Harbor Symp. op. cit. 32, 211.
29. Matsuoka, Y., G.E. Moore, Y. Yagi and D. Pressman. (1967). Production of free light chains of immunoglobulin by a hematopoietic cell line derived from a patient with multiple myeloma. Proc. Soc. Exp. Biol. Med. 125, 1246.
30. Askonas, B. and A.R. Williamson. (1967). Balanced heavy and light chain synthesis in immune tissue and disulphide bond formation in IgG assembly. in J. Killander (ed) *Gamma Globulins*, structure and control of biosynthesis. Interscience Publ. New York p. 369.
31. Scharff, M.D. (1967). The assembly of gamma globulin in relation to its synthesis and secretion. Ibid. p. 385.

AUTHOR INDEX

(Underscored numbers indicate complete papers in this volume.)

SUBJECT INDEX

GPSR Compliance

The European Union's (EU) General Product Safety Regulation (GPSR) is a set of rules that requires consumer products to be safe and our obligations to ensure this.

If you have any concerns about our products, you can contact us on ProductSafety@springernature.com

In case Publisher is established outside the EU, the EU authorized representative is:

Springer Nature Customer Service Center GmbH
Europaplatz 3
69115 Heidelberg, Germany

Batch number: 10838292

Printed by Printforce, the Netherlands